HYDRODYNAMICS OF THE EQUATORIAL OCEAN

FURTHER TITLES IN THIS SERIES

1 J.L. MERO
THE MINERAL RESOURCES OF THE SEA
2 L.M. FOMIN
THE DYNAMIC METHOD IN OCEANOGRAPHY
3 E.J.F. WOOD
MICROBIOLOGY OF OCEANS AND ESTUARIES
4 G. NEUMANN
OCEAN CURRENTS
5 N.G. JERLOV
OPTICAL OCEANOGRAPHY
6 V. VACQUIER
GEOMAGNETISM IN MARINE GEOLOGY
7 W.J. WALLACE
THE DEVELOPMENTS OF THE CHLORINITY/SALINITY CONCEPT IN OCEANOGRAPHY
8 E. LISITZIN
SEA-LEVEL CHANGES
9 R.H. PARKER
THE STUDY OF BENTHIC COMMUNITIES
10 J.C.J. NIHOUL (Editor)
MODELLING OF MARINE SYSTEMS
11 O.I. MAMAYEV
TEMPERATURE-SALINITY ANALYSIS OF WORLD OCEAN WATERS
12 E.J. FERGUSON WOOD and R.E. JOHANNES
TROPICAL MARINE POLLUTION
13 E. STEEMANN NIELSEN
MARINE PHOTOSYNTHESIS
14 N.G. JERLOV
MARINE OPTICS
15 G.P. GLASBY
MARINE MANGANESE DEPOSITS
16 V.M. KAMENKOVICH
FUNDAMENTALS OF OCEAN DYNAMICS
17 R.A. GEYER
SUBMERSIBLES AND THEIR USE IN OCEANOGRAPHY AND OCEAN ENGINEERING
18 J.W. CARUTHERS
FUNDAMENTALS OF MARINE ACOUSTICS
19 J.C.J. NIHOUL (Editor)
BOTTOM TURBULENCE
20 P.H. LEBLOND and L.A. MYSAK
WAVES IN THE OCEAN
21 C.C. VON DER BORCH (Editor)
SYNTHESIS OF DEEP-SEA DRILLING RESULTS IN THE INDIAN OCEAN
22 P. DEHLINGER
MARINE GRAVITY
23 J.C.J. NIHOUL (Editor)
HYDRODYNAMICS OF ESTUARIES AND FJORDS
24 F.T. BANNER, M.B. COLLINS and K.S. MASSIE (Editors)
THE NORTH-WEST EUROPEAN SHELF SEAS: THE SEA BED AND THE SEA IN MOTION
25 J.C.J. NIHOUL (Editor)
MARINE FORECASTING
26 H.G. RAMMING and Z. KOWALIK
NUMERICAL MODELLING MARINE HYDRODYNAMICS
27 R.A. GEYER (Editor)
MARINE ENVIRONMENTAL POLLUTION
28 J.C.J. NIHOUL (Editor)
MARINE TURBULENCE
29 M. WALDICHUK, G.B. KULLENBERG and M.J. ORREN (Editors)
MARINE POLLUTANT TRANSFER PROCESSES
30 A. VOIPIO (Editor)
THE BALTIC SEA
31 E.K. DUURSMA and R. DAWSON (Editors)
MARINE ORGANIC CHEMISTRY
32 J.C.J. NIHOUL (Editor)
ECOHYDRODYNAMICS
33 R. HEKINIAN
PETROLOGY OF THE OCEAN FLOOR
34 J.C.J. NIHOUL (Editor)
HYDRODYNAMICS OF SEMI-ENCLOSED SEAS
35 B. JOHNS (Editor)
PHYSICAL OCEANOGRAPHY OF COASTAL AND SHELF SEAS

Elsevier Oceanography Series, 36

HYDRODYNAMICS OF THE EQUATORIAL OCEAN

PROCEEDINGS OF THE 14th INTERNATIONAL LIEGE COLLOQUIUM ON OCEAN HYDRODYNAMICS

Edited by

JACQUES C.J. NIHOUL

Professor of Ocean Hydrodynamics, University of Liège, Liège, Belgium

ELSEVIER

Amsterdam – Oxford – New York 1983

ELSEVIER SCIENCE PUBLISHERS B.V.,
Molenwerf 1,
P.O. Box 211, 1000 AE Amsterdam, The Netherlands

Distributors for the United States and Canada:

ELSEVIER SCIENCE PUBLISHING COMPANY INC.
52, Vanderbilt Avenue
New York, N.Y. 10017

ISBN 0-444-42196-3 (Vol. 36)
ISBN 0-444-41623-4 (Series)

Printed in The Netherlands

The International Liège Colloquia on Ocean Hydrodynamics are organized annually. Their topics differ from one year to another and try to address, as much as possible, recent problems and incentive new subjects in physical oceanography.

Assembling a group of active and eminent scientists from different countries and often different disciplines, they provide a forum for discussion and foster a mutually beneficial exchange of information opening on to a survey of major recent discoveries, essential mechanisms, impelling questionmarks and valuable recommendations for future research.

The Scientific Organizing Committee and all the participants wish to express their gratitude to the Belgian Minister of Education, the National Science Foundation of Belgium, the University of Liège, the Intergovernmental Oceanographic Commission and the Division of Marine Sciences (UNESCO) and the Office of Naval Research for their most valuable support.

LIST OF PARTICIPANTS

ARY, S., Mr., University of Niamey, Niger.

BAH, A., Dr., Ecole Polytechnique de Conakry, Guinea.

BOUKARY, S., Mr., University of Niamey, Niger.

BOYD, J.P. Dr., Space Research Bldg., Ann Arbor, Mi., U.S.A.

BUSALACCHI, A.J., Dr., Florida State University, Tallahassee, Fl., U.S.A.

CASAS-VAZQUEZ, J., Prof., Dr., Universidad Autonoma de Barcelona, Bellaterra, Barcelona, Spain.

DELAFOSSE, G., Dr., EPSHOM, Brest, France.

DELECLUSE, P., Dr., Museum d'Histoire Naturelle, Laboratoire d'Océanographie Physique, Paris, France.

DISTECHE, A., Prof., Dr., Université de Liège, Belgium.

DJENIDI, S., Mr., 2, rue BP Cité Plaisance, Annaba, Algéria.

ERIKSEN, Ch., Prof., Dr., Massachusetts Institute of Technology, Cambridge, Mass., U.S.A.

FAHRBACH, E., Dr., Institut für Meereskunde an der Universität Kiel, Kiel, Germany.

FINE, R.A., Dr., National Science Foundation, Washington, D.C., U.S.A.

FOUMAKOYE, G., Mr., University of Niamey, Niger.

FRANKIGNOUL, C., Prof., Dr., Université Paris VI, Laboratoire de Physique et Chimie Marines, Paris, France.

GASPAR, Ph., Mr., Université de Louvain, Louvain-La-Neuve, Belgium.

GIBSON, C.H., Prof., Dr., University of California San Diego, La Jolla, Ca., U.S.A.

GONELLA, J.A., Dr., Museum d'Histoire Naturelle, Laboratoire d'Océanographie Physique, Paris, France.

HALPERN, D., Prof., Dr., Pacific Marine Environmental Laboratory, Seattle, Wa., U.S.A.

HENROTAY, P., Mr., Université de Liège, Belgium.

HISARD, Ph., Dr., Antenne ORSTOM, Centre Océanologique de Bretagne, Brest, France.

KINDLE, J.C., Dr., NORDA, Naval Ocean Research & Development Activity, NSTL Station, Ms., U.S.A.

KRAUS, E.B., Prof., Dr., Cooperative Institute for Research in Environmental Sciences, University of Colorado at Boulder, Boulder, Co., U.S.A.

LASS, H.-U., Dr., Institut für Meereskunde, Rostock-Wärnemunde, Germany.

LEBON, G., Prof., Dr., Université de Liège, Belgium.

LEVY, Cl., Melle, Museum d'Histoire Naturelle, Laboratoire d'Océanographie Physique, Paris, France.

Mc PHADEN, M.J., Dr., JISAO, University of Washington, Seattle, Wa., U.S.A.

MERLE, J., Dr., Museum d'Histoire Naturelle, Laboratoire d'Océanographie Physique, Paris, France.

MONREAL, M.A., Mme, CICESE, Ensenada, B.C., Mexico.

MOLINARI, R.L., Dr., NOAA/AOML/PhOL, Miami, Fla., U.S.A.

MOORE, D.W., Prof., Dr., JIMAR, University of Hawaii at Manoa, Honolulu, Hawaii, U.S.A.

NGENDAKUMANA, Ph., Mr., Bujumbura, Republic of Burundi.

NIHOUL, J.C.J., Prof., Dr., Université de Liège, Belgium.

OLBERS, D.J., Dr., Max-Planck-Institut für Meteorologie, Hamburg, Germany.

PANITZ, H.-J., Dr., Institut für Geophysik und Meteorologie, Universität Köln, Germany.

DU PENHOAT, Y., Dr., Antenne ORSTOM, Centre Océanologique de Bretagne, Brest, France.

PICAUT, J., Dr., Laboratoire d'Océanographie Physique, Faculté des Sciences, Brest, France.

REVERDIN, G., Dr., Museum d'Histoire Naturelle, Laboratoire d'Océanographie Physique, Paris, France.

RIPA, P., Dr., CICESE, Ensenada, B.C., Mexico.

RONDAY, F.C., Dr., Université de Liège, Belgium.

ROWE, M.A., Mr., Department of Oceanography, University of Southampton, U.K.

RUAL, P., Dr., Antenne ORSTOM, Centre Océanologique de Bretagne, Brest, France.

SALAS DE LEON, D.A., Mr., CICESE, Ensenada, B.C., Mexico.

SARACHIK, E.S., Dr., Center for Earth and Planetary Physics, Harvard University, Cambridge, Mass., U.S.A.

SERVAIN, J., Dr., Laboratoire d'Océanographie Physique, Faculté des Sciences, Brest, France.

SMITZ, J., Mr., Université de Liège, Belgium.

TRANNOY, P., Mr., Ecole Polytechnique, Palaiseau, France.

TREGUIER, A.-M., Melle, Ecole Polytechnique, Palaiseau, France.

CONTENTS

EQUATORIAL SEA SURFACE TEMPERATURE VARIATIONS ON SEASONAL TIME SCALES

M. J. MC PHADEN

The Joint Institute for the Study of the Atmosphere and Ocean
University of Washington
Seattle, Washington, 98195

1.1 INTRODUCTION

The study of sea surface temperature (SST) variability in the equatorial oceans is of great value to physical oceanographers and atmospheric scientists alike. Aside from an interest in simply documenting and describing equatorial SST, oceanographers often use SST as a tracer for subsurface dynamical phenomena that are either too difficult or too expensive to observe directly. For example, the accumulation of warm water off the coast of Peru has often been interpreted as evidence for the existence of equatorial Kelvin waves (e.g. McCreary, 1976). Likewise, Moore et al. (1978) have hypothesized that cold SST in the Gulf of Guinea during coastal upwelling season is the result of a coastal Kelvin wave, which through a complex series of events involving equatorial waves, is excited by changing winds in the western equatorial Atlantic. The physical basis for this dynamical influence on SST is that when cold thermocline water is brought closer to (further from) the surface, it is easier (harder) to mix across the base of the mixed layer. This in turn leads to colder (warmer) SST. These are only qualitative arguments however, and ignore other important physical processes such as horizontal advection by surface currents. To quantify the connection between SST and subsurface dynamics, it is therefore necessary to determine the relative significance of various terms in the upper ocean heat budget.

Variations in SST may also induce important dynamical effects in the equatorial oceans. McPhaden (1981) has shown in a steady state model that the response to zonally varying SST patterns is two orders of magnitude larger in the equatorial oceans than at mid latitudes and is comparable in magnitude and spatial structure to the wind-driven equatorial circulation. This strong dynamical effect is due to the sensitivity of geostrophic currents to small changes in horizontal thermal gradients at low latitudes. Our understanding of equatorial dynamics will therefore be limited to the extent that quanitative information on the processes which generate and maintain SST gradients is lacking.

SST is an important lower boundary condition for atmospheric general circulation models. Results from a number of recent numerical experiments based on these models have shown that equatorial SST anomalies have a larger effect on the overlying atmosphere than comparable mid-latitude anomalies (e.g. Shukla, 1975; Rowntree, 1976). Moreover, these effects can be global in nature as Bjerknes (1966, 1969) suggested in his early work on meridional teleconnections. It is clear therefore that effective climate modeling will require a better understanding of the factors that control equatorial SST than we have at present.

This study has two purposes. The first, which is discussed in Section 2, is to provide a brief description of seasonal SST variations in the equatorial oceans and the processes which affect them. The second, which is discussed in Section 3, is to present in detail the results of a case study of SST variability in the central equatorial Indian Ocean. It will be shown that in the vicinity of Gan Island (00°41'S, 73°10'E), surface heat flux accounts for most of the observed SST variations on seasonal time scales. This paper concludes in Section 4 with a summary and comments on the relevance of the Gan results to other equatorial regimes.

1.2 OBSERVED SEASONAL SST VARIATIONS

The patterns of SST variability on seasonal time scales in the equatorial oceans are well documented from historical data analyses. The most striking feature of the seasonal cycle is the development during the northern summer in the Pacific (east of the dateline) and Atlantic of a cold tongue extending from the eastern boundary (Fig. 1 and Fig. 2). The axis of this temperature minimum is generally found between 1°S - 3°S except at its westernmost limit where it is more symmetric about the equator. In both oceans, the seasonal range diminishes from 5°C in the east to $\lesssim$1°C in the west. In the Pacific, Horel (1982) has shown that the minimum temperature occurs later in the year as one progresses to the west; it is not clear whether a similar progression occurs in the Atlantic.

The Pacific, unlike the Atlantic, is subject to large interannual fluctuations in SST which are characterized by basin scale warm anomalies of $\sim$2°C (Wyrtki, 1975). Such "El Niño" episodes are a manifestation of the Southern Oscillation, a coupled ocean-atmosphere phenomenon of global scale. Theoretical arguments involving ocean basin width and the dynamic response of the equatorial ocean to wind forcing explain why the Pacific and not the Atlantic should experience such events (Philander and Pacanowski, 1981). However, we do not now have a complete detailed understanding of the generation, evolution, and decay of El Niño in terms of ocean dynamics, thermodynamics, and air-sea interaction. A better quantitative knowledge of the processes affecting the seasonal march of SST in non-El Niño years will no doubt improve our understanding of El Niño as well.

Seasonal SST variations in the equatorial Indian Ocean are dramatically different from those in the Pacific and Atlantic. To the east of 55°E, i.e. outside the influence of the east African coast, SST shows little significant spatial variation, and the annual range is typically only ∿2°C (Fig. 3). Furthermore, one does not observe an equatorial SST minimum. These differences can be rationalized in terms of the prevailing wind systems. The Indian Ocean is characterized by a monsoon circulation in which northeasterlies predominate in northern winter and southwesterlies predominate in northern summer. Westerly jets appear twice a year along the equator between the northeast and southwest monsoons. These "transition jets" of April/May and October/November are of sufficient intensity that on a yearly average winds at the equator are westerly. On the other hand, the Pacific and Atlantic, excluding limited areas in the eastern and western Pacific and eastern Atlantic that are influenced by monsoon regimes, are dominated by the easterly Trade Winds. While these wind systems may exhibit seasonal variations, they rarely reverse direction for extended periods of time. The relevance of these differences will become more apparent below.

Equatorial SST variations are difficult to interpret because there is a multiplicity of physical processes that can affect them over a wide range of space and time scales. The most important of these processes are 1) insolation, 2) evaporative (or latent) heat loss, 3) upwelling, which is the combined

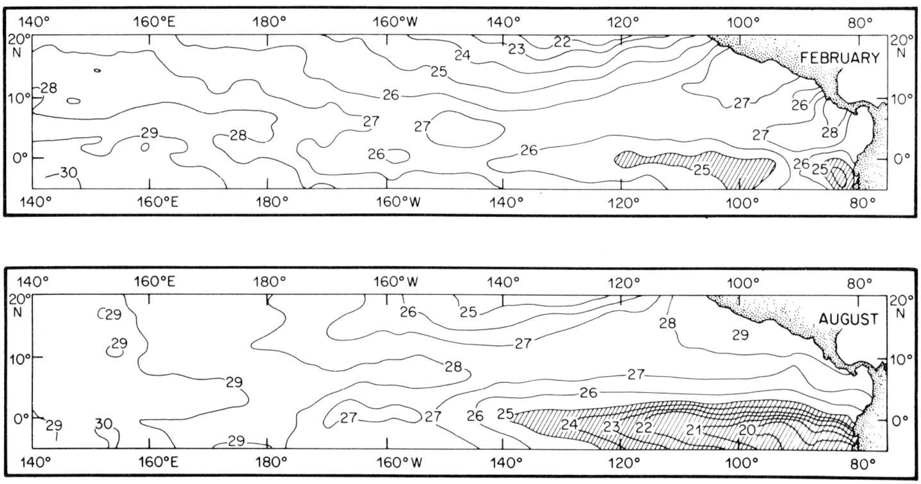

Fig. 1. SST in °C for February and August in the Pacific Ocean from Robinson (1976).

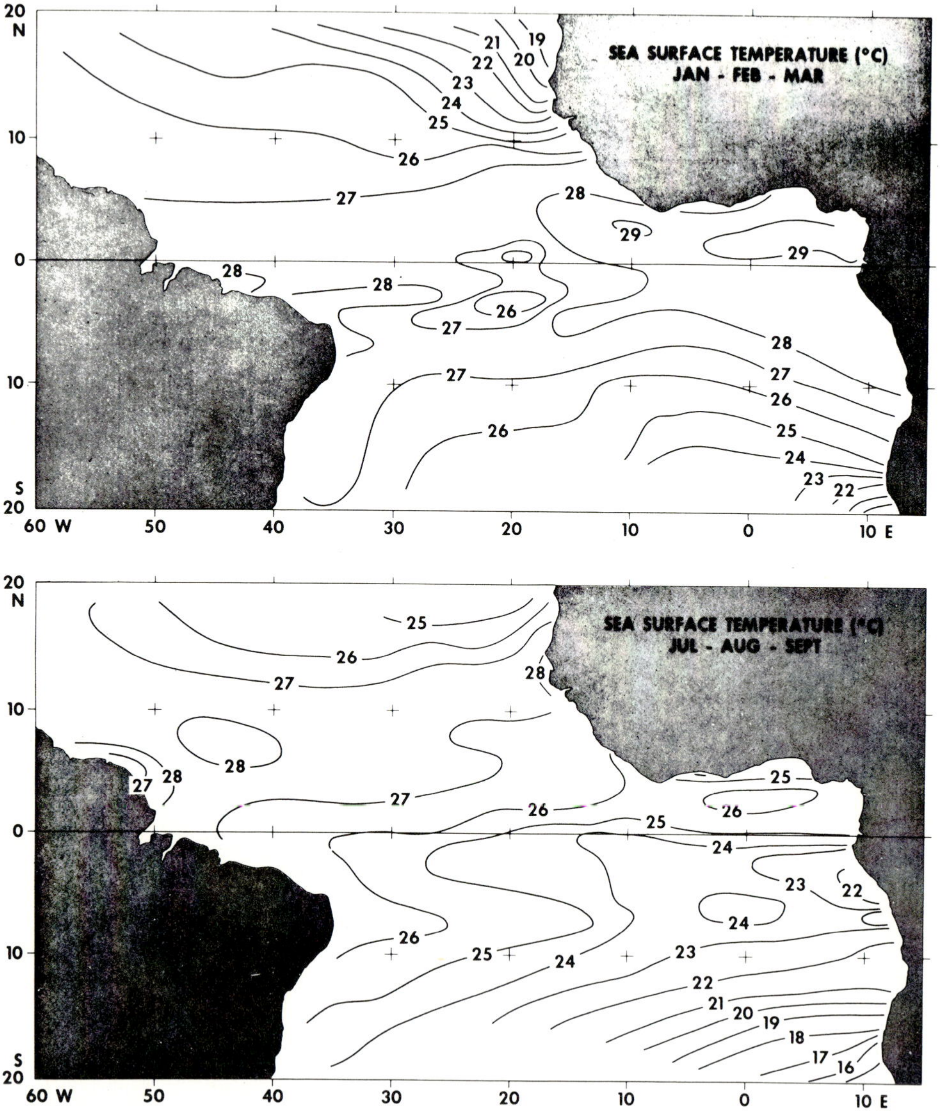

Fig. 2 SST in °C for Jan-Feb-Mar. and July-Aug-Sept. in the Atlantic Ocean from Düing et al. (1980)

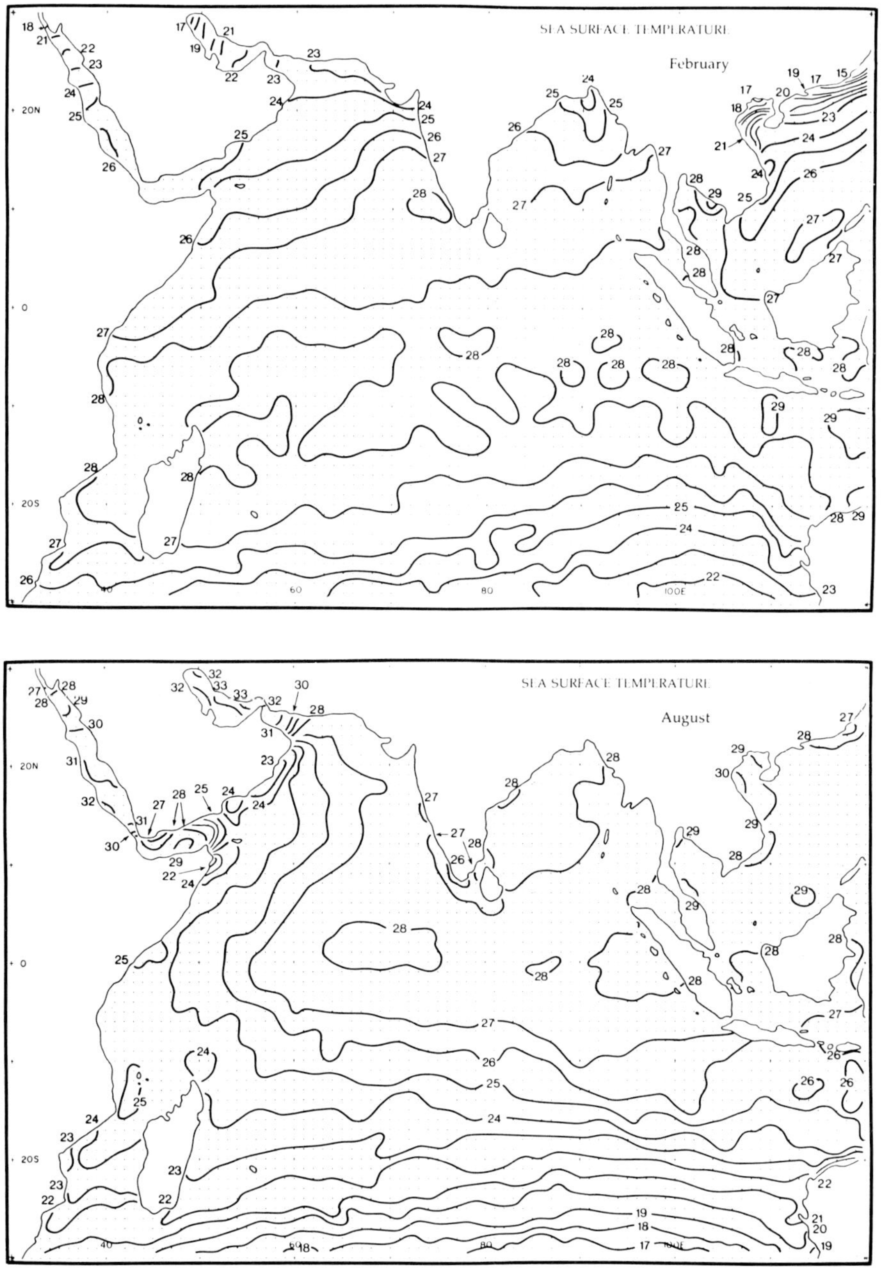

Fig. 3. SST in °C for February and August in the Indian Ocean from Hastenrath and Lamb (1979).

effect of an upward mass flux of cold thermocline water and upper ocean turbulent mixing process, and 4) horizontal advection by strong surface flows such as the South Equatorial Current (SEC). Of secondary importance in general are sensible heat exchange and long wave radiation from the sea surface, and lateral turbulent mixing. There may be, however, restricted areas where these latter phenomena assume a more significant role, e.g. turbulent lateral mixing in frontal zones of the eastern Pacific and Atlantic. Nonetheless they will be ignored in the following discussion.

One can clearly see the effects of insolation in Figure 2. In March there is a large area of water warmer than 27°C straddling the equator in the Atlantic. By September, 27°C water has migrated northward with the sun and is centered near 10°N in the vicinity of the Intertropical Convergence Zone (ITCZ). To the south, temperature gradually decreases into the southern hemisphere where it is winter. Solar heating however cannot explain the equatorial SST minimum or the zonal variations associated with that minimum in the Pacific and Atlantic. Nor can evaporative heat loss, because this tends to be negatively correlated with SST (e.g. Weare, et al., 1981). When the surface is cold, decreased evaporation anomalously warms it, and vice versa, so that evaporative heat fluxes operate to diminish SST extrema rather than to create them. By contrast, equatorial minima can be attributed to equatorial upwelling.

Equatorial upwelling is a coupled dynamic/thermodynamic phenomena driven by the easterly Trade Winds. Easterlies produce an Ekman divergence at the equator which is balanced by an upward mass flux from the thermocline. This brings cold water to the surface where it is mixed by turbulence generated in the high vertical shear zone between the Equatorial Undercurrent (EUC) and SEC. The east-west variation in upwelling intensity is due to continual easterly wind forcing which piles water up in the west and removes it from the east via the SEC. This results in a deeper thermocline in the west than in the east (Colin et al., 1971; Halpern, 1980), where because of the closer proximity of cold water to the surface, a given level of turbulent energy generation will produce more rapid cooling.

The lack of an equatorial SST minimum in the Indian Ocean can be understood within this framework, since there winds are typically westerly along the equator. This favors an Ekman convergence and downward mass flux which depresses the thermocline. Moreover, an EUC may develop during the latter part of the northeast monsoon, but it is a short-lived, transient phenomena. More often flow at all depths to 200 m is from the west with relatively little vertical shear across the base of the mixed layer (McPhaden, 1982a). Thus, the equatorial upwelling is supressed in the Indian Ocean.

One can argue that the position of the SST minimum south of the equator in the Pacific and Atlantic is the result of an upwind displacement of the

Ekman surface divergence (Cromwell, 1953) and EUC (e.g. Charney and Spiegel (1971) by the southerly wind component of the Southeast Trades. However, the axis of minimum temperature does not always coincide with the EUC axis (and therefore the axis of maximum shear induced mixing). This observational fact is consistent with recent theoretical work by McPhaden (1980, 1981), who showed that in a stratified equatorial ocean steady meridional winds drive shallow, frictional surface flows that are not capable of displacing the EUC off the equator. Together with the fact that the SEC has speeds of O(50 cm sec^{-1}) and and the SST gradients are generally very pronounced, this indicates that horizontal advection must be important as well.

While the above interpretation is qualitatively correct, there is a great deal of uncertainty concerning how, in a quantitative sense, the above processes interact in space and time to produce the observed SST patterns. In part this uncertainty is a consequence of the degree to which various processes can be correlated. Consider, for example, the hypothetical scenario in which a seasonal intensification of the Trades leads to a stronger SEC and hence stronger zonal advection of cold water from the east. Simultaneously, local surface divergence and shear between the SEC and EUC may intensify, resulting in more pronounced upwelling. Isolating the effects of these two processes on SST will be very difficult unless carefully selected measurements are made and rigorous analyses carried out. An example of one such analysis is Wyrtki's (1981) box model calculation of the combined heat, salt, and mass budgets of the equatorial Pacific. He concludes that on average upwelling is more important than zonal advection by a factor of 2 or more in maintaining the Pacific cold tongue. This is an illuminating result, but unfortunately one that is sensitive to a number of model assumptions, such as meridional extent of the equatorial box and depth of the Ekman layer. It clearly indicates, however, that progress can be made in the quantitative assessment of SST variability.

In the following section we examine in depth the results of diagnostic study from the Indian Ocean, where it is shown that surface heat fluxes account for most of the observed temperature fluctuations.

1.3 SST VARIATIONS IN THE CENTRAL EQUATORIAL INDIAN OCEAN: A CASE STUDY

1.3.1 Observed variability

Knox (1976) and McPhaden (1982a, b) have extensively discussed month to month fluctuations in oceanic and atmospheric data from the island of Gan (00°41'S, 73°10'E) in the central equatorial Indian Ocean. This section summarizes the important results of McPhaden's (1982b) heat and turbulent kinetic energy budget study for the period January 1973 to May 1975. Figure 4 shows time series of temperature and mixed layer depth (MLD), defined as the first depth below the surface at which the vertical temperature gradient exceeds

5°C/100 m. The mixed layer is extremely well mixed at this location and is on average ∿55 m deep (Fig. 5) so that the 0 - 20 m average in Fig. 4 is equivalent to SST and the 60-80 m average corresponds to upper thermocline temperature. One sees an interesting change in the spectral content of temperature between the surface and the thermocline. The former is dominated by a 1 cycle per year (cpy) variation with a maximum in May and minimum in November, wheras the latter is dominated by a 2 cpy variation. The thermocline variations are a manifestation of the zonal, geostrophically balanced currents driven by westerly transition winds (McPhaden, 1982a). It is not

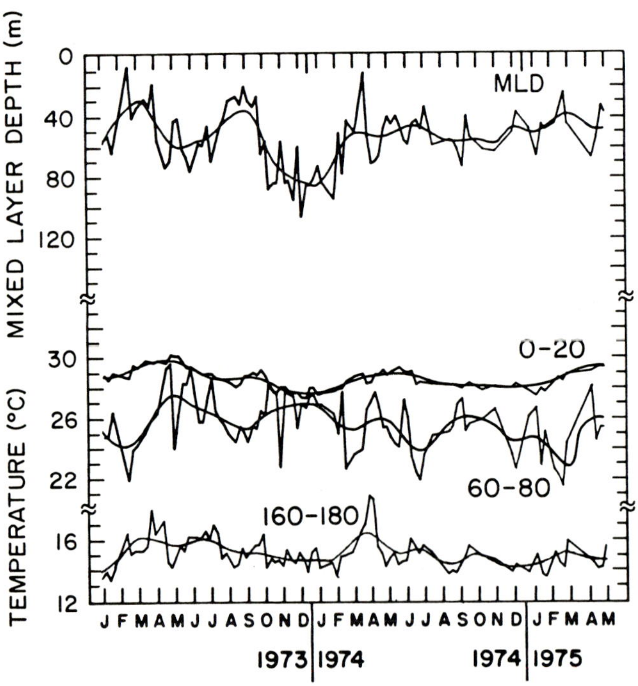

Fig. 4. Weekly estimates of mixed layer depth (MLD) and temperature averaged between 0 - 20 m, 60 - 80 m and 160 - 180 m for the period January 1973 - May 1975. Smooth lines are low passed versions consisting of monthly means.

immediately clear, however, what controls SST. If zonal advection were important, one might expect 2 cpy variations in SST; if SST were changing in response to fluctuations in the intensity of vertical mixing as the thermocline moves up and down, one might expect the time series in SST to mirror that of MLD. In fact, it will be shown below that surface heat flux is the primary controlling mechanism on SST at this location.

Components of the surface heat flux calculated from standard bulk formulae are shown in Fig. 6. Total flux, Q_0, is defined as

$$Q_0 = Q_R + Q_S + Q_L \quad (1)$$

where Q_R ($= Q_{SW} - Q_{LW}$) is the net radiative flux at the sea surface, Q_{SW} is insolation or incoming shortwave radiation, Q_{LW} is outgoing longwave radiation, Q_S is sensible heat flux, and Q_L is latent heat flux. Net shortwave minus longwave radiation (Q_R) is plotted, since Q_{LW} is fairly constant (≈ 60 W m^{-2}) over the record length. Variations in Q_R are therefore mostly due to variations in Q_{SW}. Net radiation shows a strong 1 cpy variation even though at these latitudes a 2 cpy variation dominates clear sky radiation due to the fact that cloud cover is higher during the southwest monsoon than during the

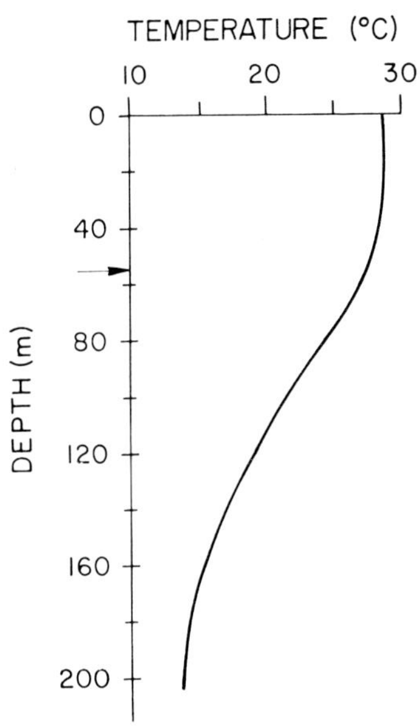

Fig. 5. Temperature as a function of depth averaged over the period January 1973 - May 1975. Arrow indicates mean mixed layer depth.

northeast monsoon. Latent heat flux exhibits a strong 2 cpy fluctuation that is clearly related to the zonal transition winds. Sensible heat flux shows a similar variation, though this component is in general at least an order of magnitude weaker than other flux terms. Total surface flux is typically into the ocean. It is highest in boreal winter when the winds are light and the skies are relatively clear; it is lowest during the transition months of April/May and October/November, when the winds are intense and the sky is overcast.

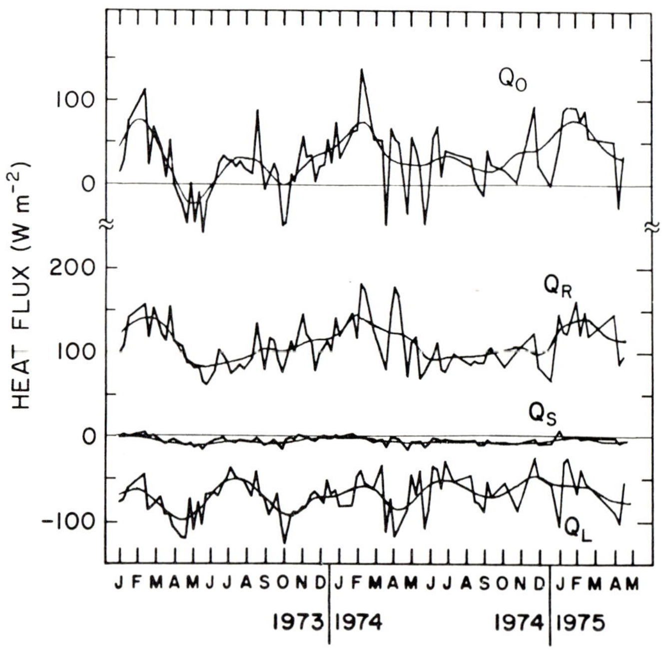

Fig. 6. Weekly estimates of latent (Q_L), sensible (Q_S), net radiative (Q_R), and total surface heat flux (Q_0) for the period January 1973 - May 1975. Smooth lines are low passed versions consisting of monthly means. Positive values indicate that the ocean is being heated.

1.3.2 Mixed layer temperature balance

Following Niiler and Kraus (1977), one can integrate the temperature conservation equation from the surface to the base of the mixed layer to get

$$\frac{\partial T}{\partial t} + \underset{\sim}{u} \cdot \nabla T = \frac{1}{h} w_e \Delta T + \frac{1}{\rho C_p h} [Q_0 - Q_{-h} - I_0 e^{-\gamma h}] \quad (2)$$

Temperature (T) in this expression is equivalent to SST because of the homogeneity of the mixed layer temperature (q.v. Fig. 5). The base of the mixed layer is located at $z = -h$, where Q_{-h} is vertical turbulent heat diffusion and ΔT is a temperature jump due to entrainment mixing. Entrainment velocity, w_e, is defined as

$$w_e = \frac{\partial h}{\partial t} - w_{-h} > 0 \quad (3)$$

in which $\partial h/\partial t$ is the time rate of change of mixed layer thickness and w_{-h} is a vertical upwelling velocity in the thermocline just below the base of the mixed layer. The inequality means that entrainment occurs only when the layer deepens relative to changes in thermocline depth. Horizontal vector velocity is designated $\underset{\sim}{u}$. The variable I_0 represents penetrative radiation which is ≈45% of incident solar flux (Ivanov, 1977); γ determines its attenuation with depth. Lateral turbulent diffusion has been neglected.

To examine the temperature balance on monthly time scales, subtract the mean from equation (2) to get

$$\frac{\partial T}{\partial t} - \frac{1}{\tau} [T(t_1) - T(t_0)] + (\underset{\sim}{u} \cdot \nabla T)' + (\frac{1}{h} w_e T)'$$
$$= \frac{1}{\rho C_p} [(\frac{Q_0}{h})' - (\frac{Q_{-h}}{h})' - (\frac{I_0}{h} e^{-\gamma h})'] \quad (4)$$

where primes denote fluctuations about the mean (denoted by an overbar), e.g. $(Q_0/h)' = Q_0/h - \overline{(Q_0/h)}$, etc., and $\tau = t_1 - t_0$ is record length. The term preceeded by $1/\tau$ corrects for trends introduced by the fact that the end points of the time series are at different temperatures. Time series of observed T, h, u, and Q_0 can now be used to estimate the relative importance of various terms in (4). For the turbulent diffusive flux term, this requires a model in which $Q_{-h} = \rho C_p K\, \partial T/\partial z$ where K is a vertical eddy diffusivity and $\partial T/\partial z$ is the temperature gradient below the base of the mixed layer. Calculation of the entrainment term requires estimates of both w_e and ΔT.

The simplest balance to test with the present data set is one in which horizontal advection, penetrative radiation, entrainment, and turbulent diffusive fluxes are neglected. Then integration of (4) yields

$$T_Q(t) = \frac{1}{\rho C_p} \int_{t_2}^{t} \left(\frac{Q_0}{h}\right)' dt + \frac{1}{\tau} [T(t_1) - T(t_0)](t - t_2) + T(t_2) \qquad (5)$$

where T_Q an estimated mixed layer temperature is derived from a single heat source, viz. the surface flux Q_0. The integration constant $T(t_2)$, where $t_0 < t_2 < t_1$, is arbitrary and chosen such that the average T_Q over the record length is equal to the average of the observed mixed layer temperature. Thus (5) predicts fluctuations and not absolute values of mixed layer temperature.

Figure 7 shows a calculation of T_Q based on a trapezoidal rule integration of observed $(Q_0/h)'$. Superimposed on this is the observed mixed layer temperature, T_{OBS}. Though there are some O(1°C) discrepancies between these two curves, most notably in April 1973, the rms temperature difference $\sigma_{\Delta T} \equiv [\overline{(T_{OBS} - T_Q)^2}]^{1/2}$, is only 0.32°C. Moreover, the coherence squared between these curves is 0.81, i.e. 81% of the variance in mixed layer temperature can be accounted for by considering only surface fluxes.

One might have expected zonal advection to dominate the surface temperature balance at Gan since the zonal currents in this region can attain speeds of ∿100 cm sec^{-1} on monthly averages. However, as is evident in Fig. 3, zonal gradients in SST in the central ocean are extremely weak. A simple scale analysis indicates that no more than 20% of the observed variance in SST could be

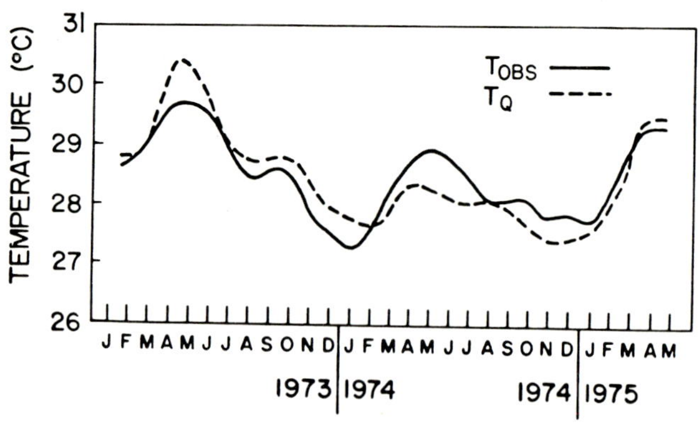

Fig. 7. A comparison of observed monthly mean mixed layer temperature (T_{OBS}) with computed temperature (T_Q) from (5).

due to this process.

Estimates of upwelling velocity, w_{-h}, from vertical motion of isotherms in the thermocline and $\partial h/\partial t$ from Fig. 4 indicate significant entrainment velocities of $O(10^{-4}$ cm sec$^{-1})$ only during the transition months of 1973. These are due to wind stirring by the transition jets, and are such that a 50 m mixed layer would cool by 0.1°C over 1 month. This would tend to reduce but not eliminate the discrepancies in Fig. 7 between T_Q and T_{OBS} in the spring and fall of 1973. For the record as a whole, however, entrainment is generally not active due to the lack of significant shear generated turbulence like that found in the Pacific and Atlantic in association with the EUC (e.g. Crawford and Osborn, 1979a, b).

One can show from a scale analysis that other physical processes neglected in (5) are small as well. An error analysis for the surface flux term indicates however an uncertainty in T_Q of $\pm$20% due to inherently inaccurate bulk parameterization schemes. This is comparable in magnitude to the supposed advection term and could plausibly account for the discrepancy between the two curves in Fig. 7 by itself.

It is important to note that in the calculation of T_Q from (5) the reference level was taken to be the time varying MLD. Had this reference level been embedded in the thermocline, say at a fixed depth of 200 m, changes in depth averaged temperature (or equivalently heat content) would be an order of magnitude larger than could be accounted for by surface fluxes. This is a consequence of the fact that vertical motion of the thermocline due to fluctuating geostrophic currents dominates the heat balance of the upper tropical ocean below the mixed layer (e.g. Merle, 1980). These fluctuations must be filtered out by a proper choice of reference level for investigations of SST variability.

1.4 CONCLUSION

Simultaneous oceanic and atmospheric measurements from January 1973 to May 1975 were used to examine the surface temperature balance near Gan. Results indicate that surface heat flux accounts for about 80% of the observed mixed layer temperature, or equivalently SST, on monthly time scales. Zonal advection is of secondary importance, and entrainment mixing is negligible most of the time, though on occasion sufficient turbulence is generated by wind stirring during the transition months to penetrate the thermocline. A major source of uncertainty in these calculations is the parameterization of surface heat flux from bulk formulae which introduces errors equivalent to ≈20% of the observed SST signal.

These results may be relevant to other regions of the equatorial oceans. For instance, oceanic conditions in the Pacific west of the dateline in many respects resemble those in the central Indian Ocean. Both regions are characterized by monsoon winds, weak horizontal SST gradients, deep mixed layers, and the absence of an equatorial SST minimum. Hence it may be that in the western Pacific, SST variations on seasonal time scales are also determined primarily by surface fluxes. In regions of the equatorial ocean influenced by Trade Winds, it is expected that though surface fluxes may not dominate the seasonal cycle, they should have a noticeable effect. Indeed, Stevenson and Niiler (1982) found from data collected during the Hawaii-Tahiti Shuttle Experiment (Wyrtki et al., 1981) that SST in the central Pacific was significantly corelated with the surface heat fluxes. In areas such as this, it would clearly be desirable to have accurate estimates of upwelling and horizontal advection to close the temperature balance.

ACKNOWLEDGMENTS

The author gratefully acknowledges the support provided by the Joint Institute for the Study of the Atmosphere and Ocean and by the School of Oceanography, University of Washington, Seattle, Washington.

REFERENCES

Bjerknes, J., 1966. A possible response of the atmospheric Hadley circulation to equatorial anomalies of ocean temperature. Tellus, 18: 820-829.

Bjerknes, J., 1969. Atmospheric teleconnections from the equatorial Pacific. Mon. Wea. Rev., 97: 163-172.

Colin, C., Henin, C., Hisard, P., and Oudot, C. Le Courant de Cromwell dans le Pacifique central en fevrier. Cah. ORSTOM Ser. Oceanogr., 9: 167-186.

Charney, J. G., and Spiegel, S. L., 1971. The structure of wind-driven equatorial currents in homogeneous oceans. J. Phys. Oceanogr., 1: 149-160.

Crawford, W. R., and Osborn, T. R., 1979a. Microstructure measuremtents in the Atlantic Equatorial Undercurrent during GATE. Deep-Sea Res., 26, GATE supplement 2: 285-308.

Crawford, W. R., and Osborn, T. R., 1979b. Energetics of the Atlantic Equatorial Undercurrent. Deep-Sea Res., 26, GATE Supplement 2: 309-323.

Cromwell, T., 1953. Circulation in a meridional plane in the equatorial Pacific. J. Mar. Res., 12: 196-213.

Düing, W., Ostopoff, F., and Merle, J. (editors), 1980. Physical Oceanography of the Tropical Atlantic During GATE. University of Miami, 117 pp.

Hastenrath, S., and Lamb, P. J., 1979. Climatic Atlas of the Indian Ocean, Part I: Surface Circulation and Climate. University of Wisconsin Press, 109 pp.

Halpern, D., 1980. A Pacific equatorial temperature section from 172°E to 110°W during winter and spring 1979. Deep-Sea Res., 27A: 931-940.

Horel, J. D., 1982. The annual cycle of the tropical Pacific atmosphere and ocean. Mon. Wea. Rev., 110: in press.

Ivanov, A., 1977. Oceanic absorption of solar radiation. Modelling and Prediction of the Upper Layers of the Ocean, E. B. Kraus (Editor). Pergamon Press, Oxford: 47-71.

Knox, R. A., 1976. On a long series of measurements of Indian Ocean equatorial currents near Addu Atoll. Deep-Sea Res., 23: 211-221.

McCreary, J. P., 1976. Eastern tropical response to changing wind systems: with application to El Nino. J. Phys. Oceanogr., 6: 632-645.

McPhaden, M. J., 1980. Models of the Equatorial Ocean Circulation. Ph. D. dissertation, Scripps Institution of Oceanography, 105 pp.

McPhaden, M. J., 1981. Continuously stratified models of the steady state equatorial ocean. J. Phys. Oceanogr., 11: 337-354.

McPhaden, M. J., 1982a. Variability in the central equatorial Indian Ocean, Part I: Ocean dynamics. J. Mar. Res., 40: 157-176.

McPhaden, M. J., 1982b. Variability in the central equatorial Indian Ocean, Part II: Oceanic heat and turbulent energy balances. J. Mar. Res., 40: 403-419.

Merle, J., 1980. Seasonal heat budget in the equatorial Atlantic Ocean. J. Phys. Oceanogr., 10: 464-469.

Moore, D. W., Hisard, P., McCreary, J. P., Merle, J., O'Brien, J. J., Picaut, J., Verstraete, J., and Wunsch, C., 1978. Equatorial adjustment in the eastern Atlantic. Geophys. Res. Lett., 5: 637-640.

Niiler, P. P., and Kraus, E. B., 1977. One-dimensional models of the upper ocean. Modelling and Prediction of the Upper Layers of the Ocean, E. B. Kraus (editor), Pergamon Press, Oxford, 143-172.

Philander, S. G. H., and Pacanowski, R. C., 1981. Response of equatorial oceans to periodic forcing. J. Geophys. Res., 86: 1903-1916.

Rowntree, P. R., 1976. Tropical forcing of atmospheric motions in a numerical model. Quart J. Roy. Meteor. Soc., 102: 583-605.

Robinson, M. K., 1976. Atlas of the North Pacific Ocean Monthly Mean Temperatures and Mean Salinities of the Surface Layer. Naval Oceanogr. Office, Ref. Pub. 2, 173 figures.

Shukla, J., 1975. Effect of Arabian sea-surface temperature anomaly on Indian summer monsoon: A numerical experiment with the GFDL model. J. Atmos. Sci., 32: 503-511

Stevenson, J., and Niiler, P. P., 1982. Tropical heat flux and storage during the FGGE Shuttle experiment. (Unpublished manuscript).

Weare, B. C., Strub, P. T., and Samuel, M. D., 1981. Annual mean surface heat fluxes in the tropical Pacific Ocean. J. Phys. Oceanogr., 11: 705-717.

Wyrtki, K., 1975. El Niño--the dynamic response of the equatorial Pacific Ocean to atmospheric forcing. J. Phys. Oceanogr., 5: 572-584.

Wyrtki, K., 1981. An estimate of equatorial upwelling in the Pacific. J. Phys. Oceanogr., 11: 1205-1214.

Wyrtki, K., Firing, E., Halpern, D., Knox, R., McNally, G. J., Patzert, W. C. Stroup, E. D., Taft, B. A., and Williams, R., 1981. The Hawaii to Tahiti Shuttle Experiment. Science, 211: 22-28.

SEASONAL OBSERVATIONS IN THE WESTERN INDIAN OCEAN

Michèle FIEUX and CLAIRE LEVY

L.O.P. Muséum Paris - LA 175 CNRS

Abstract

XBT and surface salinity observations which are part of the SINODE (Surface INdian Ocean Dynamics Experiment) program are presented to describe the variations in the surface layers of the western Indian ocean between Cape Guardafui and Reunion island. In the Somali area, the temperature structure presents a high variability related to a strong eddy structure; the surface salinities are a good indicator of the circulation. At the equator, during the two transition periods, after the two monsoons, when the eastward jet is present, the temperature structures are different : the thermocline is broader after the SW monsoon. South of the equator, the South Equatorial Current northern limit wanders between 6° S and 9° S.

1. INTRODUCTION

The "MARION DUFRESNE" who supplies the "Terres Australes et Antarctiques Françaises" in the southern Indian Ocean crosses the tropical and equatorial western region of the Indian ocean four times a year.To reach the seasonal variations, since the SINODE cruises of 1979 and 1980, this opportunity is used to make XBT sections, surface temperature and salinity measurements, ship drift estimations along the route between Cape Guardafui and la Réunion, and to launch satellite tracked drifting buoys at the equator.

Here are presented the preliminary results from the first eight SINODE cruises except the drifting buoys ones which are presented in this volume by REVERDIN et al..

The routes are somehow different from one cruise to another (Fig.1 and 2). During SINODE 1, SINODE 2 and SINODE 6 zig-zag sections were done in the Somali basin. It is only during SINODE 4, SINODE 7 and SINODE 8 that the direct meridional route had to be followed because no ship time was available. Surface temperature and salinity were recorded continuously except during SINODE 7. The ship drifts, estimated from the difference between dead reckoning and satellite positioning, were obtained only for SINODE 4, 5, 7, 8. The XBT probes (T 7 type) gave the temperature down to at least 850 meters. Due to the ship program constraints, the cruises are not evenly spread over the seasons ; the months sampled are May, June, July, October, November and December i.e. at the beginning and during the SW monsoon, during the transition period of October and during the beginning of the NE monsoon.

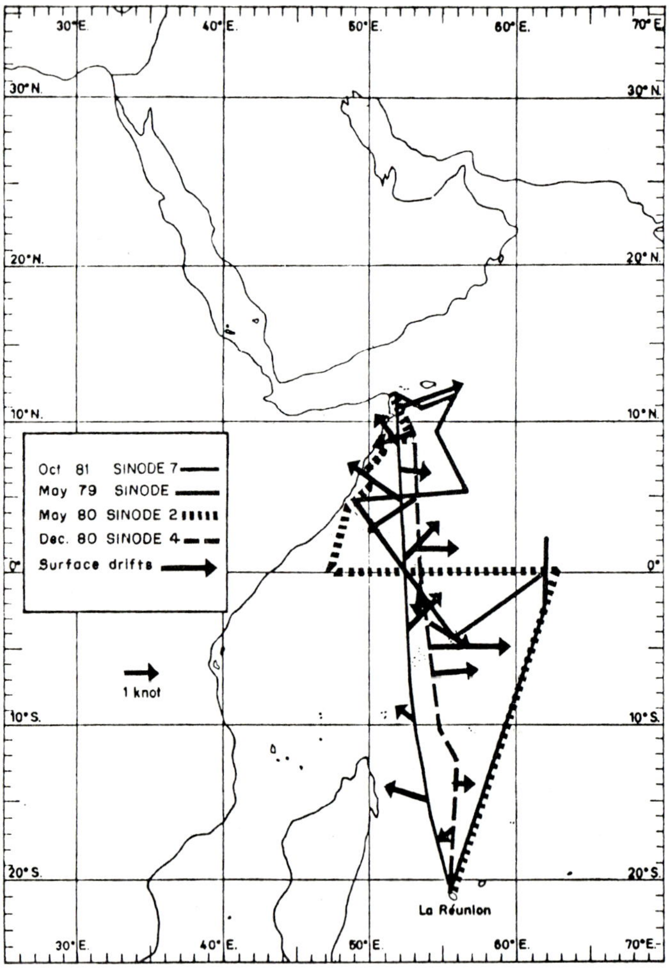

Fig. 1. Routes and ship drift of cruises : SINODE 1, SINODE 2, SINODE 4 and SINODE 7.

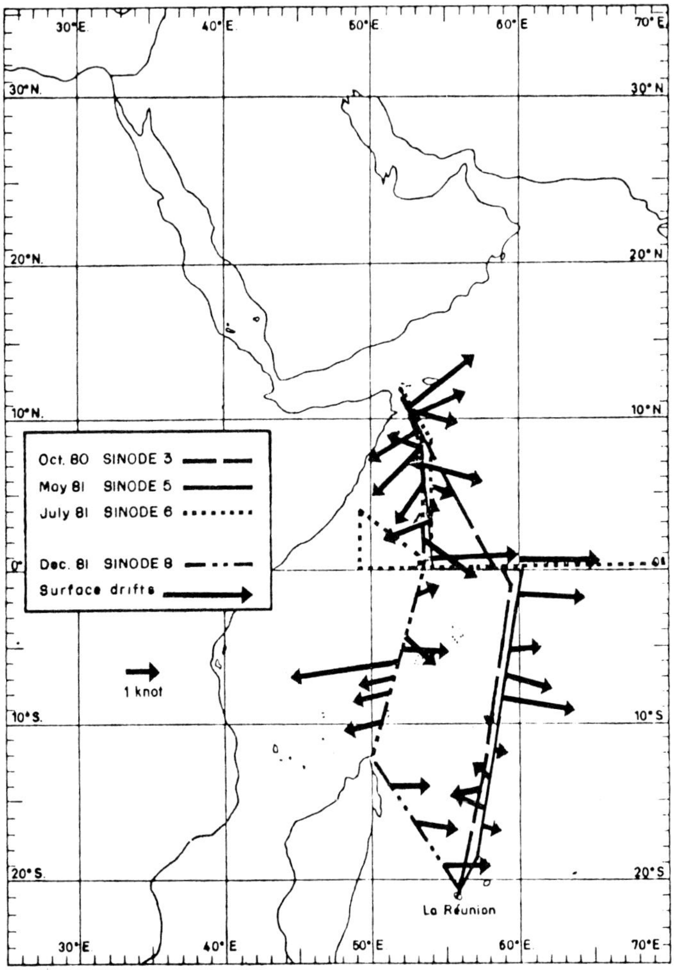

Fig. 2. Routes and ship drift of cruises : SINODE 3, SINODE 5, SINODE 6 and SINODE 8.

2. PRELIMINARY RESULTS

2.1 Somali basin

The temperature structure in that area is similar to the density structure because the density variations are more dependent on the temperature than on the salinity. The slope of the isopycnes along a section gives the component of the geostrophic current perpendicular to the section. Thus the geostrophic current perpendicular to the section can be guessed from the slope of the isotherms.

In the Somali basin the dominant feature of most of the temperature sections (except in June 1979 and May 1980) is a trough in the thermocline with a steeper front on the northern side than on the southern one (fig.3, the 20°C isotherm lies in the middle of the thermocline thus can be taken as the thermocline index (QUADFASEL, 1982)).

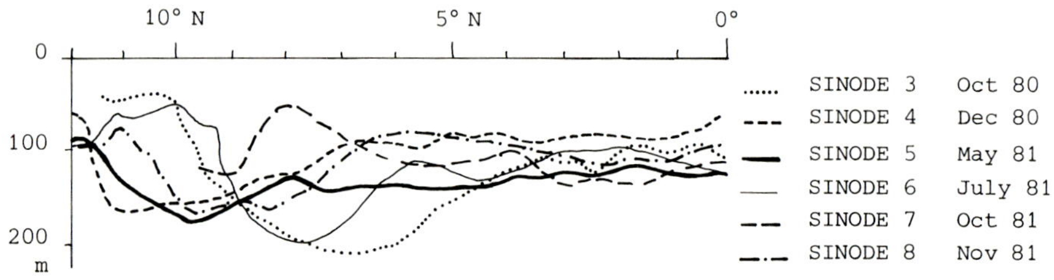

Fig. 3. 20° C isotherm depth.

This trough corresponds to an anticyclonic eddy print, the northern Somali eddy (Bruce, 1968, 1979, DÜING, 1978, SWALLOW et al. 1966, 1982) which seems to be driven by the SW monsoon wind stress curl (SCHOTT et al. to appear, ANDERSON 1980). In June 79 and May 80, the sections were made just one or two days after the onset of the SW monsoon so the eddy was not formed yet. The observations show the beginning of a cooling ($\Delta T \simeq 2°C$) in the surface layer close to the coast which is a rapid response to the onset (SCHOTT et al , 1980). In October of the same year after the cessation of the SW monsoon the trough was broad with an axis around 7° N and a northern front around 9°30 N (fig.4). In December 80, at the beginning of the NE monsoon the trough was much reduced and the northern front has migrated to around 11°30 N. The surface drifts are in agreement with the subsurface temperature structure with a strong northeastward drift in the front area and a southwestward drift from 10°N to around 4°N which brings high salinity waters from the Arabian Sea (S‰ > 36‰). In May 81, after the onset of the SW monsoon with no observation between December and May, there is the print of a northern eddy with a northern front around 11°N, less steep than in the previous December (Fig.3). The ship drifts directions (Fig.2) are coherent with the subsurface struc-

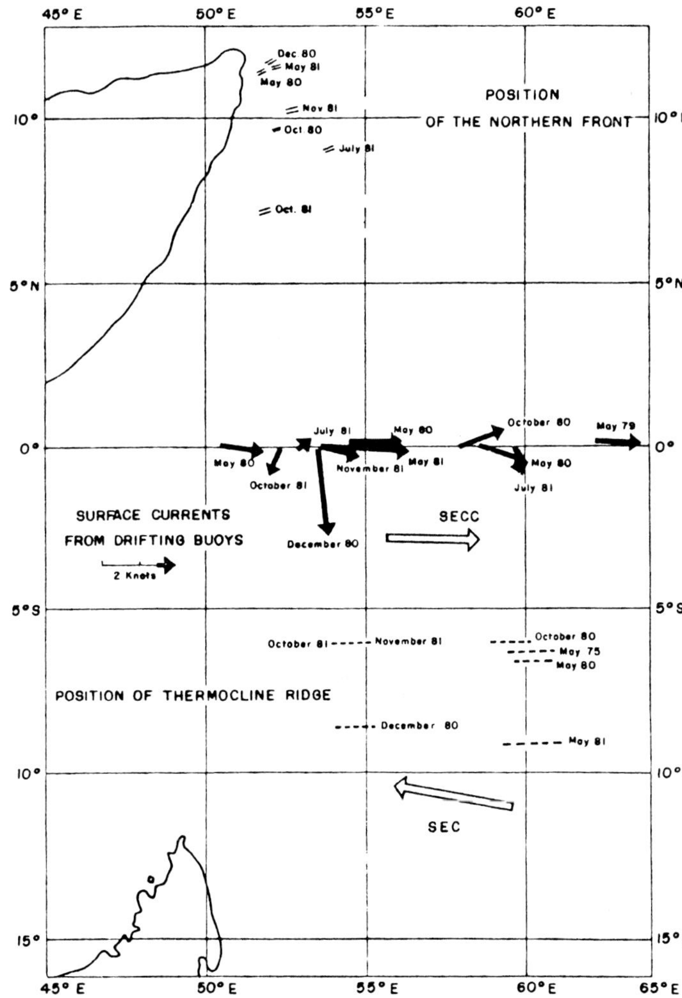

Fig. 4. Positions of the northern fronts, of the buoy drifts at the equator and of the thermocline ridge north of the South Equatorial Current, for each cruise when available.

ture ; the surface salinity is high in the eddy (S > 35.6‰) ; south of 5°N the surface salinity decreases (S < 35.3‰) which shows its southern origin (LEETMAA et al. 1972). At the beginning of July 81, the trough is much more south with the axis around 7°30N (Fig.3) and a northern front around 8°45N. There is the sign of a smaller eddy between 10°N and 12°N. The salinity is higher than 36‰ north of 10°N (water from Arabian Sea) and higher than 35.6‰ between

3°N and 10°N.

In October 81, the trace of an anticyclonic eddy is present between 3°30 N and 7°30 N with the northern front around 7°N which is quite south (fig.3). North of it, the temperature section shows the southern part of another anticyclonic eddy. The ship drift directions are in agreement with a double eddy structure (Fig.1).

Between 1°N and 2°N in May 81, 1°30 N and 2°30 N in July 81 and 1°N and 3°30 N in October 81, there are cooler and fresher waters at the surface (1° or 2° C cooler) which correspond to eastward drift directions. This can be related to the eastward extension of the southern loop which leaves the coast around 3°N and brings waters from the South Equatorial Current and East Africa coastal current with low salinity characteristics (S< 35.3‰). (LEETMAA et al., SWALLOW et al. to appear in J.P.O.). In November 81, 6 weeks later there was only one trough between 6°N and 11°N with a northern front around 10°30 N (Fig.3). Again the surface drift directions are in agreement with the subsurface structure (Fig.5) and with the surface salinity which is maximum between 7°N and 10°N corresponding to the westward drift of the southern part of the northern eddy. In contrast north of 10°N in the northeastward drift, the surface salinity is lower (35.6 < S < 36‰) similar to the values between 3°N and 7°N. The values less than 35.3‰ appear south of 0°30 N in the eastward equatorial jet.

To summarize these observations in the Somali basin, it seems that the print of the northern Somali eddy is always there after the onset of the SW monsoon with a trough as deep as 150 meters and with a steeper front in the north which could be due to the presence of the coast. The surface salinity is a good index of the circulation : westward drift in the southern branch of the eddy brings high Arabian Sea salinity waters (S> 36‰),on the contrary, low salinity waters (S < 35.3‰) are brought from the SEC by the Somali current loop. Drift, surface salinity and subsurface structure are usually in good agreement (Fig.5). The observations made on different years at the same season are not similar, for example in October 81, a sharp temperature ridge is present around 8°N, much further south than in October 80 when the structure was a single-eddy one with northern front around 9°30 N. But it seems, from these observations, that at the end of the eddy supposed life the northern subsurface front gets steeper (cf. Nov. 81 and Dec.80). which could be due to the concentration of the eddy energy in a much restricted coastal zone.

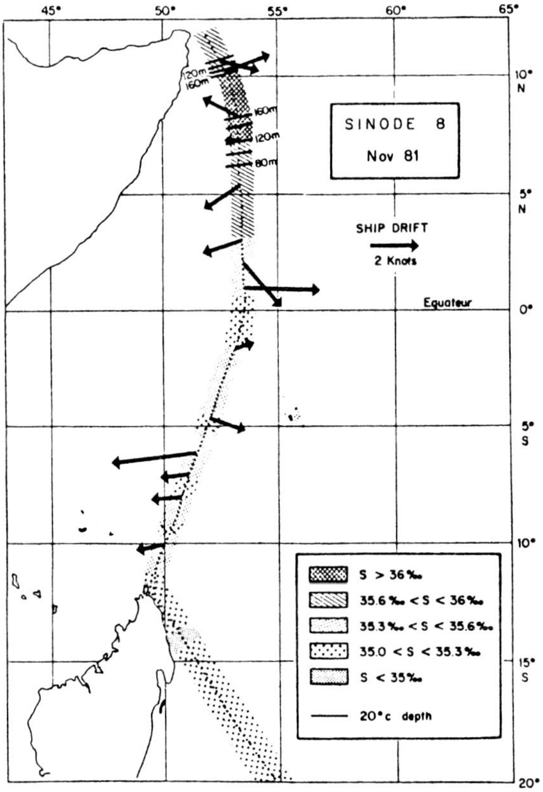

Fig. 5. Surface salinity, surface drift and depth of 20° C isotherm during SINODE 8 , November 1981.

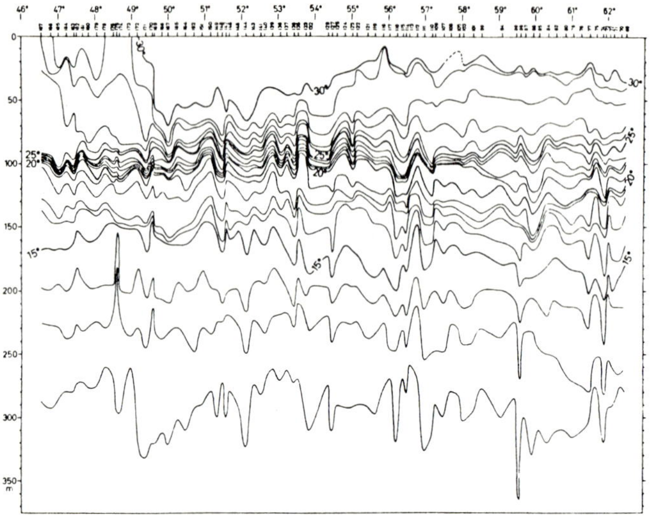

Fig. 6. Equatorial temperature section from 46° 30' E to 62° 30' E (in May 1980).

2.2 Equatorial region

At the equator the eastward jet which appears east of around 49° E - 50° E, between the monsoons, during the transition periods of April-May and October-November (WYRTKI, 1973, CRESSWELL, et al. 1981, LUYTEN, et al. 1980) was present during each crossing (with speed around 1m/s) except in July 81 during the SW monsoon and in November 81 when the jet was present further east of the section (Fig.4).

The thermocline structure does not show much variations with longitude along the equator (Fig.6,8) except west of 49°E in May 80 where the thermocline is thinner. This western equatorial area, where the jet does not exist, presents low salinity characteristics ($S < 35.20‰$) i.e. waters of southern hemisphere origin (fig.7). The salinity front extends from 0 to 90 meters and stands around 49°E. As in 1980, the 49°E July 81 temperature section presents a thinner thermocline than at 53° - 54°E ; its slope corresponds to the return of Somali current fresher flow from the coast.

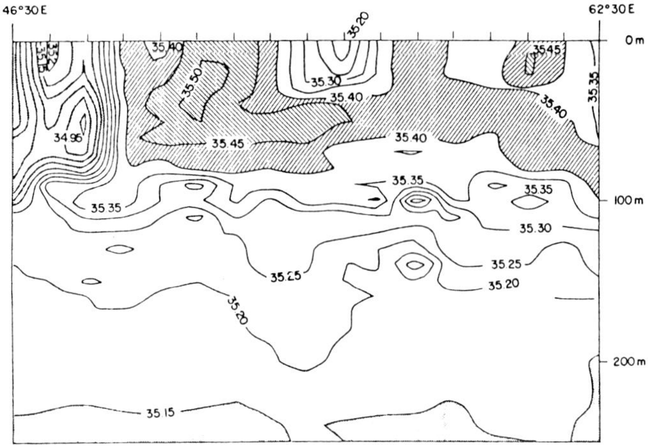

Fig. 7. Equatorial salinity section from 46° 30' E to 62° 30' E (May 1980).

Comparing now the two transition periods cross-equatorial sections it appears that the thermocline is much broader during the second period (Oct.- Nov.- Dec.) when the jet speeds are higher (cf.REVERDIN et al.) (Fig.9). This difference must reflect different processes of formation of the jet and/or different water masses involved. The two different circulation systems at the end of the two monsoons seem to affect differently the equatorial structure during the two eastward jet transition periods.

The cross-equatorial temperature sections made during the eastward jet period do not show a strong convergence. The jet must be fed partly by waters from the west and partly through horizontal convergence.

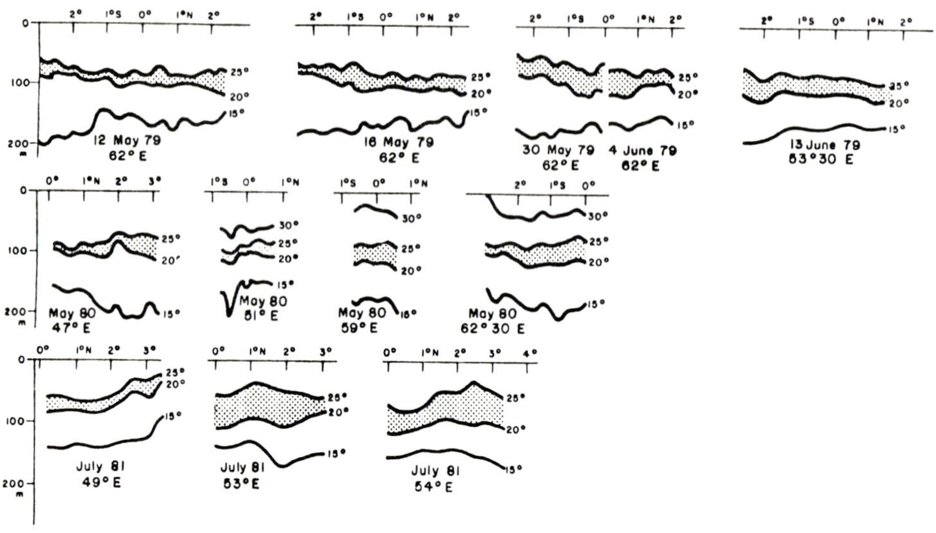

Fig. 8. Cross equatorial temperature sections at different times and longitudes during three cruises.

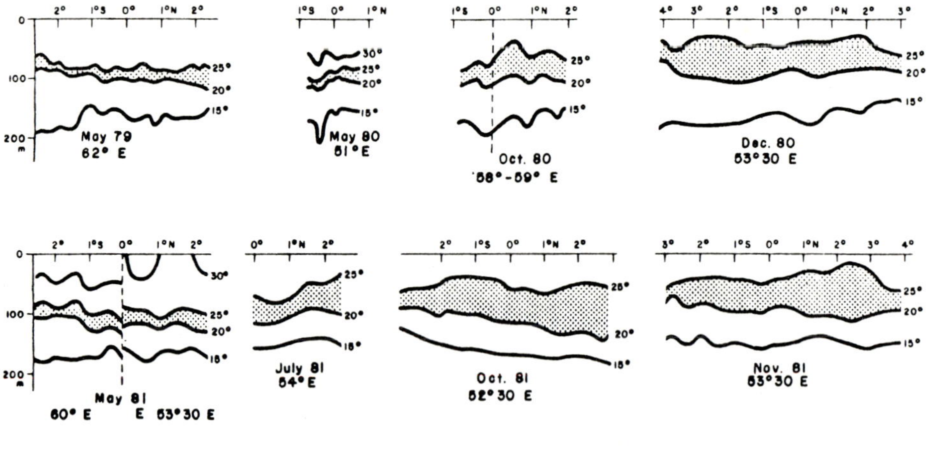

Fig. 9. Cross equatorial temperature sections at different periods of the year.

2.3 South equatorial current and counter-current region

The XBT sections have been expanded down to Reunion island. The prominent feature is the slope of the isotherms between Reunion and a thermocline ridge which varies with the cruises from 9°S to 6°S (Fig.4 and 10). This slope is the mark of the south equatorial current (SEC) which is steadier and stronger during the SW monsoon period when the SE trade winds are also stronger and when the currents are weak and variable in the region between the thermocline ridge (limit of the SEC) and the equator. During the NE monsoon the SE trade winds decrease and the SEC retreats towards the South like in December 80. Simultaneously the South Equatorial Counter current increases (between 2°S and 8°30 S in December 80 - Fig.1). Until now, due to ship program constraints, the sections have not been carried out during characteristic period (i.e. full SW or NE monsoon). During the transition periods (May and Oct - Nov) the thermocline ridge lies around 6°- 7° S except in May 81 (Fig.4). In December 80 at the beginning of the NE monsoon the ridge had retracted down to 8°30 S. At depth, in the central water, the ridge of the deep isotherms is somewhat displaced towards the south and could be as far as 11°S - 12°S. In the SECC region the depth variations of the thermocline is very small.

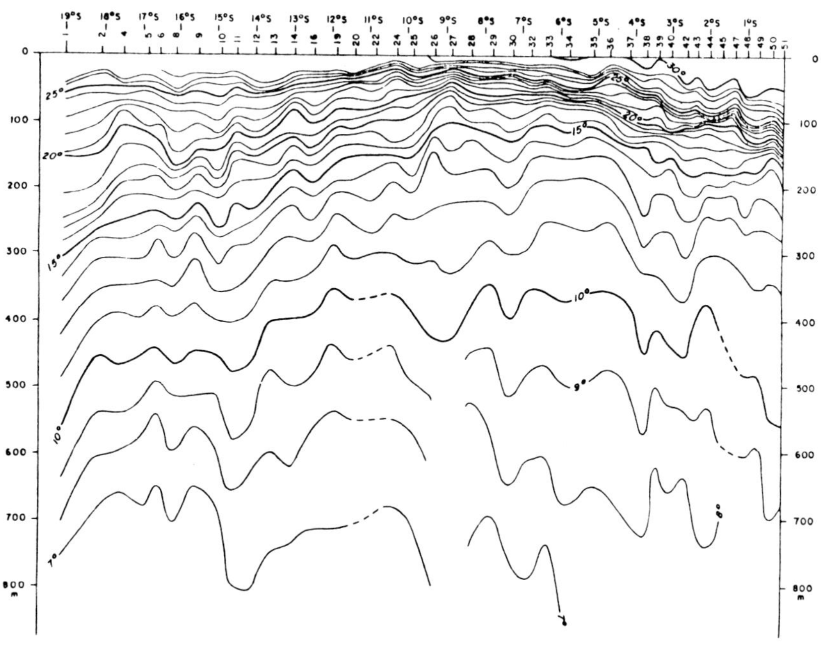

Fig. 10. South Equatorial Current sections (SINODE 5).

3. CONCLUSION

The preliminary results of the first eight SINODE XBT sections in the western Indian ocean show the high variability of the temperature structure in the Somali basin which can be related to the circulation variations. In that region, the surface salinities are a good indicator of the circulation together with the temperature sections. The northern front of the Somali eddy is always steeper than the southern one.

At the equator, during the existence of the eastward jet, the two transition periods present different temperature structures with a broader thermocline during the second transition period (Oct. Nov.Dec.) when the jet is faster.

South of the equator, the thermocline ridge associated with the northern limit of the South Equatorial Current, varies between 6°S and 9°S.

Sections during the characteristic periods (NE and SW monsoons) will permit a better description of the annual cycle.

The geostrophic calculations undertaken, using mean T-S relationships, will allow to quantify the circulation.

As the observations made on different years at the same season are not similar, a careful study of the correlation with the wind stress variations has to be done to further understand these different features.

Acknowledgments

This study is included in the SINODE program supported by the "Terres Australes et Antarctiques Françaises" (TAAF) and carried out on board their ship, the "MARION DUFRESNE", by the Laboratoire d'Océanographie Physique du Muséum National d'Histoire Naturelle, Laboratoire Associé 175 du CNRS, under the direction of Joseph GONELLA. The French "Marine Nationale" provided the 1979 XBT probes, the other ones were provided by the TAAF.

REFERENCES

Anderson,D.L.T.,1980 . The Somali Current. Ocean Modelling, No.34 (unpublished manuscript).
Bruce, J.G. 1968 . Comparison of near surface dynamic topography during the two monsoons in the western Indian Ocean. Deep-Sea Res., 15, 665-677.
Bruce, J.G., 1979 . Eddies off the Somali coast during the southwest monsoon. J. Geophys. Res., 84, 7742-7748.
Cresswell, G., M.Fieux and J.Gonella . The Wyrtki equatorial jet, May/June 1980. Tropical ocean-Atmosphere Newsletter - January 1981. Unpublished manuscript.
Düing, W.,1978 . The Somali Current. Past and recent observations. In : Review Papers of Equatorial Oceanography, Fine Workshop Proceedings. Nova/Nyit Univ. Press (unpublished manuscript).
Fieux,M., J.F. Murail, G.Soares, 1980 . Navigation et sections bathythermiques (XBT). Recueil des données collectées, campagne SINODE - MD 18 - PEMG . Vol.I. Rapp. Int. du Laboratoire d'Océanographie Physique du Muséum Paris. (Unpublished manuscript).
Fieux,M., M.Fremont, A. Kartavtseff, J.F. Murail, 1982. Navigations sections bathythermiques (XBT) et bouées dérivantes. Recueil des données collectées, campagne SINODE II-MD 22 - Vol.I. Rapp. Int. du Laboratoire d'Océanographie Physique du Muséum Paris. (Unpublished manuscript).
Leetmaa,A.and V. Truesdale, 1972. Changes in the currents in 1970 off the East African coast with the onset of the southeast monsoon. J. Geophys.Res., 77, 3281-3283.
Leetmaa, A., D.R. Quadfasel, and D.Wilson, 1982 : Development of the flow field during the onset of the Somali Current, 1979 in J. Phys. Oceanogr.
Luyten, J.R., M.Fieux, and J.Gonella, 1980 . Equatorial currents in the western Indian Ocean. Science, 209, 600-603.
Quadfasel, D.R. 1982. Low frequency variability of the 20°C isotherm topography in the western Indian Ocean. J.Geophys. Res., 87, C3, pp. 1990-1996.
Schott,F. and D.R. Quadfasel, 1980 . Development of the Northern Somali current gyre in 1979. Science. Vol, 209. Aug.1980. pp.593-595.
Schott,F., and D.R. Quadfasel, 1982. Variability of the Somali Current system during the monsoon onset, 1979 (to appear in J. Phys Oceanogr).
Swallow,J.C. and J.G. Bruce, 1966 . Current measurements off the Somali coast during the Southwest Monsoon of 1964. Deep Sea Res. 13-5, 861-888.
Swallow, J.C. and M.Fieux, 1982 . Historical evidence for two gyres in the Somali Current. J. Mar. Res. Vol.40, Suppl.pp.747-755.
Swallow, J.C., R.Molinari, J.G.Bruce, O.B. Brown. Development of near-surface flow pattern and water mass distribution in the Somali basin, in response to the southwest monsoon of 1979. To appear in J. Phys. Oceanogr.
Wyrtki, K. 1973. An equatorial jet in the Indian Ocean. Science,181 262-264.

SEASONAL VARIABILITY OF SUBSURFACE THERMAL STRUCTURE IN THE TROPICAL ATLANTIC OCEAN

J. MERLE [1]

[1]ORSTOM/LOP/MUSEUM
43 rue Cuvier 75005 PARIS

ABSTRACT

A merged data set including MBT, XBT, and NANSEN - CTD observations, made in the 20° S - 30° N tropical Atlantic region until 1978, is used to study the seasonal changes in the upper ocean thermal structure. Maximum changes in depth of the thermocline and in the 0-300 meters layer heat content are found in two regions : (i) North of the equator (until 10° N) and West of 25° W (ii) Central and Eastern equator. Annual signals of thermocline depth (and heat content) variations are not in phase in the two regions. The maximum depth of the thermocline is observed in fall in the North-West equatorial region and in Spring in the Eastern Central equatorial region. A double seasonal tilt of the thermocline is resulting of these changes in the two regions : one along the equatorial plan, the other one along a zonal pivot line situated under the mean position of the ITCZ. This double axis of rotation of the thermocline on a seasonal time scale seems to be the result of two kinds of response of the ocean to the wind forcing : (i) a remote equatorial response which mainly affects the Eastern equatorial region. (ii) A local North equatorial response associated to the seasonal migration of the ITCZ which induces a change in the sign of the curl of the wind stress and an up and down movement of the thermocline.

1) INTRODUCTION

The large seasonal and interannual changes of Sea Surface Temperature (SST) in the tropics are well known. Their effects on the atmosphere have been investigated recently by both observations and models (HOREL and WALLACE, 1981 - ROWNTRY 1976). The climatological effect of these SST variations over the continents seems also to be significant (SHUKLA 1975, MOURA and SHUKLA 1981).

The Pacific ocean draws first the attention because of the large interannual SST signal which affects its eastern basin (EL NINO) and because of the economical consequences of this phenomenon. More recently the tropical Atlantic ocean focussed also the attention because of the particularity of the SST variability which is mostly confined in the annual signal (MERLE, FIEUX, HISARD 1979). Some kind of "EL NINO" responses have been inferred, different of those of the Pacific but similar by their physical mechanisms (HISARD and MERLE 1979 ; HISARD 1980 ; MERLE 1980-a). The SST changes could be due in a first part, to the subsurface thermal structure changes. But little is known about this subsurface variability. We know that in both Pacific and Atlantic oceans there is an East-West equatorial slope of the thermocline. The mean depth of the 20°C isotherm is about 150 meters in the West and about 50 meters in the East. Salinity and density distribution present also an East-West slope (Fig. 1 a-b-c). We know that the large range of SST variability in the East of the equatorial basin is mostly due to the shallowing of the thermocline and its seasonnal surfacing. We know that in the West, even if the SST variability is small, the thermocline has the largest depth variation (MERLE 1980-b). A phase change of the annual signal of depth variation of the thermocline is observed between the West and the East of the equatorial basin suggesting a seasonal tilt of the thermocline in an equatorial plan around a pivot point situated in the middle of the basin (MERLE 1980-b).

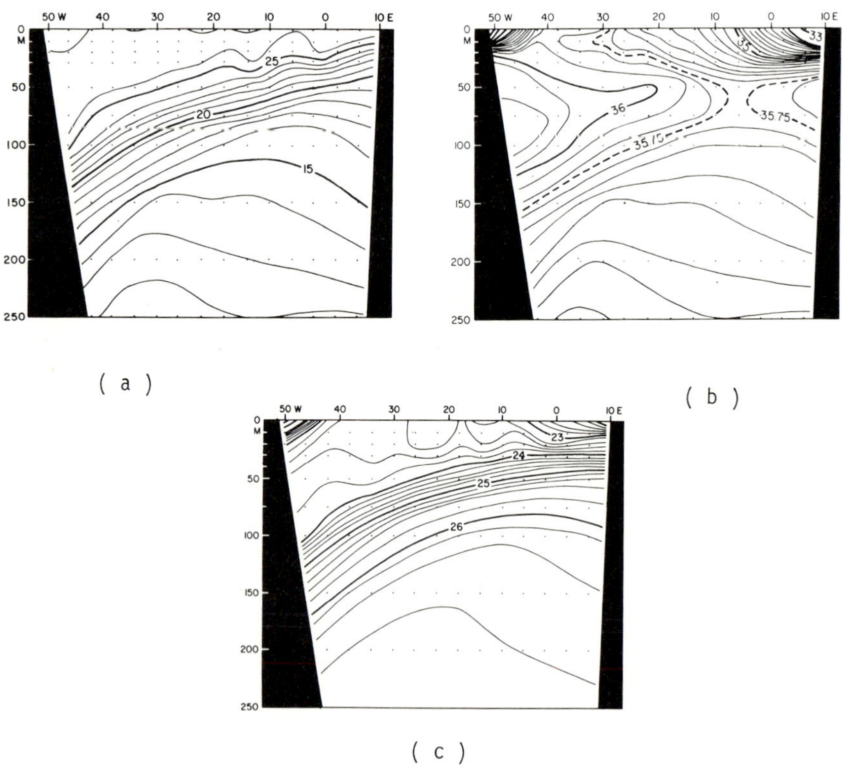

Fig. 1. Equatorial section of (a) temperature averaged from 0° to 2° S,
(b) salinity averaged from 2° N to 2° S,
(c) density (σ_t) averaged from 2° N to 2° S.

In order to understand the response of the upper tropical ocean to the atmospheric forcing, several sets of questions need to be investigated : (i) how does the thermocline vary seasonnally and interannually in the tropical ocean as a whole ? (ii) How does this thermocline variability affect the SST or does not affect it ? (iii) What is the relative importance of the different forcing mechanisms on SST and depth of the thermocline ? namely a) the local wind forcing ; b) the remote wind forcing ; c) the local thermodynamic forcing ?

The purpose of this study is to bring some new facts to the question : how does the thermocline vary seasonnally in the tropical Atlantic ocean ? We use a data file including all the MBT, XBT, and NANSEN observations available until 1978. These data and their processings are examined in section 2. Section 3 describes the mean thermal structure. Section 4 describes the seasonal variability of the depth of the thermocline (20° C isotherm). Section 5 presents some consequences of this thermocline variability upon heat content and heat budget. In Section 6, some preliminary conclusions of these facts are presented.

2) DATA PROCESSING

The data base used in this study is a merged file of MBT, XBT, and NANSEN temperature data. About 140,000 temperature profiles are considered between 20° S, 20° N, African coast and 80° W. Most of the profiles come from the NODC but additionnal XBT from the French Navy and NANSEN - CTD cast from the oceanographic vessel CAPRICORNE are also included. The file includes the oldest cruises like the METEOR in 1924, the ATLANTIS in 1931, the DISCOVERY in 1935, the CROWFORD in 1957 and the EQUALANT expedition in 1962-1963. Most of the GATE data are in the file but the file ends in 1978 and does not include the FGGE data. The data distribution is irregular in time and space. Nevertheless for the purpose of the large scale study presented here, a monthly mean with a large grid spacing (4° in longitude and 2° in latitude) brings a sufficient number of observations in each rectangle (usually over 30). The confidence interval is thus reduced to about.2°C. A weighted interpollation from adjacent squares and months has been applied in order to reduce the noise level in some square with scanty data or no data at all. Prior to this smoothing procedure, various screening routines have been applied to eliminate conspicuous data.

The monthly mean temperature is calculated in each box at each standard level. Each type of data (MBT, XBT, NANSEN) are weighted by their own standard deviation and frequency. The depth of a given isotherm is interpollated linearly. Heat content and rate of heat storage are calculated as following : the relative heat content at a depth Zn and with reference to 0° C, is caculated by :

$$HS(Z_N) = \frac{1}{10} J\bar{D}C_p \sum_{i=1}^{N} (T_i - T_{i-1})(Z_i - Z_{i-1})$$

with Ti, the temperature at each standard level in degrees centigrades ; Zi, the depth of standard levels in meters ;
Cp, the heat capacity of sea-water ;
$\bar{D}$, the mean density of the water column.
$J\bar{D}Cp$ has been taken equal to 0.4096. Then $HS(Z_N)$ is given in Joule per square meter. In this study $Z_N \doteq 300$ m or 50 m. The rate of heat storage HS for a given month, is obtained through the difference of heat content between two adjacent months. For example, HS July = (HS August - HS June) / 2 ; HS is converted into watt per square meter.

3) MEAN THERMAL STRUCTURES

Dynamic topography of the surface (Fig. 2a), heat content into the 0-300 meters layer (Fig. 2b), and depth of the thermocline of the 20° C isotherm (Fig. 2c) present similar feature in the equatorial Atlantic ocean. In the regions where dynamic topography is low, as it is along the North equatorial trough near 10° N or along the equator, the heat content is low and the thermocline is shallow. Between these two lows a North equatorial dynamic ridge along 3-4° N is characterized by a maximum of heat content and by a trough in the thermocline. Furthermore thermal structures present a zonal slope. Dynamic height and heat content are higher in the West where the thermocline is deeper than in the East.

These features indicate clearly that the heat content is directly function of the depth of the thermocline. The same remark is valid for oceanic topography (and consequently the geostrophic circulation) of the upper layers. This is obviously a result of the sharp and thin thermocline layer that is almost permanent in the tropics. This thermocline is separating a well mixed and warm superficial layer from the deep cold waters layer. The result is an almost perfect two-layers ocean (Fig. 3).

This peculiar aspect of the near surface thermal structures of the tropical oceans simplifies considerably the study of the response of the superficial ocean to the atmospheric forcing. Characteristics of this response can be divided in two parameters : (i) the depth of the thermocline, (ii) the temperature of the mixed layer (or of the Sea Surface).These two paramaters are probably not

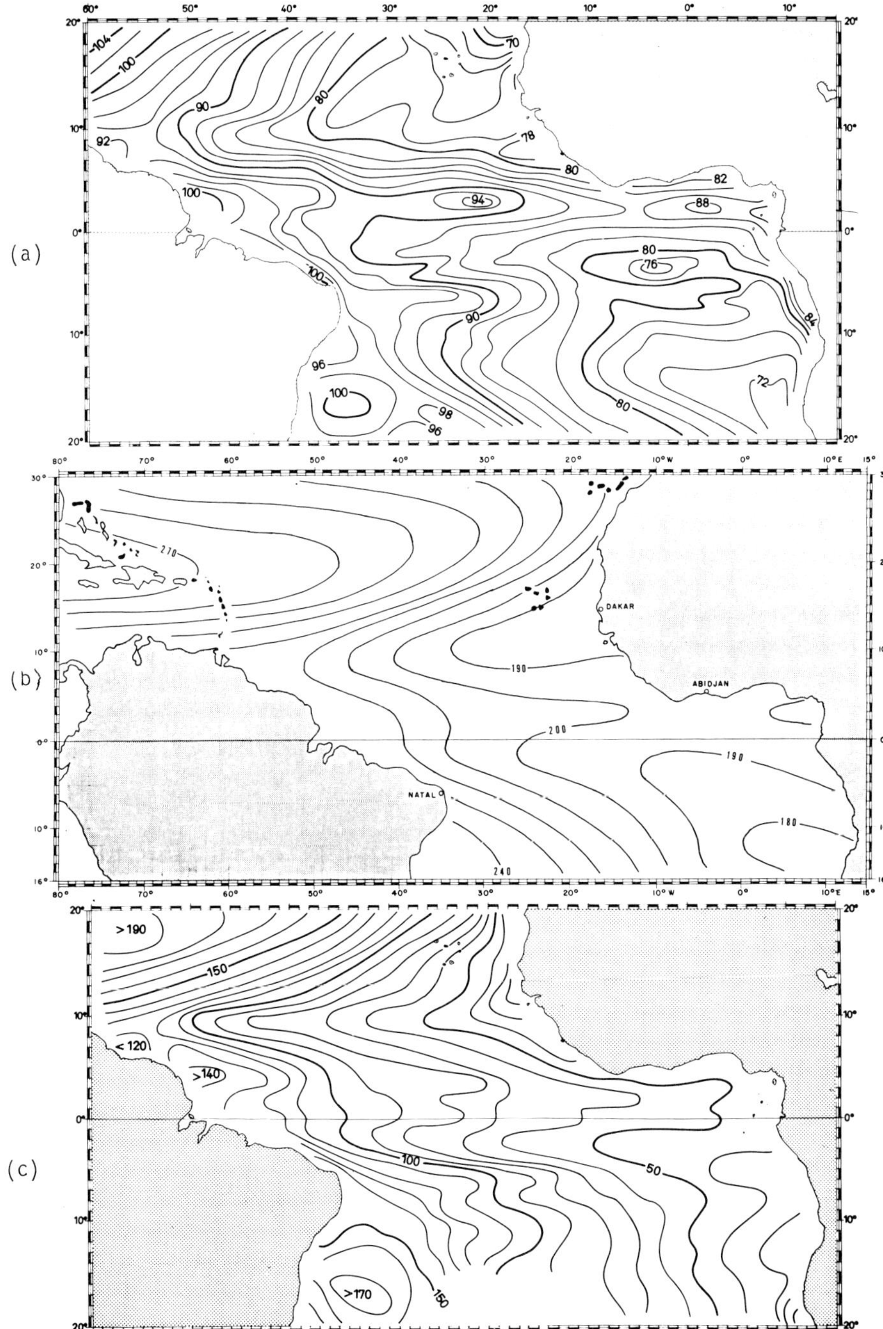

Fig. 2 : a) Surface dynamic topography relative to 500 DB.
b) Mean annual heat content from 0 to 300 meters depth (in 10^8 J/m2).
c) Mean annual depth of the 20° C isotherm.

totally independent especially in the upwelling regions. In a first approximation, the depth of the thermocline is a direct response of the ocean to the wind forcing (either local, remote or global). The temperature of the mixed layer is the result of the response of the superficial ocean to the local thermodynamic forcing of the atmosphere. These simplified views apply especially in the western regions where the thermocline is deep. In the East and during the upwelling seasons the thermocline reaches the surface. This is creating particular surface conditions (cold water) that in turn modify the conditions of the local thermodynamic forcing (the ocean absorbs heat from the atmosphere). Here the two forcings and the two corresponding responses of the ocean (depth of the thermocline and temperature of the mixed layer) are clearly not independent.

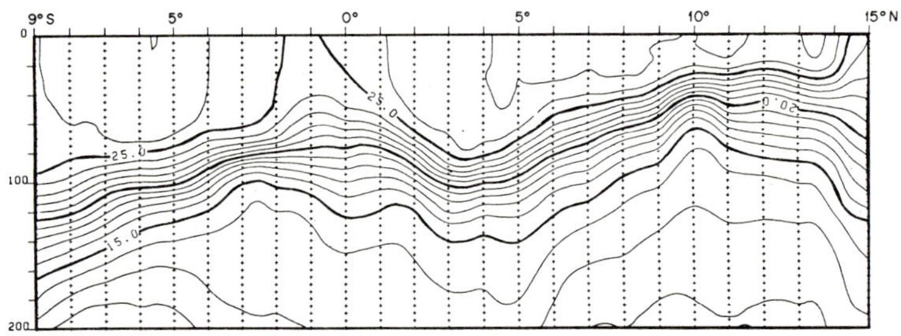

Fig. 3 : Meridional temperature section along 23.5° W during GATE (June-September 1974).

In this study we will consider only the variability of the depth of the thermocline on a seasonal time scale. We assimilate this depth to the depth of the 20° C isotherm which is approximatively in the middle of the thermocline (see Fig. 1 a and Fig. 3). We assume that the interannual variability of the subsurface thermal structures is weak comparatively to the seasonal variability. If this interannual variability is large we are in serious danger of important biases due to the unequal yearly distribution of the data. But we know that the SST interannual variability is about 1 to 5 times the seasonal variability (MERLE et al. 1979). The consistency of the results presented in the following sections gives confidence in their significance.

4) SEASONAL VARIABILITY OF THE 20° C ISOTHERM TOPOGRAPHY

Fig. 4 a and 4 b show the topography of the 20° C isotherm during March and August. During March the structure is simpler than during August. The 20° C isotherm is deep in the North West, reaching 200 meters in a trough around 20° N. The sloping isotherm moves upward slowly until 40° W and more rapidly further east to reach the surface near the African coast. Further South, from 10° N to 15° S, the topography does not show any particular pattern. There is a general East-West slope and a slight doming extending to the south of the equator from the South-East to the West and underlining the Benguela Current and the South-Equatorial Current.

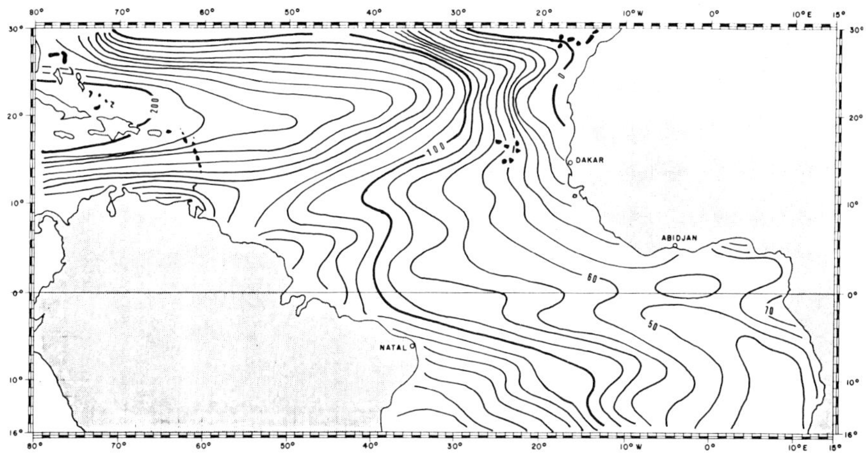

Fig. 4 a : Depth of the 20° C isotherm in March (meters).

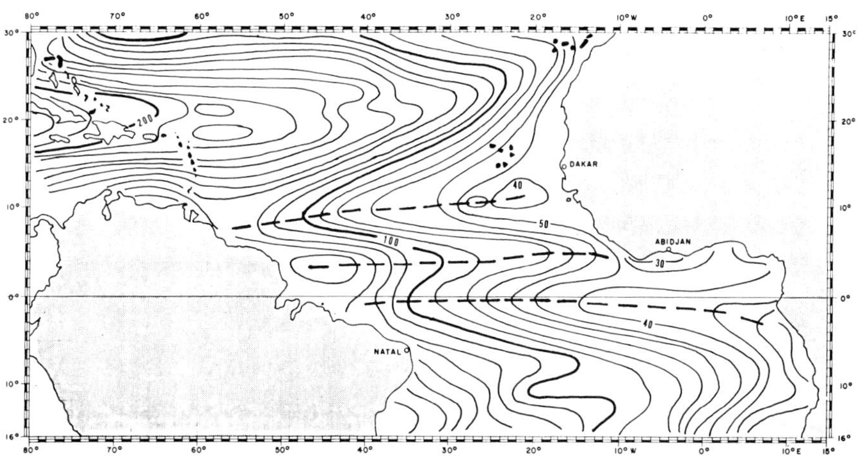

Fig. 4 b : Depth of the 20° C isotherm in August (meters).

In August the pattern is totaly different. Three axes of deformation appear. A central trough is extending at about 3-4 degrees North. This trough is bordered by two ridges. The North ridge, centered at about 10° N, is to the South of the North Equatorial Current (NEC). The South ridge is lying along the equator. Between the North ridge and the 3-4° N trough flows the North Equatorial Counter-Current (NECC). And between the 3-4° N trough and the equatorial ridge flows the northern branch of the South Equatorial Current (SEC).

March and August are the two extreme months of the seasonal variability but progressive changes are observed between these two periods of the year. The North equatorial trough which is the dominant feature of the Northern Summer in the 20° C isotherm topography, begins to appear in June and ends in November (maps are not shown here).

A Fourier analysis has been performed to derive maps of the maximum change of depth and phase of the Annual Component (Fig. 5 a et b). The depth change of the 20° C isotherm shows a large maximum at the North-West of the equator between 3° - 8° North and 25° - 45° West. The difference of depth is over 60 meters. At the other end of the ocean, South of the equator and close to the African coast, another maximum appears (over 50 meters depth change). These two maxima are associated along the equator with a band of relatively large depth variation (over 40 meters). North of 20° N and South of 10° S the change of depth is also increasing rapidly. But it is a different phenomenon in regions where the thermocline is less marked and where the 20° C isotherm could even reach the surface.

The 15° N - 10° S band is marked by important phase changes. A first phase change is observed along the equator from the Gulf of Guinea in the East to the North Brasil in the West. In the Gulf of Guinea the maximum depth of the 20° C isotherm is observed during the first months of the year (from the end of February near the African coast to April-May at 10° W). Between 15° W and 20° W the phase change is very rapid (about 180 days change over 5° of longitude). Further West until the Bresilian coast, the maximum depth of the 20° C isotherm is found in October-November. Thus a tilt of the thermocline with a pivot point around 25° W confirms a previous result obtained with a different and smaller data set (MERLE 1980 b).

The second phase change is observed along a line standing roughly from 3 - 4° N in the West to about 10° N in the East. This phase change is also very rapid and large (about 180 days). This suggests a meridional tilt of the thermocline along a pivot line approximatively under the mean position of the ITCZ. We will discuss this point later on.

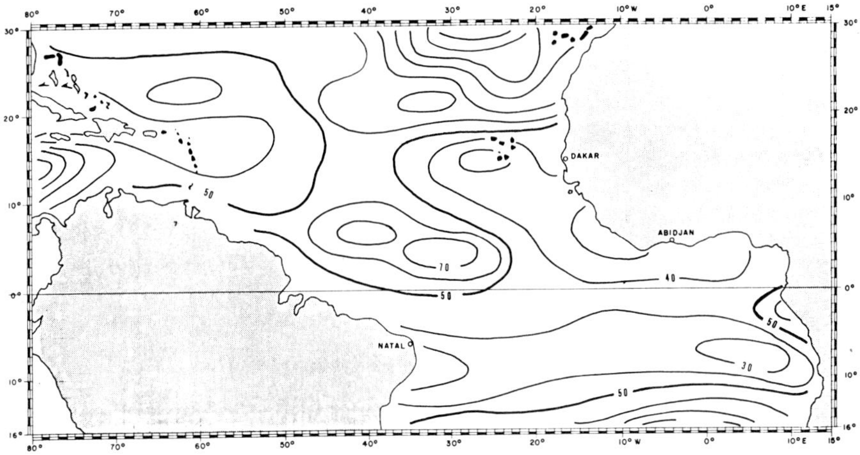

Fig. 5 a : Depth change in meters of the 20° C isotherm (approximatively twice the amplitude of the annual component of a Fourier analysis).

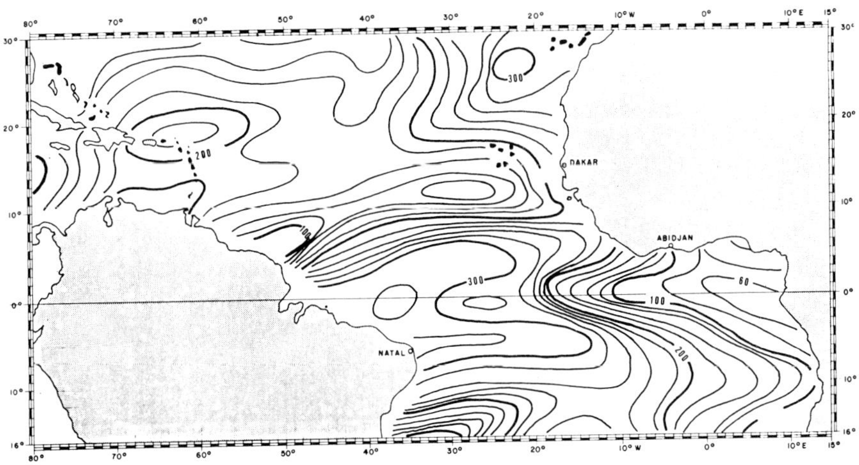

Fig. 5 b : Phase (in days) of the annual component of the 20° C isotherm depth in a Fourier analysis .

5) SEASONAL CHANGES OF HEAT CONTENT

A previous study of heat budget in the equatorial Atlantic ocean (MERLE 1980 b) showed that seasonal variations of heat content in the West and in the East of the equatorial basin are different (Fig. 6 a and b). In the West the 0-300 meters heat content signal is larger than the one of the East. More over the two signals are not in phase. Conversely, the 0-50 meters heat content signal is smaller in the West than in the East but they are in phase. These results indicate that in the West, and at least at the equator, the subsurface thermal structures vary seasonally, independently of the overall surface conditions.

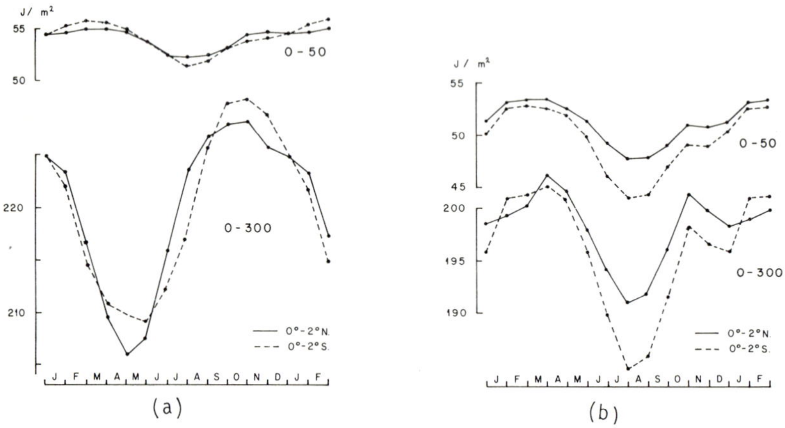

Fig. 6 a) : Seasonal variation of heat content in the West equatorial Atlantic 0-2° N and 0-2° S from 0 to 300 meters and 0 to 50 meters (from MERLE 1980).
b) : Seasonal variation of heat content in the East equatorial Atlantic (0-2° N and 0-2° S) from 0 to 300 meters and 0 to 50 meters (from MERLE 1980).

Amplitudes and phases of the annual component of heat content for both the 0-50 meters and 0-300 meters layers indicate that in the Western part of the whole intertropical region (20° N - 15° S) the subsurface (0-300 meters) thermal conditions are independent of the near surface (0-50 meters) thermal conditions (Fig. 7 a-b and Fig. 8 a-b).

The annual variation of the 0-300 meters heat content layer and of the depth of the 20° C isotherm are very similar when amplitude and phase are concerned. The 0-300 meters heat content shows a large maximum of amplitude in the West and at the North of the equator with values close to 6.10^8 J/m2. An eastern equatorial maximum is also observed with values over 10.10^8 J/m2 at the African coast and south of the equator. The phase map shows also a phase difference between the Eastern and the Western part of the equatorial band with a transition point at about 15° W. Another change phase is observed along an almost zonal line near 6-8° N which is very similar to the one described for the 20° C isotherm depth.

The annual variation of the 0-50 meters heat content (Fig. 8 a - b) is similar to the SST annual variation in both amplitude and phase (Fig. 9 a - b). The maximum of amplitude is found just South of the equator in the Gulf of Guinea with a maximum over 5 10^8 J/m2 near the African coast. A zone of minimum (less than 2 10^8J/m2) is lying under the mean position of the ITCZ. The phase map shows a slight increase in the date of the maximum of heat content from the Gulf of

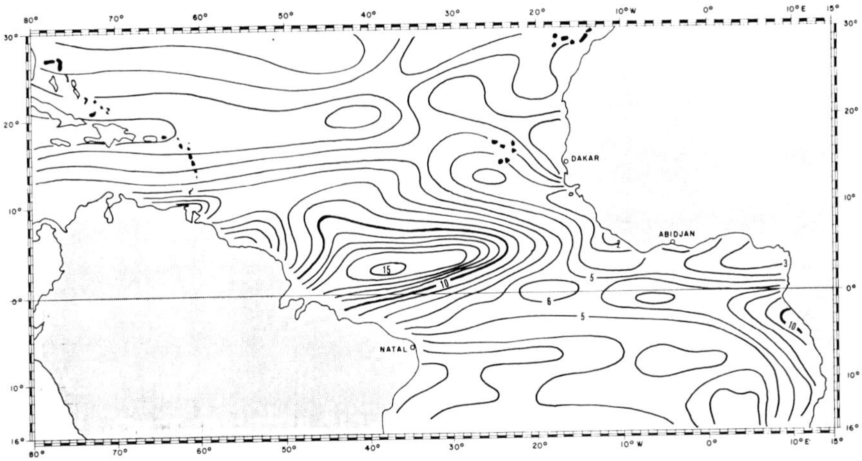

Fig. 7 a : Amplitude of the annual component of heat content from 0 to 300 meters (in 10^8 J/m2).

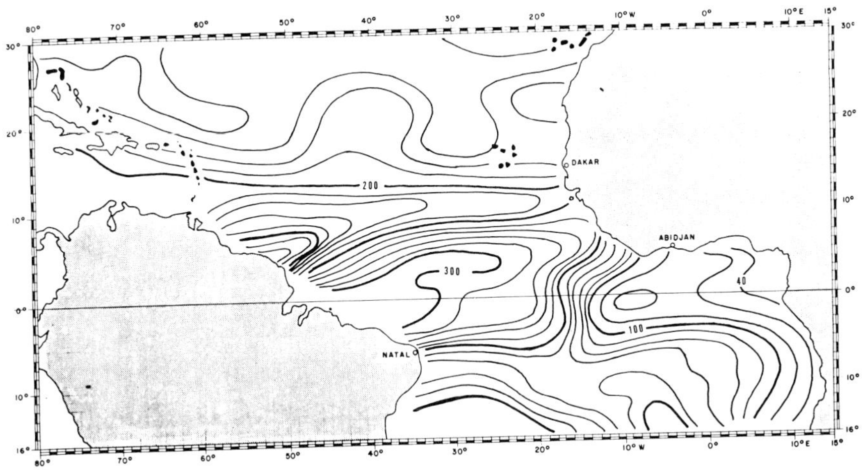

Fig. 7 b : Phase of the annual component of heat content from 0 to 300 meters (in days).

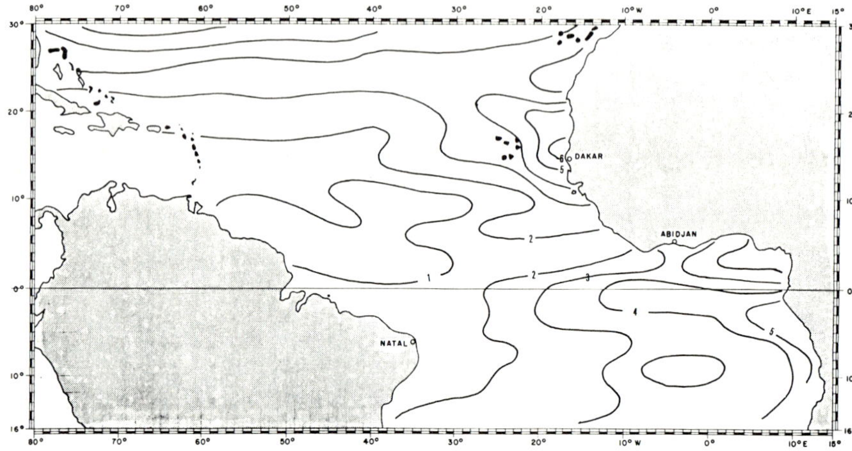

Fig. 8 a : Amplitude of the annual component of heat content from 0 to 50 meters (in 10^8 J/m2).

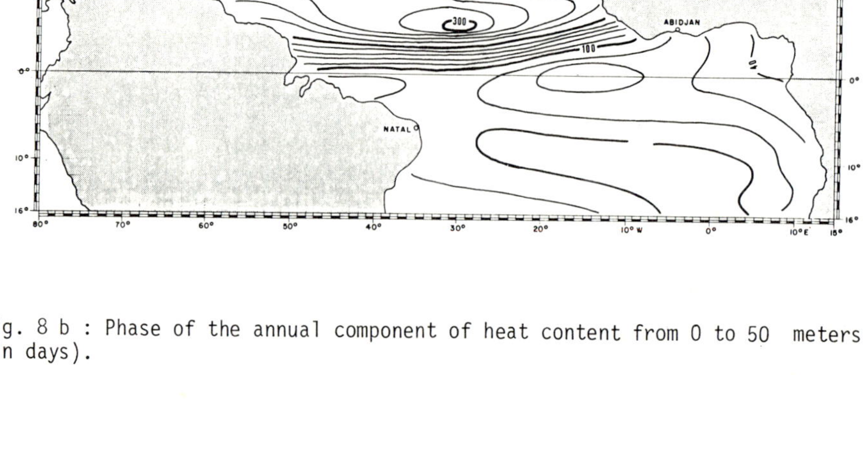

Fig. 8 b : Phase of the annual component of heat content from 0 to 50 meters (in days).

Guinea toward the West, at the equator. The phase change is observed under the mean position of the ITCZ as for the SST phase change (Fig. 9 b).

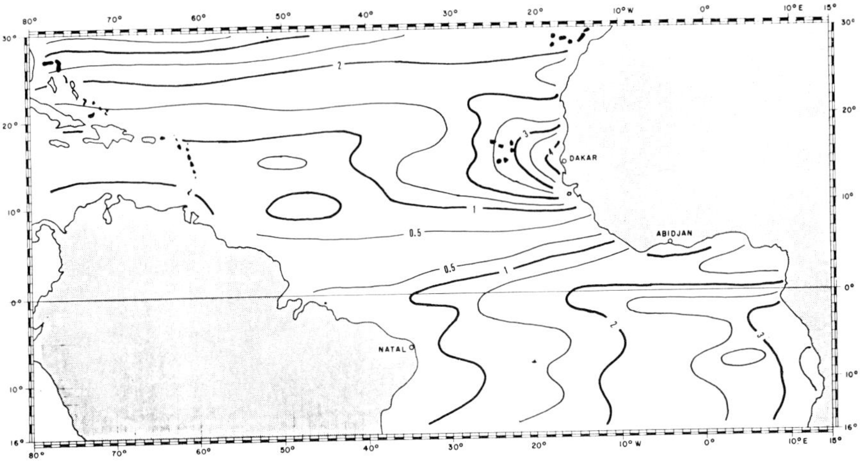

Fig. 9 a : Amplitude of the annual component of Sea Surface Temperature (in degrees C)

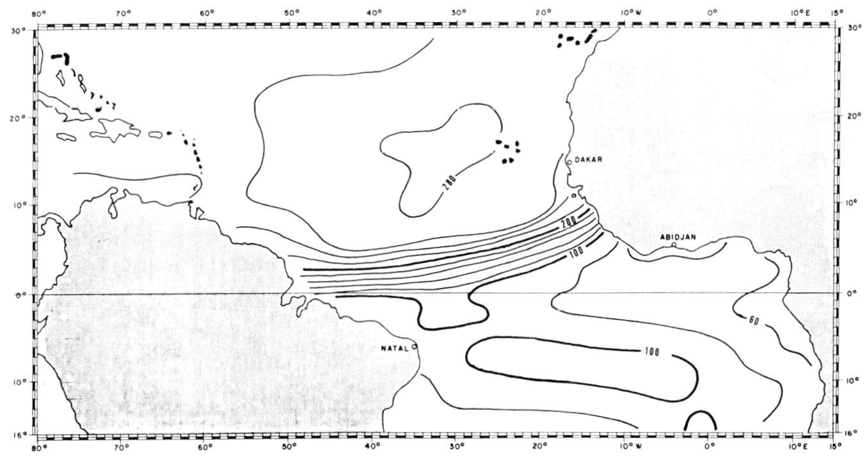

Fig. 9 b: Phase of the annual component of Sea Surface Temperature (in days).

In order to differentiate the surface and subsurface oceanic energy exchanges, we computed the rate of heat content change for two layers: the layer including the thermocline (0-300 meters) and the surface layer (0-50 meters). We selected two months : June and September. These two months frame the equatorial upwelling which is the major event of the tropical ocean and peaks in July-August. Fig.

10 a presents the rate of heat storage (change of heat content) in the 0-300 meters layers during June. Positive values extend North of the equator where the thermocline is depressed, negative values are covering the whole Gulf of Guinea where the thermocline is upwelling. The extrema are over 300 W/m2. Comparing these numbers with the net heat exchange at the surface during the same month obtained by HASTENRATH and LAMB (1978), we see that the rate of heat storage of the 0-300 meters layer is about ten times the heat exchange at the surface. (The maximum value of heat gain obtained by HASTENRATH and LAMB (1978) is confined along a zonal band south of the equator with a value close to 50 W/m2). In September (Fig. 10 b) the situation is reversed. Positive values (over 300 W/m2) extend along the equator in the Gulf of Guinea where the thermocline is downwelling.

In the 0-50 meters layer (Fig. 11 a - b) the features are different : the changes in heat content are confined in the East (Gulf of Guinea and Dakar region). In June, the rate of heat storage is negative with extrema over 180 W/m2 in the Gulf of Guinea. The Dakar region, in reverse shows positive values (over 160 W/m2). It is interesting to note that, in the Dakar region, the rate of heat storage is larger in the 0-50 meters layer than in the 0-300 meters layer and of opposite sign. When the sun is heating the mixed layer, the deep cold waters are upwelling. In September the rate of heat storage in the 0-50 meters layer is positive in the Gulf of Guinea (over 160 W/m2 at the African coast and South of the equator), and negative in the Dakar region (less than 60 W/m2).

6) CONCLUSION

The seasonal signal of the subsurface thermal variability of the tropical Atlantic ocean presents a maximum amplitude in the region North of the equator (3 - 8° N) and West of 30° W. This maximum variability appears in the depth of the 20° C isotherm and in the heat content of the 0-300 meters layer. A secondary zone of maximum seasonal variability extends all along the equator in the central and East part of the basin with a net increase in the East near the African coast.

The seasonnal signals in these two regions (North-West of equator and South East of equator) are not in phase. The maximum depth of the thermocline (and maximum of heat content) is observed during Spring time (February-April) in the Eastern equatorial region, and in Fall (October-December) in the North-West equatorial region. Thus a phase difference of about 6-8 months separates the two events.

The resulting seasonal movement of the thermocline induces a double tilt of the thermocline : one in an equatorial plan with a pivot point around 25° W ; another one, along a pivot line approximatively parallel to the equator situated

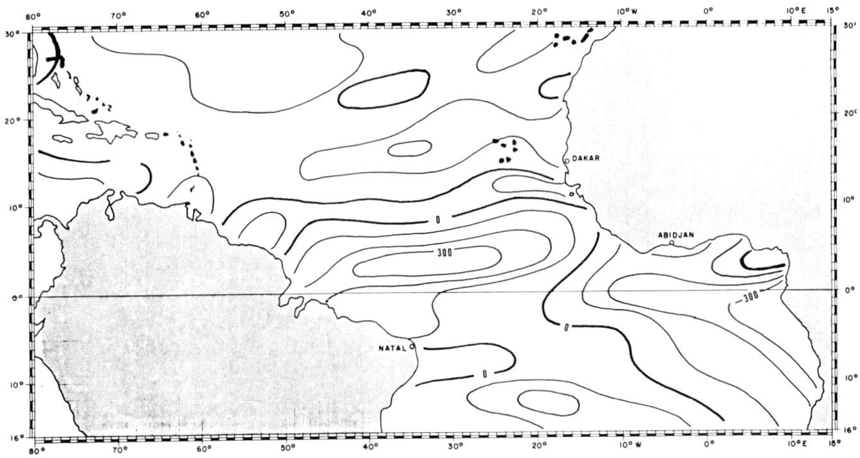

Fig. 10 a : Rate of change of heat content from 0 to 300 meters in June (in W/m2).

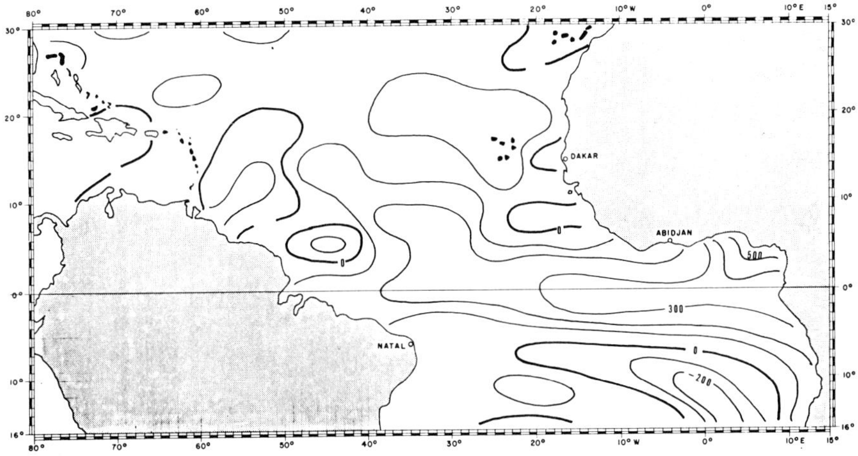

Fig. 10 b : Rate of change of heat content from 0 to 300 meters in September (in W/m2).

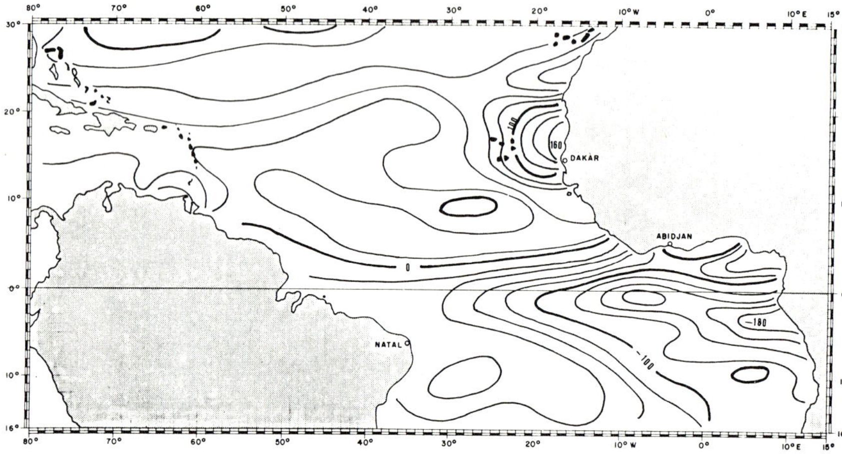

Fig. 11 a : Rate of change of heat content from 0 to 50 meters in June (in W/m2)

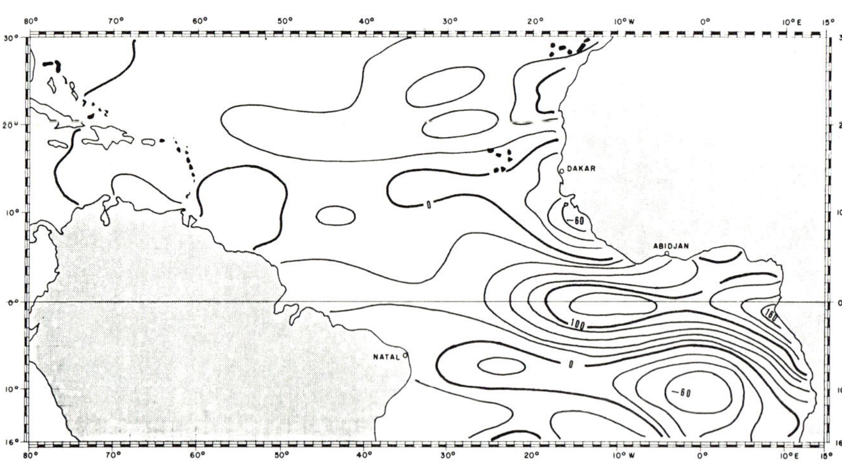

Fig. 11 b : Rate of change of heat content from 0 to 50 meters in September (in W/m2).

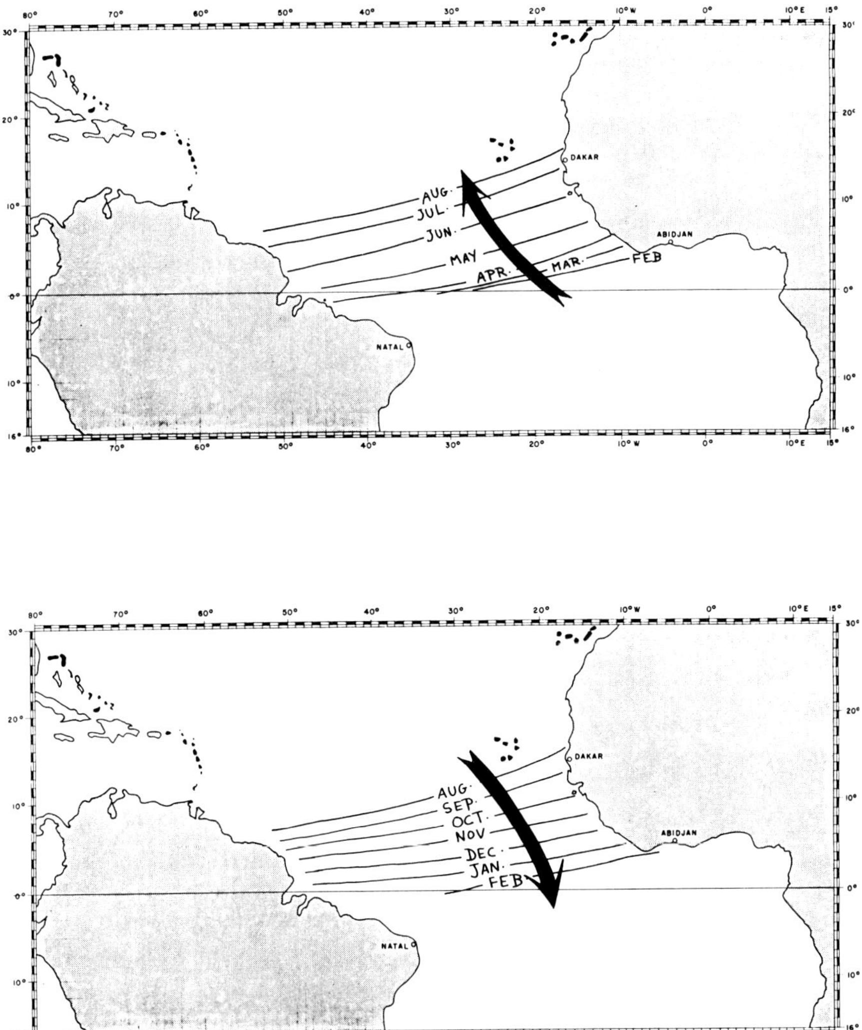

Fig. 12 a and b : Seasonal migration of the ITCZ obtained from the atlas of HASTENRATH and LAMB (1977).

between 3 and 8° N. The zonal tilt of the thermocline at the equator was already known (MERLE 1980b) and confirmed by equatorial models (CANE and SARACHIK 1980 PHILANDER and PACANOWSKY 1981). The meridional North equatorial tilt of the thermocline is a new fact brought up by this study.

This meridional seasonal oscillation of the thermocline is situated approximatively under the mean position of the Intertropical Convergence Zone (ITCZ). The ITCZ position varies seasonally over a large latitudinal band which extends roughly from the equator to 10-15° North (Fig. 12 a - b) and covers the oceanic regions where maximum amplitude and phase changes in the depth of the thermocline are observed. The curl of the wind stress changes sign North and South of the ITCZ. Thus, we can easely speculate that this seasonal up and down movement of the thermocline associated to the latitudinal migration of the ITCZ results directly from the change in the sign of the curl of the wind stress. Recent calculations using a monthly mean wind data set confirm this hypothesis (DELCROIX, personnal communication). A linear two layer model using the same wind data set provides also an amazing similarity with our results in both phase and amplitude changes of the annual component of the thermocline depth variation (PICAUT and BUSALLACHI 1982).

Thus it appears that the seasonal variation of the upper thermal structures in the tropical Atlantic ocean are mostly due to the combination of two events : (i) an equatorial perturbation which affects the Eastern equatorial region ;(ii) a North equatorial perturbation due to the local response of the subsurface ocean to the migration of the ITCZ.

ACKNOWLEDGEMENTS

This study belongs to the FOCAL program and has been partly supported by CNRS /PNEDC research contract. We thank the BNDO (CNEXO) for part of treatment. Yves TOURRE and Thierry DELCROIX have provided helpful comments and corrections.

REFERENCES

Cane M.A., and E.S. Sarachik, 1981 : The response of a linear baroclinic equatorial ocean to periodic forcing. J. Mar. Res., 39, 651-693.

Hastenrath S. and Lamb P.J., 1977 : Climatic atlas of the tropical Atlantic and Eastern Pacific oceans. The University Wisconsin Press, 112 p.

Hastenrath S. and Lamb P.J., 1978. Heat budget Atlas of the Tropical Atlantic and Eastern Pacific Ocean. The University Wisconsin Press, 140 p.

Hisard P. and MERLE J., 1979. Onset of summer surface cooling in the Gulf of Guinea during GATE, Deep Sea Res., GATE, Suppl. II to V, 26, 325-342.

Hisard P., 1980 : Observation de réponses de type "El Nino" dans l'Atlantique tropical oriental-Golfe de Guinée. Ocean. Acta, 3, 69-78.

Horel J.D., and J.M. Wallace, 1981 : Planetary-scale atmospheric phenomena associated with the Southern Oscillation. Mon. Wea. Rev., 109, 813-829.

Merle J., M. Fieux and P. Hisard, 1979 : Annual signal and interannual anomalies of sea-surface temperatures in the eastern equatorial Atlantic ocean. Deep-Sea Res., GATE Suppl.II to V, 26,77-102.

Merle J., 1980a : Variabilité thermique annuelle et interannuelle de l'océan. Atlantique équatorial Est. L'hypothèse d'un "El Nino" Atlantique. Ocean. Acta, 3, 209-220.

Merle J., 1980b : Seasonal heat-budget in the equatorial Atlantic Ocean, J. Phys. Oceanogr. 10, 3, 464-469.

Moura A.D., and J. Shukla, 1981 : On the dynamics of droughts in northeast Brazil : Observations, theory and numerical experiments with a General Circulation Model. J. Atmos. Sci., 38, (to appear).

Picaut J. and A.J. Busalacchi, 1983 : Seasonal variability from a model of the tropical Atlantic. (To appear in J. Phys. Oceanogr.).

Philander S.G and R.C Pacanowski, 1981 : Response of equatorial oceans to periodic forcing. J. Geophys. Res., 86, 1903-1916.

Rowntree P.R., 1976 : Response of the atmosphere to a tropical Atlantic Ocean temperature anomaly. Q.J.R. Meteorolo. Soc., 102, 607-625.

Shukla J., 1975 : Effect of Arabian sea-surface temperature anomaly on Indian summer monsoon : A numerical experiment with the GFDL model. J. Atmos. Sci., 32, 503-511.

THE VARIABILITY OF LOCAL WINDS AT 22°W AND THEIR INFLUENCE ON THE OCEANIC SYSTEM AT THE EQUATOR IN THE ATLANTIC DURING FGGE

P. SPETH and H.-J. PANITZ
Institut für Geophysik und Meteorologie
der Universität zu Köln
Kerpenerstr. 13
D-5000 Köln 41
Federal Republic of Germany

ABSTRACT

Local wind forcing is a possible mechanism which drives fluctuations of the oceanic system in the equatorial region. Local winds in the equatorial belt between 1°S and 3°N along 22°W are analysed by spectral methods for two periods in 1979. For the first period from February to March the wind was strongly influenced by the ITCZ. For the second period from May to June we found fluctuations of the zonal wind component in the 7 day range. These oscillations were meridionally coherent. The meridional wind showed fluctuations in the 3-4 day range which could be identified as easterly waves. We found a coherent relationship between the zonal wind and the zonal currents at the 7 day period at the Equator. A similar relationship could not be found between the atmospheric easterly waves and oceanic oscillations of the same period. The upwelling events, one in the February/March period and one at the end of May, are not connected with the local wind.

1 INTRODUCTION

The upper equatorial Atlantic reacts relatively quick to changes in atmospheric forcing. In the past local wind forcing was made responsible for some observed oceanic oscillations by various authors. We have used measurements along 22°W between 1°S and 3°N during the First GARP (Global Atmospheric Research Program) Global Weather Experiment (FGGE) in 1979 to investigate such a relationship. In the following section we summarize what is known about the equatorial current system in the Atlantic in so far as a wind forcing is concerned. The data are described in Section 3 and in Section 4 results of spectral analysis follow. A conclusion is given in Section 5.

2 THE EQUATORIAL CURRENT SYSTEM IN THE ATLANTIC OCEAN

Since Neumann (1960) and Voigt (1961) the Equatorial Undercurrent (EUC) is a well-known feature of the Atlantic Ocean. As described by Katz et al. (1981) the EUC is an ever present eastward zonal flow, about 2oo km wide and centered within 100 km of the Equator, which flows counter to the usually prevailing trade winds. Though not always a subsurface flow, the core of the EUC is generally to be found between 50 m and 125 m depth in the upper part of the thermocline. The EUC carries high salinity water with a maximum above the velocity core (e.g. Cornus and Meincke, 1979). The mean depths of the velocity and salinity maxima decrease monotonically from the western part of the Atlantic to the east. Only on the average the EUC is overlaid by the westward flowing South Equatorial Current (SEC).

From recent investigations, which are mainly results of the FGGE, it seems reasonable to us to conclude that the western equatorial Atlantic is in close equilibrium with the seasonally varying winds (Katz et al., 1981; Lass et al., 1982), which is in contrast to the Pacific. To demonstrate this Katz and Garzoli (1982) compared observations from FGGE directly to predictions derived from a zonally wind-forced nonlinear model of Philander and Pacanowski (1980). Katz and Garzoli (1982) conclude that the oceanic cycle of the region can be reasonably described by two sequential events: the response to a westward zonal wind which begins in mid-May and the response to the cessation of that wind after January. With the onset of the wind the thermocline deepens by about 40 m. Connected with this is an initially westward, but soon near-zero, surface current and an EUC which rapidly levels off. With the relaxation of the wind the thermocline rises and the surface flow becomes eastward, and the undercurrent transport first increases to its annual maximum in March and then decreases to its minimum just after the wind resumes again. The minimum undercurrent transport lags the minimum zonal wind stress by about two months (Katz et al., 1981).

For the eastern Atlantic (Gulf of Guinea) several mechanisms have been proposed to explain the equatorial and coastal upwelling. Servain et al. (1982) analysed historical sea surface temperatures (SST) and surface winds. Their results support the remote forcing mechanism proposed by Moore et al. (1978). According to this hypothesis increasing easterly winds in the western equatorial Atlantic excite an internal equatorial Kelvin wave that propagates into the Gulf of Guinea. Servain et al. (1982) point out, that an anomaly of the zonal wind stress near the north Brazilian coast is followed by a SST anomaly in the Gulf of Guinea about one month later, which implies a wave propagation speed of $\sim 1\ ms^{-1}$.

In the equatorial belt low frequency energetic fluctuations with time scales from a few days to a few weeks are superimposed on the processes as mentioned above (e.g. Moore and Philander, 1977; Weisberg et al., 1979). The most prominent of these wave-like disturbances are inertia-gravity waves and a north-south meandering of the EUC. For the former Garzoli and Katz (1981) find for the western tropical Atlantic variable spectral density levels in the band of 2-5 days centered at 3.75 days. The most likely interpretation of this spectral peak is an atmospherically forced, vertically propagating inertia-gravity wave with a vertical wavelength of $\sim$730 m. The westward propagating meander of the EUC with a period of $\sim$16 days, a phase speed of $\sim 1.7\ ms^{-1}$ (Düing et al., 1975) and a wavelength around 2,000 km (Düing et al., 1975: 2,600 km; Düing and Hallock, 1979: 1,800 km) could also be observed during FGGE (Lass et al., 1980). Additionally Düing and Hallock (1979) observed, that these long oscillations of the EUC have a lag of $\sim$4 days relative to the surface currents. This meander is believed to be caused by instabilities of the equatorial current system (Philander, 1976; Philander, 1978) or by wind-generated equatorially trapped waves (Hallock, 1979).

The seasonally varying belt of relatively low SST along the Equator (Merle, 1978; Hastenrath and Lamb, 1977) are often attributed to upwelling caused by the equatorial f-divergence, which we consider to be a more local effect. On the other hand these low SST could be caused by a second mechanism: the meridionally increased vertical mixing due to the high current shear between EUC and SEC. But Fahrbach and Bauerfeind (1982) demonstrated that this mechanism is not responsible for local temperature decrease - at least at 22°W at the Equator.

3 DATA

In the preceeding section we mentioned the possible relationship between the atmospheric flow and the current system in the equatorial Atlantic. We intended to find out which of these results can be verified by *in situ* measurements of combined oceanic and meteorological measurements. For that purpose it was planned to deploy 5 surface moorings by R.V. METEOR over an entire 6 month period along 22°W between 1°S and 3°N. However, because of technical problems, we got measurements only of two separate time periods. The times of surface wind recordings are given in Table 1. For that reason we have to restrict ourselves to an examination of relatively short period fluctuations and we will begin with a discussion of the data (compare also Garzoli et al., 1982).

TABLE 1

Times of surface wind recording during 1979 at 22°W.

3°N	01.02. - 29.03.	
2°N	01.02. - 09.03.	05.05. - 17.06.
1°N	31.01. 30.03.	
Eq.	18.02. - 07.03.	07.05. - 16.06.
1°S	30.01. - 28.03.	

The oceanographic equipment consisted of Vector Averaging Current Meters (VACM: sampling interval 7.5 min.) at 15 m depth at 1°S, Equator, 1°N and 3°N for the first period and at the Equator for the second period. Below the depth of 15 m Aanderaa current meters were used down to 600 m (sampling interval 30 min.). For 2°N there were no current measurements in the near surface layer for both periods. Additionally some moorings were equipped with thermistor chains.

Wind speed and wind direction were registrated by Aanderaa datalogger, with sampling intervals of 10 minutes or 20 minutes, which depends on the equipment. From these data hourly averages for the zonal and meridional wind component were calculated. The zonal (u: positive to the east) and the meridional (v: positive to the north) wind component along 22°W at 1°S, Equator, 1°N, 2°N and 3°N for the February/March period are shown in Figure 1.

Figure 3 gives a similar representation for the Equator and 2°N for the May/June period. Hourly averages are displayed.
Figures 2 and 4 show the daily variation of the magnitude of the mean wind vector, $(\bar{u}^2+\bar{v}^2)^{1/2}$, to the mean magnitude, $\overline{(u^2+v^2)^{1/2}}$. An overbar represents a daily average. The steadiness is expressed as a percentage; it equals zero for winds which shift randomly, and one hundred for winds that always blow in the same direction.

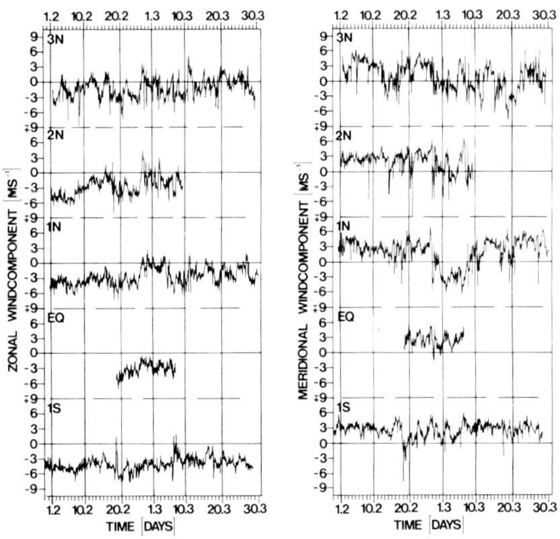

Fig. 1.
Time series of zonal (left) and meridional (right) wind components (ms^{-1}) along 22°W between 1°S and 3°N for February/March 1979.

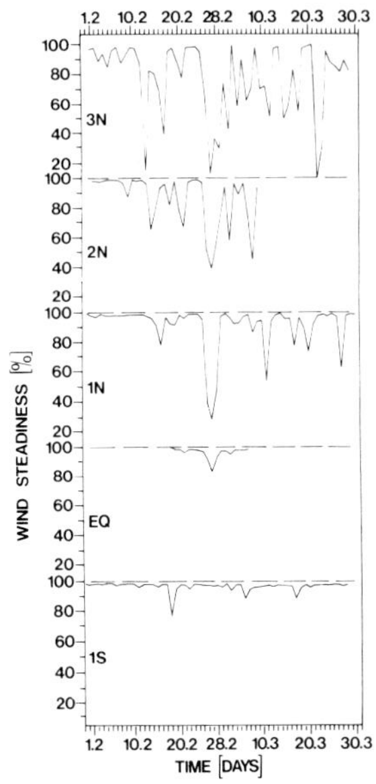

Fig. 2.
Wind steadiness (%) for February/March 1979. For further explanation see text.

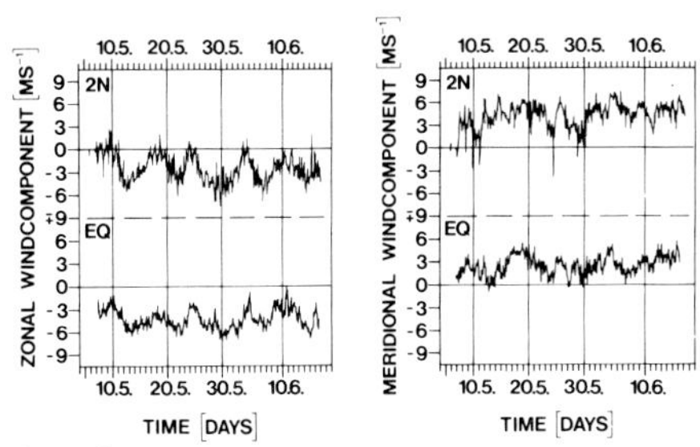

Fig. 3.
Time series of zonal (left) and meridional (right) wind components (ms^{-1}) at the Equator and at 2°N for May/June 1979.

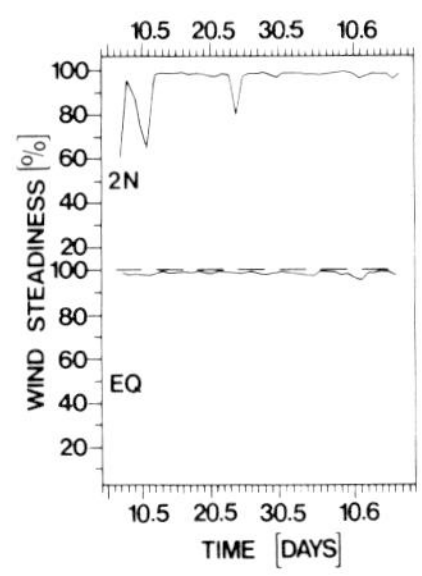

Fig. 4.
Wind steadiness (%) for May/June 1979.

We have related our wind measurements to the daily position of the Intertropical Convergence Zone (ITCZ), which was subjectively determined from METEOSAT cloud images (Köhne and Speth, 1982). The band of maximum cloudiness was defined as the position of the ITCZ using Nicholson's (1975) criteria to distinguish the intensity. The mean monthly position of the ITCZ over the Atlantic from January to November 1979 is shown in Figure 5.

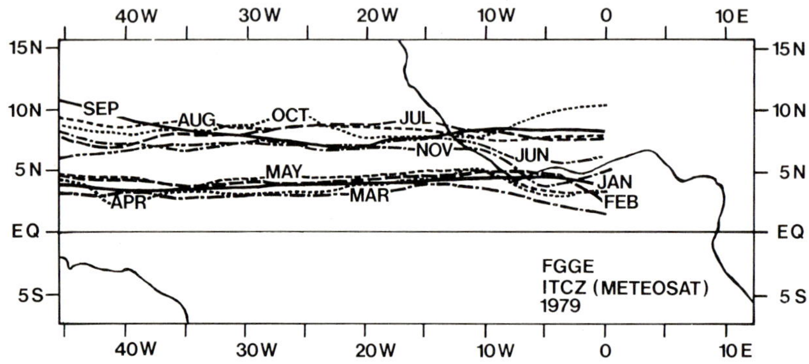

Fig. 5.
Mean monthly position of the ITCZ over the Atlantic from January to November 1979.

From January through May the ITCZ was found around 4°N with its southernmost position in March. Beginning in June the ITCZ migrated to the north with a mean position around 7°- 8°N. However, the band of maximum cloudiness was not necessarily identical with the surface flow discontinuity. Sadler (1975) and Ramage (1974) suggested that the cloudiness maximum was separated from the zone of trade wind confluence. The climatological mean position of the maxima of cloudiness, precipitation frequency and convergence lie ∿350 km south of the surface confluence in the eastern Atlantic (Hastenrath and Lamb, 1977). We found the same for our FGGE data. To determine the synoptic-scale surface flow we have used the objective analysis of the European Centre for Medium Range Weather Forecasts in Reading, UK. For the western Atlantic the maximum cloud zone was north of the confluence zone, which is in good agreement with the results of Sadler (1975). At 22°W for February/March and May/June 1979 the band of maximum cloudiness was close to the surface flow discontinuity, generally not more than one degree north of it.

We see from Figures 1 and 2 that for February/March the wind was in general from south-east, as expected for a trade wind region. The steadiness was relatively low in the north and increased southward. We can see from Figure 5 and especially from the daily position of the ITCZ (not shown here) that a low steadiness was always connected with a southward movement of the ITCZ. Between February 26 and March 6 the ITCZ was far to the south. So we can identify from Figure 1 a strong confluence between 1°N and the Equator. This was confirmed by synoptic surface observations which were performed frequently aboard R.V. METEOR. In contrast to these features the trade-winds are very stable and stronger from May to June (Figures 3 and 4). These results are in qualitative agreement with historical data (e.g. Hellerman, 1980).

Our principal aim was to investigate the influence of the atmospheric motion on the oceanic circulation. So far we described the available meteorological data sets. The figures 6 and 7 represent a selection of the oceanographic measurements, which we need for discussion in section 4. Those are temperatures in 15 m depth (Figure 6). Our interpretation of Figure 6 is, that we find in each of the considered periods an upwelling event at the Equator. It starts with January 17 and May 29 respectively.

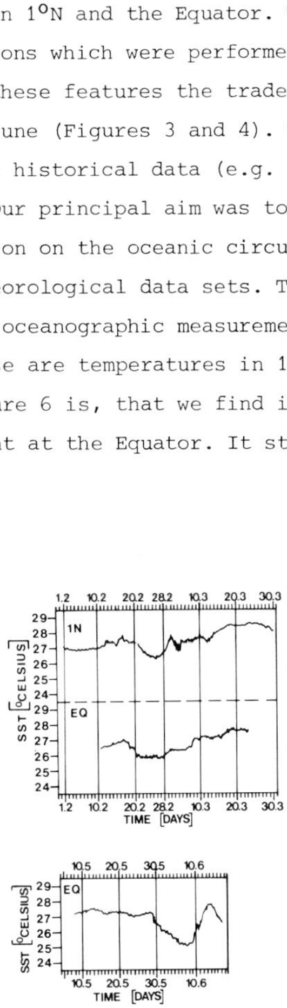

Fig. 6.
Time-series of SST (°C) at the Equator and at 1°N for February/March 1979 (top) and at the Equator for May/June 1979 (bottom).

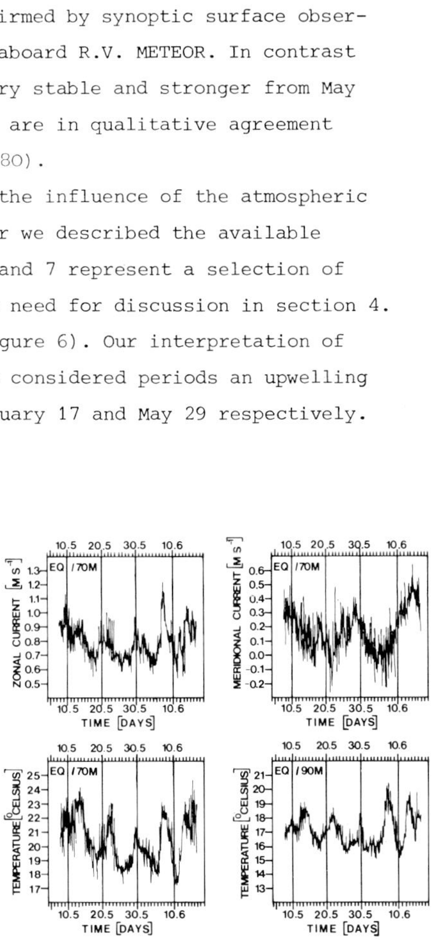

Fig. 7
Time-series of the zonal and meridional current component (ms^{-1}) (top) at the Equator at 70 m depth and time-series of water-temperatures (°C) bottom) at 70 m and 90 m at the Equator for May/June 1979.

For the first period the upwelling event can also be identified in 1°N. Additionally, we need currents and temperatures of the second period (Figure 7). At the Equator we find the EUC in 70 m, which is the mean depth of its core for that period. In the meridional current component we find a low frequency variability, which we attribute to the 16-days meander, as described by Düing et al. (1975). A similar meander was inferred by Lass et al. (1980) from current profile measurements across the Equator along 28°40'W between May and June 1979. They found that the meander was week at the beginning of the observations and amplified simultaneously with the amplification of the SEC after May 20. The temperatures in 70 m depth and also in 90 m depth (Figure 7, bottom) show similar fluctuations.

4 SPECTRAL ANALYSIS

To come close to our goal we have applied spectral methods to our data. Wind spectra for the February/March observations barely showed any significant features and are not reproduced here. Spectra for the May/June period are shown in Figure 8 in which the following representation was chosen.

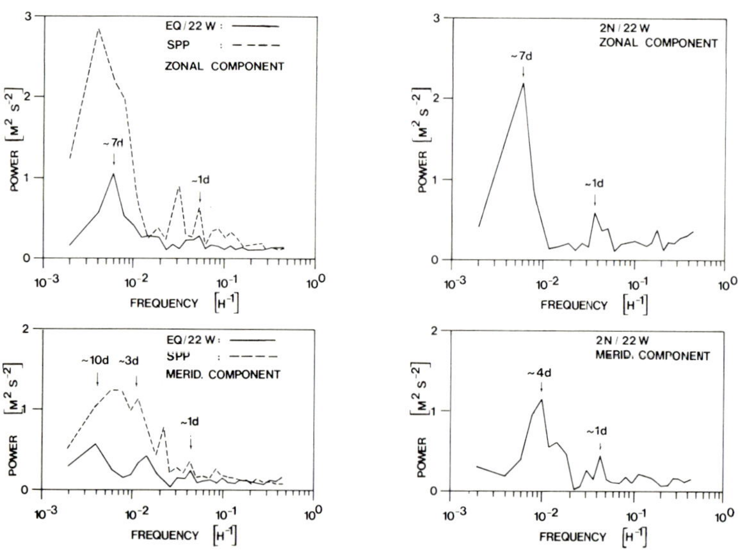

Fig. 8.
Power spectral estimates (m^2s^{-2}) of the zonal and meridional wind components at the Equator (left) and at 2°N (right), indicated by solid lines. Additionally, power spectral estimates of the wind components at St. Peter and St. Paul Rocks (SSP) are presented by dashed lines. All spectra are computed for May/June 1979.

The abscissa gives the period in a logarithmic scale, while the ordinate is the power density multiplied by the frequency. This kind of representation has the advantage that the area under the curves is proportional to the total variance. Significant peaks in the spectra are marked by arrows. The solid curves represent spectra at the Equator (left) and at 2°N (right) for the zonal and meridional wind components. The spectral estimates of each time series with zero mean value and zero least-square linear trend were obtained by dividing the time series into segments with a length of 512 hours. We allowed a 50 % overlapping and applied a sinus-shaped Hanning window. After a direct Fast Fourier Transformation the frequency spectra were averaged over all segments.

It is seen from Figure 8 that maxima do exist for periods of ∿1 day in each spectrum. The meridional wind component exhibits additional maxima for ∿3 to 4 days in both latitudes and 10 days on the Equator. This is in good agreement with Krishnamurti and Krishnamurti (1979) who found spectral peaks in the ∿4 to 6 days range and around 11 days period in the south east trades during GATE (Garp Atlantic Tropical Experiment). The zonal wind component, however, has maxima for ∿7 days, which is most expressed in 2°N. Co-spectra, not represented here, show that the zonal wind components between the Equator and 2°N are coherent for this period; the zonal wind component at 2°N lags that on the Equator by ∿7 hours. The meridional wind components exhibit a high coherence in the 3 to 4 range. The meridional wind component at the Equator leads with ∿5 hours. The latter result is consistent with a westward travelling disturbance with an axis directed from southwest to northeast. This is in agreement with Burpee (1974) who found the same features for easterly waves (wave length ∿3,800 km) of a period of ∿4.5 days over West-Africa at the 700 mb level. He found an inclination of the axis of ∿480 km per 5 degrees latitude which corresponds to a lag of ∿5.5 hours per 2 degrees latitude. This is comparable to our result, but with a tendency that the trough axis is a little more inclined (see Reed et al., 1977).

For the period February 2, 1979 to January 23, 1980 wind records at St. Peter and St. Paul Rocks (0°55.13'N, 29°20.60'W) are available from Garzoli et al. (1982). We have computed spectra from these data for the same periods as for the measurements from 22°W and included in Figure 8. These spectra show essentially the same structure as those for 22°W, however, the variance of relative low frequencies is much larger. This effect is larger than it could be expected from the different height of the wind recorders (St. Peter and St. Paul Rocks: 21.3 m; 22°W: 1.5 m). Coherence spectra for the wind components at 22°W and 29°W were computed by Garzoli et al. (1982). From these it is inferred that the wind components are coherent for the periods of ∿3-4 days and 7 days. The phase between the wind components at 22°W and 29°W (not shown) suggests a westward wave propagation with a wavelength of $\sim$5,000 km. We argue that also this result is consistent with Burpee (1974). We are aware of the fact that the two positions are 800 km apart and the data intervals are short, but the absence of any previous simultaneous *in situ* wind records from the Central Atlantic moves us to comment on what we have.

To investigate how far the oceanic oscillations at 22°N are forced by the local wind we have computed co-spectra and empirical orthogonal functions of the wind and oceanic parameters. For periods of $\sim$7 days for the May/June campaign, we could find a coherence between the zonal wind and the zonal current component in depths of 15 m, 70 m and 90 m at the Equator[1]. The connection is most pronounced at 70 m and is shown in Figure 9. The signal is first to be seen at 15 m and has at 70 m a lag of 3.5 days (not shown) against the wind.

The temperature is in phase with the zonal current component (not shown, but compare the time series in Figure 7). We did not find a similar relationship for 2°N.

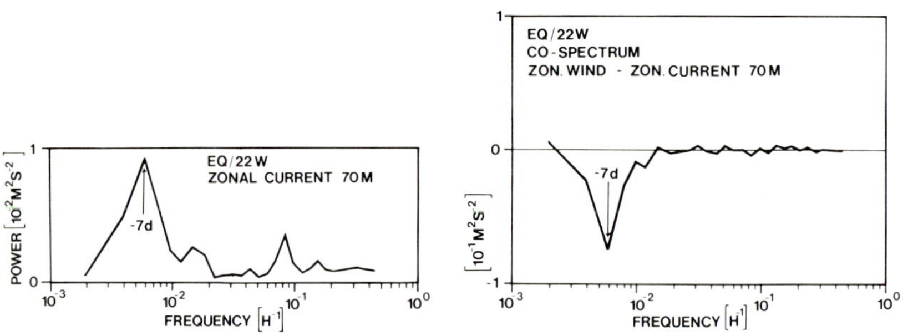

Fig. 9.
Auto-spectral estimates ($10^{-2}m^2s^{-2}$) of the zonal current component at 70 m at the Equator (left) and its co-spectrum ($10^{-1}m^2s^{-2}$) with the zonal wind component (right). Both spectral presentation are calculated for May/June 1979.

Garzoli and Katz (1981) studied observations from inverted echo sounders near the Equator in the western Atlantic. They found an increase in the variance centered at 3.5 days, which they interpreted as an atmospherically forced and equatorially confined, vertically propagating inertia-gravity wave. Because our spectra for the meridional wind component exhibit peaks for periods of $\sim$3-4 days we expected to find coherent oscilla-

---

1) Current meters were developed in the following depths: 15 m, 70 m, 90 m, 110 m, 255 m, 590 m (Equator) and 45 m, 65 m, 85 m, 105 m, 255 m, 590 m (2°N).

tions of the currents. However from our data we could not confirm the existence of such coherent relationships with the local wind at 22°W. Moreover, we could not find any other coherent oscillations with the local wind, with the exception we discussed in the penultimate paragraph.

5 CONCLUSION

It was our aim to investigate the atmospheric forcing of currents in the equatorial Atlantic at 22°W. We had data available for two periods: February/March and May/June of 1979. For that reason we had to restrict ourselves to an examination of relatively short period fluctuations.

During the first period the steadiness of south-east trades was relatively low in the north and increased southward; the wind spectra barely showed any significant characteristics. We attribute both features to the fact that during the February/March period the measurements were made pretty close to the ITCZ. Towards May/June the ITCZ migrates to the north, the wind steadiness is high, and in the zonal wind spectra we find increased variance for periods of ∿7 days. The zonal wind component at 2°N lags that on the Equator by ∿7 hours. The meridional wind components exhibit increased variance for periods of ∿10 days on the Equator and ∿3 to 4 days on the Equator and 2°N. We attribute the latter periods to easterly waves with an axis inclined from southwest to northeast.

From these characteristics of the wind field it was not surprising to us that for February/March we could not find any coherent oscillations between the local wind and the ocean. For May/June we could detect a significant coherence for periods of ∿7 days between the zonal wind and the zonal current only for the Equator. This relationship is most pronounced in the mean depth of the core of the EUC, where the current has a lag of ∿3.5 days. We expected that the increased variance in the meridional wind components for periods of ∿3-4 days would be favourable to excite inertia-gravity waves in the ocean. However, we could not confirm such a relationship from our data.

Also the two upwelling events, starting February 17 and May 29, which were observed in the time series of temperature in 15 m depth are not coherent with the local wind. Because Fahrbach and Bauerfeind (1982) demonstrated that these low SST are not a result of increased vertical mixing due to a strong current shear between EUC and SEC, we conclude that these upwelling events could be caused by the remote forcing mechanism proposed by Moore et al. (1978). The necessary assumption for this, namely increasing easterly winds in the western equatorial Atlantic, did exist for the February/March as well as for the May/June period (compare Figures 4 and 9 of Garzoli et al., 1982). However, we cannot exclude the possibility of an upwelling event caused by the f-divergence under permanent easterly trade winds as Wyrtki and Eldin (1982) found for the Central Pacific. They conclude that the upwelling becomes evident at the sea surface 10-20 days after the onset of a strong zonal wind. We could not infer such a relationship because of the shortness of our time-series.

For the meridional current component at the Equator for the May/June period we found a low frequency variability, which we attribute for the 16 - days meander. We could deduce this only from the time series, because of their relatively short length. This was also the reason that we made no attempt to look for their possible generation by the wind.

REFERENCES

Burpee, R.W., 1974. Characteristics of North African easterly waves during the Summer of 1968 and 1969. J.Atmos.Sci., 31, 1556-157o.
Cornus, H.P. and J. Meincke, 1979. Observation of near-surface layer changes related to the Atlantic equatorial undercurrent. Deep-Sea Res., 26A, 1291-1299.
Düing, W., P. Hisard, E. Katz, J. Meincke, L. Miller, K.V. Moroshkin, G. Philander, A.A. Rybnikov, K. Voigt and R. Weisberg, 1975. Meanders and long waves in the equatorial Atlantic. Nature, 257, 28o-284.
Düing, W. and Z. Hallock, 1979. Equatorial waves in the upper central Atlantic. Deep-Sea Res., GATE Suppl. II to Vol. 26, 161-178.
Fahrbach, E. and E. Bauerfeind, 1982. The variability of equatorial thermocline spreading as an indication of equatorial upwelling. Submitted to: Océanographie Tropicale.
Garzoli, S.L. and E.J. Katz, 1981. Observations of inertia-gravity waves in the Atlantic from inverted echo sounders during FGGE. J.Phys.Ocean., 11, 1463-1473.
Garzoli, S.L., E.J. Katz, H.-J. Panitz, P. Speth, 1982. In situ wind measurements in the equatorial Atlantic during 1979. Ocean.Acta, 5, 281-288.
Hallock, Z., 1979. On wind-excited, equatorially trapped waves in the presence of mean currents. Deep-Sea Res., Suppl. II to Vol. 26, 261-284.
Hastenrath, S. and P.J. Lamb, 1977. Some aspects of circulation and climate over the eastern tropical Atlantic. Mon.Wea.Rev., 1o5, 1o19-1o23.
Hastenrath, S. and P.J. Lamb, 1978. Climatic Atlas of the tropical Atlantic and eastern Pacific oceans. The University of Wisconsin Press, Madison, Wisc. 537o1.
Hellerman , S., 1979. Charts of the variability of the wind stress over the tropical Atlantic. Deep-Sea Res., Suppl. II to Vol. 26, 63-75.
Katz, E.J., R.L. Molinari, D.E. Cartwright, P. Hisard, H.U. Lass, A. de Mesquita, 1981. The seasonal transport of the equatorial undercurrent in the western Atlantic (during the Global Weather Experiment), Ocean.Acta, 4, 445-45o.
Katz, E.J. and S.L. Garzoli, 1982. Response of the western equatorial Atlantic ocean to an annual wind cycle. Manuscript.
Köhne, A. and P. Speth, 1982. Variation of the ITCZ in the Atlantic during FGGE. Tropical Ocean Atmosphere Newsletter, 9, 7.
Krishnamurti, T.N., R. Krishnamurti, 1979. Surface meteorology over the GATE A-scale. Deep-Sea Res., GATE Suppl. II to Vol. 26, 29-61.
Lass, H.U., W. Fennel, R. Helm, F. Möckel, M. Sturm, K.-H. Till, H. Wiechert and H. Will, 198o. Vorläufige Ergebnisse der Expedition des Forschungsschiffes "A.v. Humboldt" in dem äquatorialen Atlantik während des Globalen Wetterexperimentes (FGGE) SOP II 1979. Beitr.Meeresk., 44/45, 89-1o7.
Lass, H.U., V. Bubnov, J.M. Huthnance, E.J. Katz, J. Meincke, A. de Mesquita, F. Ostapoff, B. Voituriez, 1982. Seasonal changes in the zonal pressure gradient in the equatorial Atlantic west of 1o°W during the FGGE year. Submitted to Ocean.Acta.
Merle, J., 1978. Atlas hydrologique saisonnier de l'océan atlantique intertropical. Travaux et documents de l'O.R.S.T.O.M., No. 82.
Moore, D.W. and S.G.H. Philander, 1977. Modelling of the tropical oceanic circulation, in: The Sea: Ideas and observations on progress in study of the seas, Vol. 6, J. Wiley and Sons Inc.
Moore, D.W., P. Hisard, J. McCreary, J. Merle, J.J. O'Brien, J. Picaut, J.M. Verstraete and C. Wunsch, 1978. Equatorial adjustment in the eastern Atlantic, Geophys.Res.Lett., 5, 637-64o.

Neumann, G., 196o. Evidence for an equatorial undercurrent in the Atlantic Ocean. Deep-Sea Res., 6, 328-334.
Nicholson, S.E., 1975. Variation of the Intertropical Convergence Zone during phases 1, 2 and 3 of GATE experiment. GATE-Report No. 14, Vol. 1, 169-175.
Philander, S.G.H., 1976. Instabilities of zonal equatorial currents. Journ.Geophys.Res., 81, 3725-3735.
Philander, S.G.H., 1978. Instabilities of zonal equatorial current, II. Journ. Geophys.Res., 83 (C7), 3679-3682.
Philander, S.G.H. and R.C. Pacanowski, 198o. The generation of equatorial currents. Journ.Geophys.Res., 85, 1123-1136.
Ramage, C.S., 1974. Structure of an oceanic near-equatorial trough deduced from research aircraft traverses. Mon.Wea.Rev., 1o2, 754-759.
Reed, J.R., D.C. Norquist, E.R. Recker, 1977. The structure and properties of African wave disturbances as observed during Phase II of GATE. Mon.Wea.Rev., 1o5, 317-333.
Sadler, J.C., 1975. The monsoon circulation and cloudiness over the GATE area. Mon.Wea.Rev., 1o3, 369-387.
Servain, J., J. Picaut and J. Merle, 1982. Evidence of remote wind forcing in the equatorial Atlantic ocean. J.Phys.Ocean., 12, 457-463.
Voigt, K., 1961. Äquatoriale Unterströmung auch im Atlantik. (Ergebnisse von Strömungsmessungen auf einer atlantischen Ankerstation der "Michail Lomonossov" am Äquator im Mai 1959). Beitr.Meeresk., 1, 56-6o.
Weisberg, R.H., L. Miller, A. Horigan and J.A. Knauss, 1979. Velocity observations in the equatorial thermocline during GATE. Deep-Sea Res., GATE Suppl. II to Vol. 26, 217-248.
Wyrtki, K. and G. Eldin, 1982. Equatorial upwelling in the Central Pacific. J.Phys.Ocean., 12, 984-938.

NEAR SURFACE TEMPERATURE OBSERVATIONS OBTAINED IN THE EQUATORIAL ATLANTIC OCEAN DURING FGGE (1979)

Robert L. Molinari*, National Oceanic and Atmospheric Administration, Atlantic Oceanographic and Meteorological Laboratories, Miami, Florida, U.S.A.

Eli Katz, Lamont-Doherty Geological Observatory, Columbia University, Palisades, New York, U.S.A.

Eberhard Fahrbach, Institute of Marine Research, University of Kiel, Kiel, Federal Republic of Germany

Hans Ulrich Lass, Academy of Sciences of GDR, Institute of Marine Research, Rostock-Warnemunde, German Democratic Republic

Bruno Voituriez, Antenne ORSTOM-Oceanographic Center of Brest, Brest, France

ABSTRACT

Temperature and surface wind data collected across the equatorial Atlantic Ocean during the First GARP Global Experiment (FGGE), 1979, have been compiled to study the seasonal evolution of the near-surface temperature field. The development of the large-scale, sea-surface temperature (SST) field is characterized by the appearance during boreal summer of a tongue of cold water which extends from the eastern to western basin, on and south of the equator. During 1979, the cold water first appears at 4°W and 28°W during early May and at 22°W some four weeks later. East of about 20°W, the thermocline rises and the mixed layer becomes shallower simultaneously with the lowering of SST's. West of about 30°W, the thermocline and mixed layer deepen at this time. Below average temperatures are observed through October, as the thermocline redeepens in the east and continues to deepen in the west. The 1979 SST and thermocline distributions along the equator have been compared to climatological distributions derived from historical data. The FGGE year fields are qualitatively similar to the climatological fields. Surface wind data collected during FGGE indicate that the surface cooling occurs within several days of an increase in both components of wind at 4°W and 28°W and within several weeks of the wind increase at 22°W.

---

*Prepared while visiting scientist at:

Museum National d'Historie Naturelle
Oceanographie Physique, L.A. 175, CNRS
Paris, France

Changes in heat content over various time intervals were computed from the FGGE data for a layer 50 m thick. These changes have been compared to changes which would be induced by energy fluxes through the sea surface, determined climatologically. The comparison shows that surface energy fluxes can not account for the initiation of the cold water tongue. A box model centered on the equator and extending from 10°W to 30°W was constructed to determine what processes may be active during the early stages of the development of the tongue. Cooling within the box is estimated at a rate of -3.0°C/60 days from the first of May through the first of July. This is approximately balanced by warming due to surface energy fluxes (+1.5°C/60 days) and cooling due to advection by South Equatorial Current (-1.6°C/60 days), vertical mixing (-1.2°C/60 days) and upwelling (-1.2°C/60 days). The observations further suggest that mixing and upwelling are important during the early stages of the development of the tongue, while advection plays a more important role at later times.

1. INTRODUCTION

In both the Atlantic and Pacific Oceans, a tongue of cold water extends on and south of the equator from the eastern to the central basins during northern hemisphere summer but not northern hemisphere winter (Hastenrath and Lamb, 1977). In the Atlantic, the tongue extends farthest to the west along the equator during August and September. According to Hastenrath (1980), the tongues play an important role in the global heat balance, as they serve as sinks for the excess radiation received by the tropics relative to higher latitudes.

Physical oceanographic data were collected in all the tropical oceans during the First GARP Global Experiment (FGGE), also called the Global Weather Experiment. A portion of those data collected in the equatorial Atlantic Ocean by various investigators and institutions have been combined and will be reviewed, with particular emphasis on the cold water tongue, for the following purposes:

(1) to describe the temporal and spatial evolution of the near-surface temperature field of the equatorial Atlantic Ocean during an extended period encompassing FGGE, January 1979 through March 1980;

(2) to determine if the FGGE-year temperature distributions were, in terms of long-term averages, "normal" or anomalous; and
(3) to determine what atmospheric and oceanic factors contribute to the evolution of the temperature field.

Some guidance for planning future efforts in the Atlantic should result as a consequence of the analyses performed to meet these objectives.

The analyses and results presented herein are qualitative and preliminary in nature, as not all the FGGE data are available at this time and those data available are not in a form suitable for detailed quantitative analyses. Efforts are now underway to obtain the entire FGGE Atlantic data-set, in particular, expendable bathythermograph (XBT) data and surface temperature and wind observations. More quantitative analyses are planned for the expanded data-set. However, the present discussion will provide the framework for the future effort.

2. DATA AND PROCESSING

The ships involved and their operating areas and times are listed in Table 1. The majority of the data were collected by Conductivity Depth Temperature (CTD) units, but some XBT data are also used. No ship-to-ship intercalibrations of the temperature sensors were performed. In general, calibrated CTD temperature data are accurate to better than .05°C.

TABLE 1
Participating ships in the Equatorial Atlantic Ocean during FGGE*

	Operating Area		
Ship	Latitude	Longitude	Approximate Dates
BESNARD	2°S - 2°N	28°W - 40°W	Jan/Feb/Mar/Jul 1979
CAPRICORNE	4°S - 4°N	4°W	Jul 1978 Jan/Apr/Jun Oct 1979 Feb 1980
CONRAD	2°S - 2°N	28°W - 40°W	Sep/Oct 1979
HUMBOLDT	2°S - 5°N	28.7°N	May/Jun 1979
KURCHATOV	2°S - 2°N	18.5°W - 22.5°W	Mar/Apr 1979
METEOR	2°S - 3°N	22°W	Feb-Jun 1979
OCEANUS	2°S - 2°N	28°W - 40°W	Feb/Mar 1980
RESEARCHER	10°S - 1°N	11°W - 27.5°W	Jul/Aug 1978 Jan/Feb/Jul 1979 Feb/Mar 1980

*Includes only those ships whose observations are considered in the text.

Several forms of data presentations are used to describe the evolution of the near-surface temperature field. In these presentations, when several stations per day are available at one location, the data were averaged over two-day periods to reduce some of the higher frequency variability. Spatially, averaging was typically performed through subjective contouring of the fields. Objective mapping is planned when the larger data-set is available.

In the following discussion, the first depth at which the temperature is 0.5°C less than the sea-surface temperature (SST) is used to define the mixed layer depth. Using a more stringent definition, e.g., 0.2°C less, typically, did not alter the mixed layer depth distributions because of the intense thermocline located just below the mixed layer in the region.

The depth of the 23°C isotherm is taken to represent the depth of the thermocline. This definition permits comparison with previous studies. The 23°C isotherm is generally located within the region of maximum vertical temperature gradient as shown by a comparison between the average depth of this isotherm and the average depth of maximum gradient for several cruises (Table 2). The only significant difference occurs during summer in the eastern Atlantic, when the 23°C isotherm sometimes breaks the surface. The average maximum gradient depth during these cruises is about 30 m, while the average 23°C isotherm depth is about 15 m. Although these depths are different, both representations indicate that the thermocline is shallower at this time than during other cruises.

TABLE 2

Comparison of average thermocline depths (m) by cruise as given by the depth of the 23°C isotherm and by the depth of the maximum vertical temperature gradient. Also given in parenthesis are standard deviations (m).

	Operating Area		Number of	Average Depth	
Month/Year	Latitude	Longitude	Data Points	23°C isotherm	Gradient
Jul 1978	4.5°S - 2°N	1°E		11	33
Jan 1979	5°S - 5°N	4°W	53	41(9)	37(11)
Apr 1979	6°S - 5°N	4°W	33	42(8)	40(10)
Jun 1979	5°S - 3°N	1°E - 4°W	17	21(13)	30(10)
Oct 1979	5°S - 5°N	4°W	43	43(7)	46(10)
Jan 1980	5°S - 5°N	4°W	18	48(10)	49(14)
Jan 1979	4°S - 3°N	27.5°W	16	78(7)	86(7)
May 1979	2°S - 4°N	28.7°W	41	72(8)	78(12)
Jun 1979	2°S - 2°N	28°W - 38.5°W	30	83(14)	90(20)
Oct 1979	2°S - 2°N	28°W - 40°W	39	102(11)	112(17)
Feb 1980	0° - 2°N	28°W - 40°W	27	89(8)	91(11)
Mar 1979	4°S - 1°N	27.5°W	6	88(6)	88(13)

3. BASIN-WIDE TEMPERATURE DISTRIBUTIONS

Sea-surface temperature distributions have been constructed for five quasi-synoptic time intervals, each approximately one-month long. Isotherms from the mean-monthly SST charts of Hastenrath and Lamb (1977) are also shown on the synoptic SST distributions (Figs. 1, 2, and 3). The predominant feature in the January-February 1979 distribution is an enclosed patch of colder water centered somewhat south of the equator at about 16°W (Fig. 1). SST's along the equator during this period range from less than 26.2°C within the patch, to greater than 28.5°C to the east of the patch and greater than 27.2°C to the west.

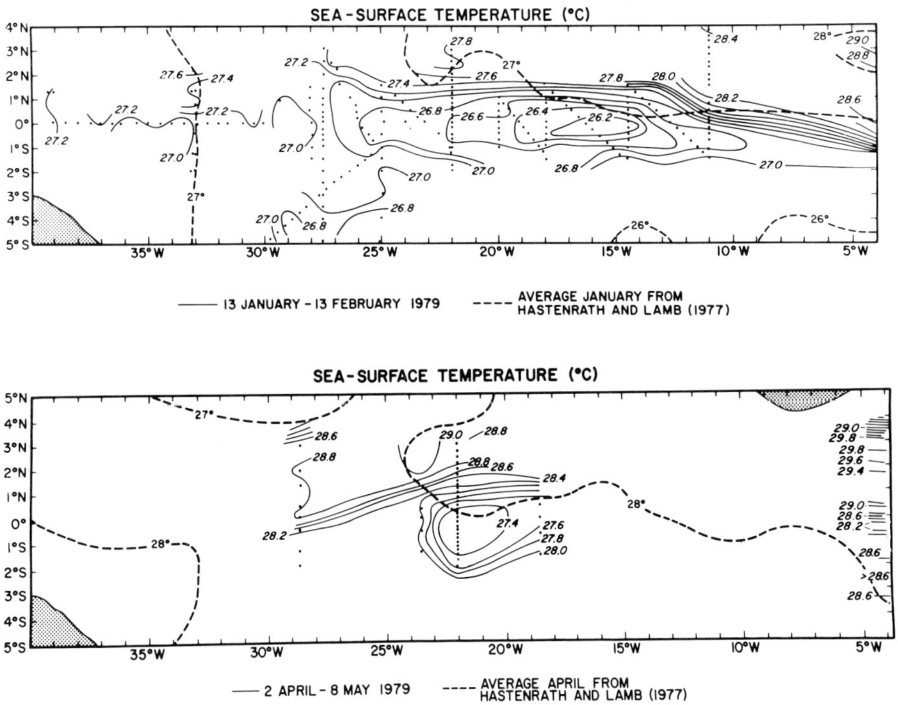

Fig. 1. Sea-surface temperature, SST (°C), distributions for the times indicated (solid lines). Dashed lines represent isotherms from the appropriate mean monthly SST chart of Hastenrath and Lamb (1977). Dots represent station positions.

Net heating at 4°W and between 18.5°W and 28.7°W has occurred between January-February and April-May 1979 (Fig. 1). Maximum temperatures in the east are greater than 29.0°C and in the west, 28.2°C. Somewhat colder waters are still observed in the central Atlantic. By June-July, a tongue of cold water, with coldest waters in the east, appears on the equator (Fig. 2). The synoptic pattern appears very similar to the climatological pattern given by the Hastenrath and Lamb (1977) isotherms. The few data available during October show that temperatures remained relatively low in the extreme western Atlantic through the early fall (Fig. 2). By February-March 1980, SST's had increased (Fig. 3). Although not well defined, there is some suggestion of a cold water patch centered at about the same location as during boreal winter 1979.

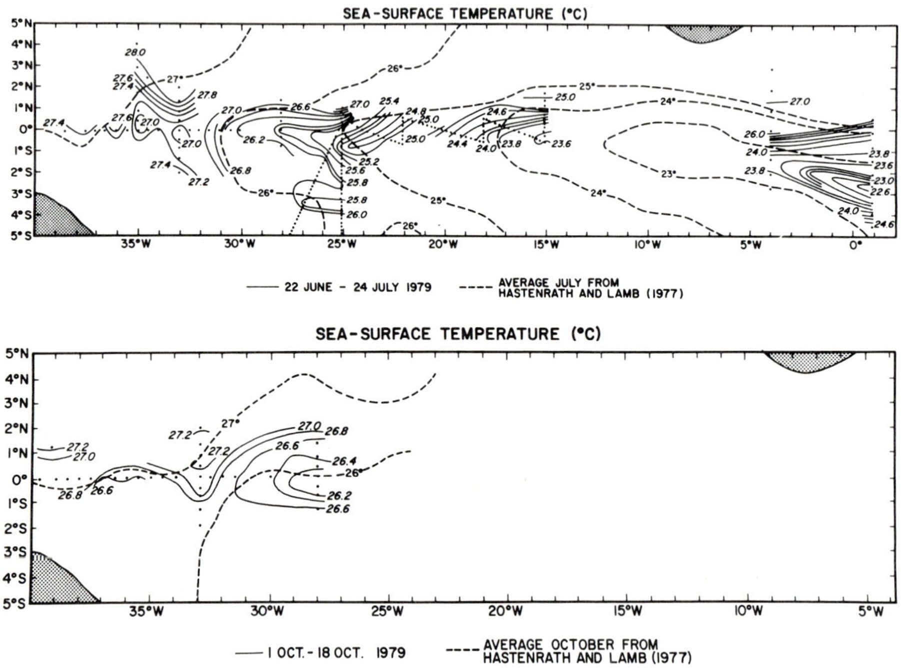

Fig. 2. Same as Figure 1.

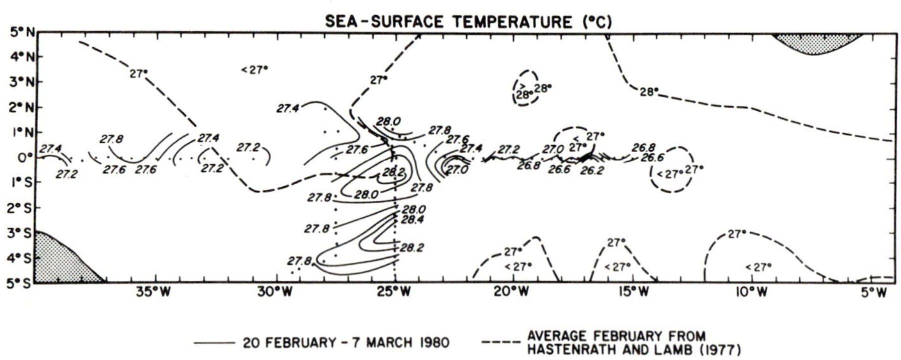

Fig. 3. Same as Figure 1.

Time-longitude plots of SST, 23°C isotherm depth and mixed layer depth have been constructed from data collected along the equator. Similar plots have been constructed from the mean-monthly SST charts of Hastenrath and Lamb (1977) and the seasonal 23°C isotherm depths of Merle (1980).

The 1979-1980 equatorial SST distributions and the climatological SST distributions are qualitatively quite similar (Fig. 4). In both representations, temperatures range from a minimum of less than 23°C, occuring in July-August, to a maximum of greater than 28°C, occuring sometime in April-May. Although these data are not adequate to describe precisely the evolution of the synoptic SST field, maximum cooling in the east occurs between late April and late June, as it does in the climatological data-set. It is difficult to resolve the timing of the cooling west of 30°W, but coldest temperatures are observed somewhat later during September-October in both representations.

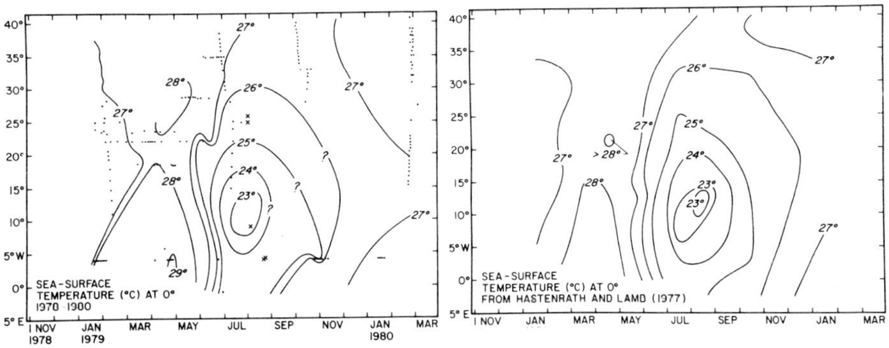

Fig. 4. Time longitude plots along the equatorial plane of sea-surface temperature, SST (°C), distributions observed during 1979 (left panel) and taken from the mean monthly SST charts of Hastenrath and Lamb (1977) (right panel). Dots represent station locations.

As was the case for the SST distribution, the 1979-1980 23°C isotherm depth distribution and the climatological distribution appear quite similar (Fig. 5). Minimum isotherm depths, in both cases, occur between 5°W and 10°W during July and August. Also shown on Fig. 5, are monthly averaged thermocline depths inferred from an inverted echo sounder deployed at 40°W. All three representations indicate that deepest thermocline depths in the west occur sometime during early fall and are greater than 120 m. Although the synoptic data are inconclusive, there is some suggestion that temporal changes during 1979-1980, particularly during the spring in the east, are similar to changes during the climatological year.

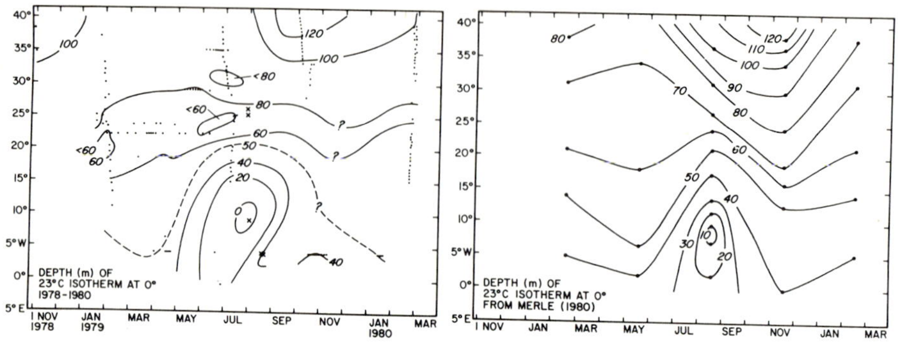

Fig. 5. Time longitude plots along the equatorial plane of the depth distribution (m) of the 23°C isotherm observed during 1979 (left panel) and from the mean seasonal distribution of Merle (1980) (right panel). Dots represent station positions.

The evolution of the equatorial mixed layer field is shown in Fig. 6. The data suggest that the cooling of SST in the east is associated initially with a shallower mixed layer. Cooling in the west, particularly during October, appears to be associated with a deeper mixed layer.

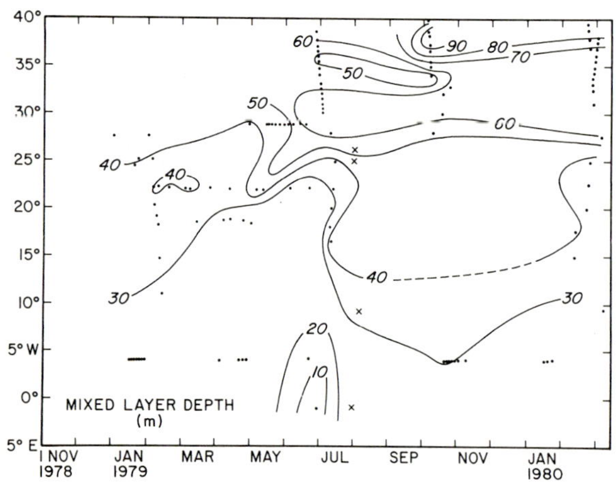

Fig. 6. Time longitude plot along the equatorial plane of mixed layer depth (m) distribution observed during 1979. Dots represent station positions.

4. EVOLUTION OF EQUATORIAL SURFACE WIND AND TEMPERATURE DISTRIBUTIONS ALONG SEVERAL MERIDIANS

Surface wind and oceanic temperature data collected simultaneously at or near the same equatorial position (4°W, 22°W, and 28°W) are compared to consider possible phase relations between atmospheric and oceanic phenomena. Wind data collected at 0°, 4°W have been kindly supplied by P. Rual, ORSTOM,

Brest, France and those at 0°, 22°W by H. J. Panitz, University of Cologne, Cologne, Federal Republic of Germany. Zonal wind components obtained at 18.5°W from wind stress data given by Bubnov (1981) are used to fill-in some gaps in the 22°W time series. Finally, Katz et al. (1981) present a time series of wind data collected at 0.9°N, 29.35°W which is used to represent equatorial conditions at 28°W.

At 4°W, in contrast to 22°W and 28°W, the meridional component of the wind is larger than the zonal component (Figs. 7, 8, and 9). Observed wind components at all three longitudes increase by at least a factor of two during May. Simultaneously, SST's decrease at the three positions.

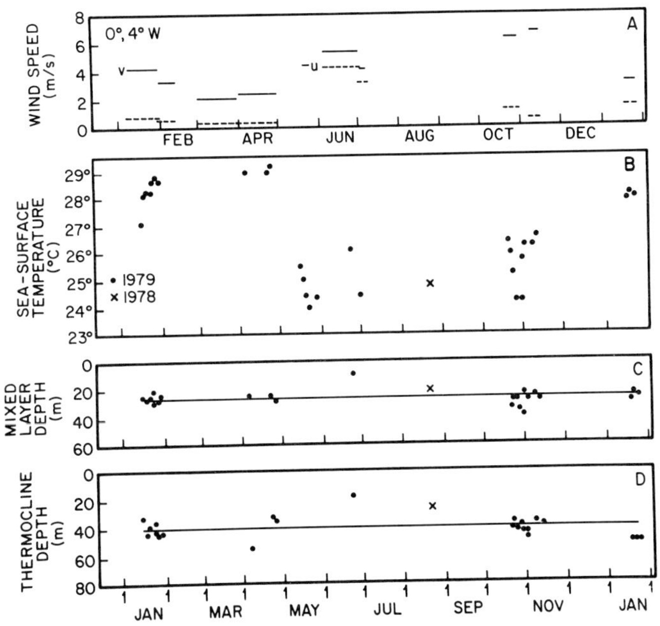

Fig. 7. Temporal evolution along 4°W of the following variables (data collected during 1979, except where indicated): panel A; average wind speed components, panel B; sea surface temperature, panel C; mixed layer depth and panel D; thermocline depth. The mean for each temperature variable is given as a straight line.

The winds increase at 4°W some time between 3 May and 5 June (Fig. 7). SST decreases over 5°C between 24 April and 16 May, rises some 2°C and then falls again during June. The data available between 24 April and 16 May were obtained from a surface current meter mooring and subsurface temperature data were not given (Jarrige and Rual, 1981). Both the synoptic and climatological depth distributions indicate that the thermocline rises concurrently with the cooling (Fig. 5), although as in the case of surface winds, gaps in the subsurface temperature records preclude the definition of precise phase relations. Similarly, the mixed layer depth becomes shallower at this time.

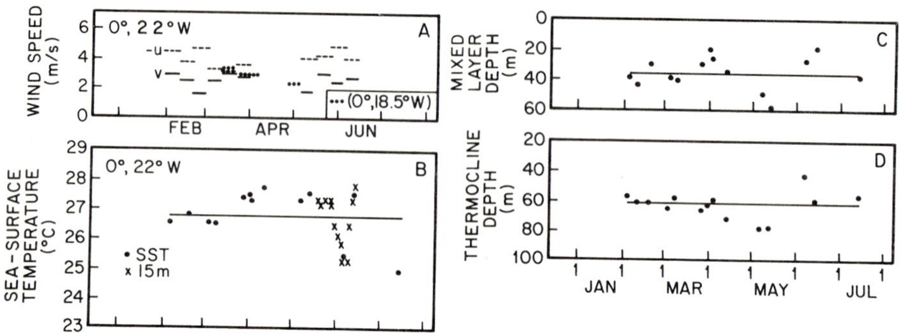

Fig. 8. Same as Figure 7, except for 22°W.

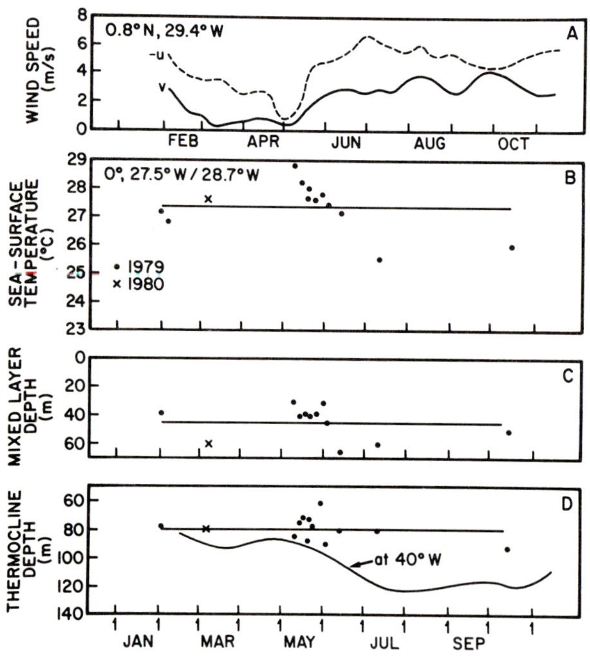

Fig. 9. Same as Figure 7, except for 27.5°W/28°W. The thermocline depth curve from 40°W is taken from mean monthly inverted echo sounder data.

Temperatures below the record average, as well as another 2°C cooling event, are also observed at 4°W during October (Fig. 7). However, the thermocline and mixed layer are deeper than during June. Deepest mixed layers are associated with the lowest temperatures during October (Fig. 10). There are no obvious systematic changes in thermocline depth at this time. Warming begins concurrently with a change in zonal wind component from easterly to westerly and a decrease in total wind speed (Fig. 10).

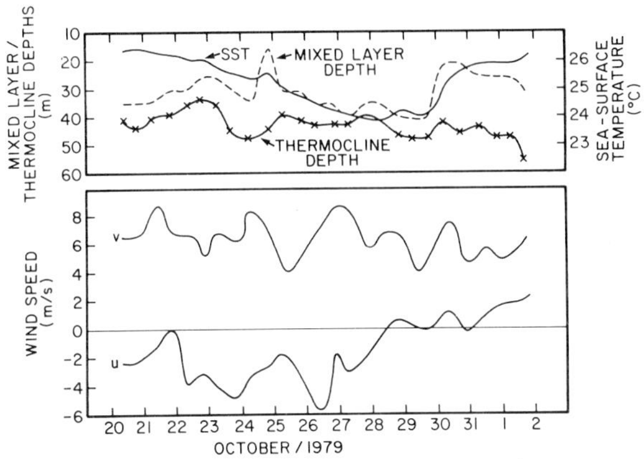

Fig. 10. Time series of sea-surface temperature, mixed layer depth, thermocline depth (upper panel) and wind speed components (lower panel) derived from twice daily observations at 0°, 4°W.

At 22°W, at least the zonal wind component of wind accelerates after the first of May (Fig. 8). Surface layers begin to cool some four weeks later, and as at 4°W, cooling does not occur monotonically. At least two periods of cooling are observed; one occurs between 28 May and 6 June and the other, which is not as well resolved, between 10 June and 13 July. Shallower mixed layer and thermocline depths are associated with the lower temperatures observed during June (Fig. 8). However, it is not clear whether the shallower thermocline depths are related to seasonal or higher frequency variability. Temperature data collected simultaneously on the equator at 70 m show large amplitude, periodic changes in the temperature (Fig. 11). Assuming an average thermocline gradient of about 8°C/50 m, these temperature changes imply changes in thermocline depth of some 30 to 40 m.

At 28°W, both increases in wind and decreases in SST begin some time during the first week of May (Fig. 9). Temperatures have fallen some 3°C by early July. During June, there is a 20 m deepening of the mixed layer, but no systematic changes in thermocline depth. Only a few data are available at 35°W, where between 8 June and 5 July temperatures fall some 1°C (not shown). Further west at 40°W, increases in thermocline depth occur simultaneously with increases in wind intensity at 29°W (Fig. 9), as described by Katz and Garzoli (1982).

Within the temporal resolution of the data, weakest winds are observed during March-April at the three locations. Highest SST's occur at this time.

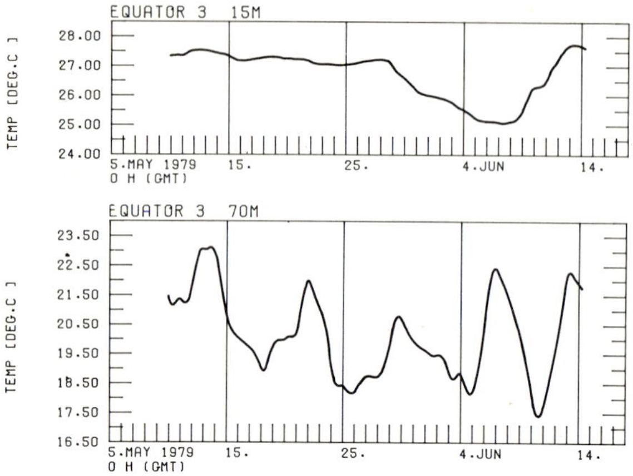

Fig. 11. Time series of temperature at 15 m (upper panel) and 70 m (lower panel) obtained from current meter moorings at 0°, 22°W.

5. HEAT BUDGET COMPUTATIONS

Some order-of-magnitude estimates of what terms can be important in the local heat budget have been made. First, it is shown that energy fluxes through the sea-surface are not responsible for the cooling observed during the summer. Changes in heat content of the mixed layer at the three meridians have been estimated from the temperature data for time periods which are functions of data availability. Net heat gain through the sea-surface has been estimated from the mean monthly heat gain charts of Hastenrath and Lamb (1978). These charts are contoured at a 40 W/m^2 contour interval and thus, only an approximate range of heat gain can be obtained. Changes in heat content are expressed in terms of changes in temperature over an average mixed layer (Figs. 7, 8, and 9). Surface heat gain values are converted to changes in mixed layer temperatures for comparison purposes in Table 3.

At all three locations, the ocean receives throughout the year a net input of energy. Thus, as found by Merle (1980) from climatological data, the summer cooling is driven by processes internal to the ocean, rather than by external atmospheric processes. Furthermore, at 22°W and 28°W, the warming observed at other times of the year is less than would be caused by surface energy exchanges. Thus, even at times of warming, either some internal oceanic processes serve to retard heating of surface layers or significant interannual variability (of the order 50 W/m^2) occurs in surface heat fluxes. The rapid warming observed during June at 22°W may be related to southward advection of warmer water after the initial cooling.

TABLE 3

Observed temperature changes and average mixed layer for the time periods given and the range of temperature changes estimated for the same periods using mean monthly net heat gains through the sea-surface from the atlas of Hastenrath and Lamb (1978).

Latitude	Julian Days (from 1 Jan 1979)	Average Mixed Layer Depth (m)	Observed Temperature Change (°C)	Range of Estimated Temperature Change (°C)
4°W	21-121	25	0.7	0 - 2.8
	121-174		-3.0	0.9 - 1.8
	174-299		-0.7	4.2 - 6.3
	299-382		2.6	2.0 - 2.8
22°W	42- 64	30	-0.1	1.2 - 1.4
	64- 93		0.8	1.2 - 1.8
	93-130		0.1	1.0 - 2.0
	130-156		-2.0	0.7 - 1.4
	156-163		2.1	0.2 - 0.4
	163-194		-2.6	0.9 - 1.8
28°W	31-133	45	1.2	3.7 - 4.2
	133-152		-0.8	0.3 - 0.5
	152-191		-1.9	0.7 - 1.0
	191-276		0.4	2.6 - 3.9
	276-431		1.6	2.9 - 5.8

Processes which can cool the surface layers include advection of cold water from the east by the South Equatorial Current (SEC), upwelling from below and vertical mixing. Wyrtki (1981) considers these processes in a model of the Pacific Ocean. We adopt his model for application to the Atlantic Ocean, but consider the initiation of the cold water tongue rather than its maintenance, as did Wyrtki (1981). For these computations, a combination of climatological and synoptic data are used. However, when all the FGGE-year data are available, the computation will be repeated using only those data.

A box model will be used which extends from 5°S to 5°N (so that reasonable estimates of Ekman transport are possible), from 10°W to 33°W (to avoid possible lateral boundary effects) and from the sea-surface to 50 m (an average mixed layer depth for early boreal summer, Fig. 5). The heat budget of the box can be written

$$\frac{\partial T}{\partial t} = Q_S + Q_{ADV} + Q_{UP} + Q_{MIX} \qquad (1)$$

where $\partial T/\partial t$ = average temperature change within the box,

Q_S = temperature change caused by heat-flux through the sea-surface,

Q_{ADV} = temperature change caused by horizontal advection,

Q_{UP} = temperature change caused by upwelling, and

Q_{MIX} = temperature change caused by vertical mixing.

A least squares fit to the five summer cooling rates given in Table 3 indicates an average cooling rate of about -3°C/60 days between May and July. Temperature increase due to heat gain through the sea-surface is estimated from the boreal summer net oceanic heat gain charts of Hastenrath and Lamb (1978) at + 1.5°C/60 days.

Temperature change caused by advection of colder waters through the eastern boundary by the SEC can be expressed as

$$Q_{ADV} = U(T_{33°W} - T_{10°W})/V$$

where U is the volume transport of the SEC, $T_{10°W}$ is the temperature of SEC waters at 10°W, $T_{33°W}$, at 33°W and V, the volume of the box. Bubnov and Egorhikin (1979) observed an average transport of the SEC during boreal summer 1974 of 15 x 10^6 m^3/s. Both the synoptic and climatological representations of SST for boreal summer (Fig. 2), indicate $T_{10°W}$ is about 24°C and $T_{33°W}$ is about 27°C. The resultant cooling due to advection is -1.6°C/60 days.

The change in temperature within the box due to upwelling will be estimated following the approach of Wyrtki (1981). In this approach,

$$Q_{UP} = W(T_1 - T_o)/V$$

where W is the upwelling rate and is computed as a difference between Ekman divergence and geostrophic convergence at the equator, T_1 is the temperture of the water leaving the box through the meridional boundary and T_o is the temperature of the upwelled water at 50 m. Ekman transport out of the box is estimated by assuming the transport through 5° is proportional to

$$\tau_x/\rho f$$

where τ_x is the zonal wind stress component, ρ is density and f is the Coriolis parameter. Climatological wind stress data given in Duing et al. (1980) are used to determine τ_x. For the summer months of July, August, and September the southward Ekman transport through 5°S is 16 x 10^6 m^3/s. The northward transport through 5°N is 9 x 10^6 m^3/s.

However, as noted by Wyrtki (1981), the divergence of Ekman flux is not equal to the rate of upwelling because of geostrophic transport towards the equator which occurs on a different vertical length scale than does the former flux. The climatological seasonal dynamic topography chart for boreal summer of Merle (1978) shows that geostrophic flow at both 5°S and 5°N is towards the equator. Geostrophic currents are approximately constant in the upper 50 m and decrease to about 0 at 100 m. Geostrophic transports computed from this chart are 8×10^6 m^3/s in the upper 50 m and 4×10^6 m^3/s between 50 m and 100 m across both the northern and southern boundaries. Thus, the upwelling transport, W, into the box is about 8×10^6 m^3/s. Temperatures at 50 m are about 22°C and at the meridional boundaries about 26°C (Fig. 2). With these estimates the change in temperature within the box due to equatorial upwelling is -1.2°C/60 days.

Temperature changes due to vertical mixing can be expressed as

$$Q_{MIX} = (AK\ \partial T/\partial z)/V$$

where A is horizontal area, K is vertical turbulent diffusion coefficient and $\partial T/\partial z$ is vertical temperature gradient. Wyrtki (1981) assumes a constant value of 0.1×10^{-4} m^2/s for the diffusion coefficient, although he notes that it is probably higher in the region of the Undercurrent. We assume a value of 2×10^{-4} m^2/s for a region 3° wide encompassing the Undercurrent (following the findings of Fahrbach and Bauerfeind, 1982) and 0.1×10^{-4} m^2/s elsewhere. The vertical temperature gradient within the area of the Undercurrent is taken as 8°C/50 m and elsewhere as 12°C/50 m. Temperature change due to vertical mixing is thus estimated as -1.2°C/60 days.

The balance given in (1) can now be written

$$-3.0°C/60\ days = 1.5°C/60\ days - 1.6°C/60\ days - 1.2°C/60\ days - 1.2°C/60\ days$$

$$-3.0°C/60\ days = -2.5°C/60\ days$$

These estimates are obviously qualitative and suffer from being based on combinations of synoptic and climatological data sets. However, they do suggest that, in contrast to the Pacific where the effects of vertical diffusion are small (Wyrtki, 1981), all three processes, horizontal advection, upwelling and diffusion can contribute equally in the Atlantic to the initiation of the cold water tongue.

6. SUMMARY AND DISCUSSION

The qualitative similarities between the climatological and FGGE-year representations of the equatorial SST and thermocline depth fields (Figs. 4 and 5) suggest that the upper layer thermal structure observed during FGGE can be considered typical of an average year. This representativeness implies that phase relationships between atmospheric and oceanic phenomena derived from the FGGE data describe average rather than anomalous conditions.

In summary, the following phase relationships between surface wind and oceanic temperature structure were observed during FGGE. Cooling of surface layers starts coincidentally with the early-May acceleration of the surface wind at 28°W and probably at 4°W (Figs. 9 and 7, respectively) and some four weeks later at 22°W (Fig. 8). Although obvious, it is worth noting that cooling starts later at 22°W than to the west or to the east. The thermocline deepens west of about 35°W (Fig. 5) coincidentally with the wind increase at 29°W (Fig. 9). At 4°W, the thermocline rises and both components of wind increase some time during the same six-week period in boreal spring (Fig. 7). Although from the ship observations it appears that the thermocline rises at 22°W when the wind increases (Fig. 8), temperature data from current meter observations indicate part or all of this rise may be related to higher frequency perturbations of the thermocline (Fig. 11).

The meridional component of the wind remains high at least through October at 4°W, while both components remain high at 28°W (Figs. 7 and 9). SST's are below the record mean at these times and the thermocline has deepened since June at both locations. Winds decrease during March and April when highest temperatures are observed.

The heat budget computations suggest that the initiation of the cold water tongue between May and July is related to upwelling, diffusion and advection. Previous numerical modelling and observational studies suggest how these processes can be related to the increase in wind observed during May.

The magnitude of cooling caused by upwelling is partially a function of the depth of the thermocline (i.e., the closer the thermocline is to the surface, the colder the upwelled water for the same Ekman and geostrophic transports). Previous observational (Katz et al., 1977, for instance) and numerical modelling efforts (Philander and Pacanowski, 1980, for instance) show that the thermocline of the equatorial Atlantic should respond in phase to basin-wide increases in the zonal wind component (i.e., the oceanic response is in equilibrium with the atmospheric forcing). Although the spatial and temporal resolution of the data are limited, there is a strong suggestion of such an equilibrium response in the observations (Fig. 5).

Furthermore, westward winds are also conducive to equatorial upwelling. Thus, as the westward component of the wind increases at all three positions during May (Figs. 7, 8 and 9) upwelling is likely to play a larger role in cooling the surface waters than previously.

Philander and Pacanowski (1981) show that increased meridional winds can produce, through advection, a cold water tongue with spatial characteristics similar to those observed (i.e., asymmetric with respect to the equator, Fig. 2). The meridional component of wind also increases during FGGE. Katz and Garzoli (1982) find that surface speeds of the SEC between 2°S and 2°N have a significant westward component (.25 to .50 m/s) only during July. Thus, advection is likely to play a significant role only at later stages of the development of the cold water tongue. The fact that cooling occurs earlier at 28°W than 22°W also suggests that advection may not play a crucial role in cooling early in the summer.

The thermocline continues deepening at 28°W and redeepens at 4°W through October. However, SST's remain less than average, while at 4°W the meridional component and at 28°W both components of wind remain high (Figs. 7 and 9). In the east, Voituriez (1981) suggests that horizontal divergence at the sea surface south of the equator, induced by the meridional component of the wind, can serve to maintain the tongue. In the west, Molinari (1982), using drifting buoy data, suggests that advection by the SEC plays a large role in maintaining the tongue later in the summer.

As the winds decrease along the equator during boreal winter and spring, SST's increase, but at a slower rate than would be predicted from heat gains through the sea surface (Table 3). Fahrbach and Bauerfeind (1982) show that, at 22°W, during February-March 1979 vertical mixing acts continuously to cause spreading of the isotherms and thus, cooling of the surface layers. They also show that during certain upwelling events vertical advection will enhance the cooling of the surface layers. One such event occurred during 17-20 February 1979 on the western edge of the patch of cold water centered at 16°W during January-February 1979 (Fig. 1). Thus, upwelling and mixing can serve to retard warming of the surface layers and can contribute to the maintenance of the cold water patch observed during boreal winter.

ACKNOWLEDGMENTS

The authors would like to thank the dedicated officers and crew of the research vessels involved in FGGE. The first author was partially supported by the Centre National de la Recherche Scientifique of France, the Department of Energy and the NOAA Special Research Programs Office.

REFERENCES

Bubnov, V. A. (1981). Some features of synoptic and seasonal variability of the Lomonosov Current. In: Recent Progress in Equatorial Oceanography: A Report of the Final Meeting of SCOR Working Group 47 in Venice, Italy, April 27-30, 1981. Edited by J. P. McCreary, D. W. Moore and J. M. Witte. Nova University/N.Y.I.T. Press, 87-96.

Bubnov, V. A. and V. D. Egorihkin (1979). Study of water circulation in the tropical Atlantic. Deep-Sea Res., 26, Suppl. 2, 125-136.

Duing, W., F. Ostapoff, and J. Merle (1980). Physical Oceanography of the Tropical Atlantic during GATE. I.O.C., UNESCO, Paris, 117 pp.

Fahrbach, E. and E. Bauerfeind (1982). The variability of equatorial thermocline spreading as an indication of equatorial upwelling. To appear in Oceanographia Tropicale.

Hastenrath, S. (1980). Heat budget of tropical ocean and atmosphere. J. Phys. Oceanogr., 10, 159-170.

Hastenrath, S. and P. Lamb (1977). Climatic Atlas of the Tropical Atlantic and Eastern Pacific Oceans. University of Wisconsin Press, 105 pp.

Hastenrath, S. and P. Lamb (1978). Heat Budget Atlas of the Tropical Atlantic and Eastern Pacific Oceans. University of Wisconsin Press, 104 pp.

Jarrige, F. and P. Rual (1981). Measurements in the Equatorial Current of the Atlantic and Pacific Oceans. In: Recent Progress in Equatorial Oceanography: A report of the Final Meeting of SCOR Working Group 47 in Venice, Italy, April 27-30, 1981. Edited by J. P. McCreary, D. W. Moore, and J. M. Witte. Nova University, N.Y.I.T. Press, 111-120.

Katz, E. J., R. Belevich, J. Bruce, V. Bubnov, J. Cochrane, W. Duing, P. Hisard, H. U. Lass, J. Meincke, A. deMesquita, L. Miller, and A. Rybnikov. (1977). Zonal pressure gradient along the equatorial Atlantic. J. Mar. Res., 35, 293-307.

Katz, E. J., R. L. Molinari, D. E. Cartwright, P. Hisard, H. U. Lass, and A. deMesquita (1981). The seasonal transport of the Equatorial Undercurrent in the western Atlantic (during the Global Weather Experiment). Oceanologica Acta, IV, 445-450.

Katz, E. J. and S. Garzoli (1982). Response of the western equatorial Atlantic Ocean to an annual wind cycle. To appear in Deep-Sea Res.

Merle, J. (1978). Atlas hydrologique saisonnier de l'ocean Atlantique intertropical. Trav. Doc. ORSTOM, No. 82, 184 pp.

Merle, J. (1980). Seasonal heat budget in the equatorial Atlantic Ocean. J. Phys. Oceanogr., 10, 464-469.

Molinari, R. L. (1982). Sea-surface temperature and dynamic height distributions in the central tropical South Atlantic Ocean. To appear in Oceanologica Acta.

Philander, S. G. H. and R. C. Pacanowski (1980). The generation of equatorial currents. J. Geophys. Res., 85, 1123-1136.

Philander, S. G. H. and R. C. Pacanowski (1981). The oceanic response to cross-equatorial winds (with applications to coastal upwelling in low latitudes). Tellus, 33, 201-210.

Voituriez, B. (1981). The equatorial upwelling in the eastern Atlantic Ocean. In: Recent Progress in Equatorial Oceanography: A Report of the Final Meeting of SCOR Working Group 47 in Venice, Italy, April 27-30, 1981. Edited by J. P. McCreary, D. W. Moore, and J. M. Witte. Nova University/N.Y.I.T. Press, 229-248.

Wyrtki, K. (1981). An estimate of equatorial upwelling in the Pacific. J. Phys. Oceanogr., 11, 1205-1214.

ON THE VARIATION OF THE HEAT CONTENT IN VARIOUS VERTICAL LAYERS IN THE CENTRAL EQUATORIAL ATLANTIC

EBERHARD FAHRBACH

ABSTRACT

From January to June 1979 R.V. "Meteor" working in the FGGE programme surveyed the central equatorial Atlantic on a section along 22° W from 3° N to 2° S. Ten hydrographic sections and moored current meter data are used to study the variations in the heat content of the upper 600 m of the ocean. The changes within four vertical layers defined by fixed isotherms are investigated. The observed changes are attributed to the variations of the equatorial current system. It is concluded that advection by the South Equatorial Current in the surface mixed layer and by the North and South Equatorial Undercurrents in the thermostad and main thermocline determine the observed variations of the heat content which are an order of magnitude higher than the heat flux through the surface.

1. INTRODUCTION

One of the aims of the "First GARP Global Experiment" (FGGE) was the investigation of the role of the ocean in the global heat budget. The meridional heat flux in the North Atlantic Ocean contributes 40 % of the total heat transport. This transport balances the net heat loss north of 30° N (Oort and Vonder Haar, 1976). The tropical Atlantic provides an annual mean net heat gain between 6° N and 6° S of about 50 W m^{-2} (Hastenrath and Lamb, 1978). If one calculates a seasonal time rate of change of the heat content within the upper 300 m, one finds that the rates are an order of magnitude higher than the net heat gain (Merle, 1980). It must be concluded that an important redistribution of heat occurs within the interior of the ocean.

This study discusses the redistribution of heat within a watercolumn in the equatorial Atlantic. The heat content is related to physically defined vertical layers. The interaction of the layers shall give insight to the redistribution processes. The relation between the layers and the atmospheric driving forces shall explain the dynamical processes.

The thermocline separates the surface mixed layer, which is exposed to direct contact with the heat sources, from the interior of the ocean. In the equatorial Atlantic a jet-like undercurrent is confined to the thermocline. The high vertical current shear induces significant vertical mixing, and meridional circulation cells are associated with equatorial upwelling. Both processes are responsible for vertical transports and are visible as equatorial thermocline spreading (e.g. Fahrbach and Bauerfeind, 1982). The inclination of isopycnals create horizontal pressure gradients which induce meridional accelerations.

2. THE DATA

From 27 January 1979 to 23 June 1979, R.V. "Meteor", working in the FGGE program, surveyed the central equatorial Atlantic on a section along 22° W from 3° N to 2° S (Fig. 1). During this time five moorings were deployed on the meridian.

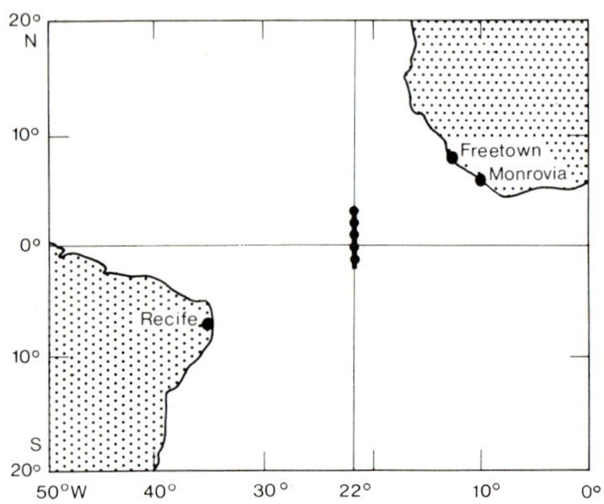

Fig. 1. The area of observation by R.V. "Meteor" during the "First GARP Global Experiment". The dots represent locations of the moorings, the line represents a ten-times repeated hydrographic section.

Taut wire techniques with surface buoys were used in order to measure wind and near surface currents. The equatorial mooring released itself after four days and was reinstalled on 10 February 1979. Because of wire corrosion all moorings had to be removed at the end of March. At 1° S only one instrument could be recovered. Two moorings could be reinstalled from May to June at 0° and 2° N. During the observation period, a hydrographic section down to 600 m was repeated ten times with a continuous Howaldt-Bathysonde CTD and a rosette sampler. The salinity data were recalibrated to fit the salinity samples which were determined with a Guildline salinometer. The stations laid between 10 and 15 nm apart. Comparisons between the conventional CTD data with those obtained by a towed undulating CTD-system, with a profiling distance of one to two nautical miles, shows that the conventional data is strongly aliased (Sy and Meincke, 1981). It is found that averaging over 30 nms remedies this problem. Therefore the data used herein has been averaged over this distance.

In order to illustrate the distribution of the sections in time and to show the temporal aliasing in the CTD data, the temperature record of the equatorial current meter at 15 m is shown along with corresponding CTD temperatures (Fig. 2).

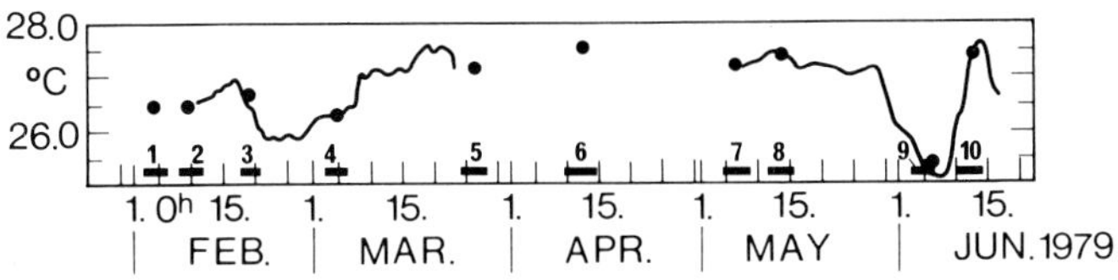

Fig. 2. The time distribution of the ten hydrographic sections carried out during FGGE in a continuous temperature record at 15 m on the equator. The dots represent temperatures measured with the CTD, the bars show the time interval of the sections.

It becomes evident that a linear interpolation between two consecutive sections is not adequate. Another problem with the interpretation of the data is the high advection rates. The South Equatorial Current (SEC) is present at 15 m with an average of 25 cm s^{-1} to the west, and the Equatorial Undercurrent (EUC) flows in 75 m with 100 cm s^{-1} in the opposite direction. It follows that variations which are observed between two consecutive sections with a time interval of 10 days might be the consequences of local processes separated by as much as 900 km.

3. THE LAYERS

In order to study the redistribution of heat within the water column I defined vertical layers and calculated the heat content within each layer at each station. I defined the limits of each of four layers by an upper and lower isotherm (Fig. 3).:

- Surface mixed layer (M) $t > 25$ °C
- Thermocline (T) 25 °C $> t >$ 16 °C
- Thermostad (TD) 16 °C $> t >$ 11 °C
- Main thermocline (MT) 11 °C $> t >$ 9 °C

In contrast to the definition of the transition from the surface mixed layer to the thermocline by a fixed temperature, a given temperature difference to the surface (e.g. ΔT = 0.5 K, Molinari et al., 1983) or a local vertical gradient can be used. In figure 4 the meridional distribution of the mixed layer depth is shown with various definitions for the average over the ten sections. With a definition of ΔT = 0.2 K or 0.5 K a clear minimum in the mixed layer depth appears at 0°30' N, which is not found with the 25 °C definition. The agreement would be better with a 26 °C definition, but at times when there is a cold mixed layer, the 26 °C isotherm lies well in a zone of very small vertical gradients which cannot be assigned to the thermocline. Even if an upward spreading of the

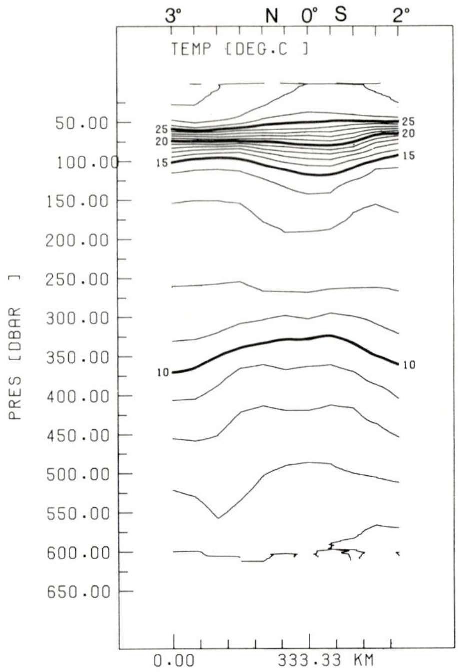

Fig. 3. Mean temperature section along 22° W calculated from ten sections between January and June 1979.

thermocline is better reproduced with the ΔT = 0.5 K definition, I rejected it, because the diurnal temperature variations which are known from the moored current meters fall well into this range. With an average from only two or three stations this can lead to serious deformations of the mixed layer depth. As the aim of this work is the variability of the layers, I compared the time change of variously defined layers at the equator (Fig. 5). As the time changes agree rather well, I accepted the 25 °C definition.

The definition of the transition from the mixed layer to the thermocline by the vertical gradient was rejected because strong vertical smoothing was necessary to avoid spurious results. In consequence the effect of the time changes was highly reduced. This is true for the deeper layers as well.

The thermocline depth is defined by other authors (e.g. Merle, 1980) by the depth of the 23 °C isotherms. This corresponds well with our upper limit of the thermocline, defined by 25 °C (Figs. 4 and 5).

For each individual station I calculated the heat content of each layer as the vertical integral over the temperature multiplied with the mean density and the specific heat.

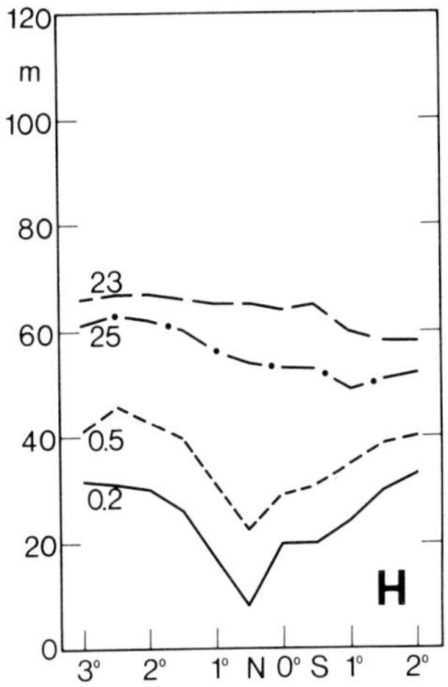

Fig. 4. The depth of the average surface mixed layer and thermocline defined in different ways. The curves denoted by 0.2, 0.5 and 25 stand for mixed layers defined by a difference of 0.2 K or 0.5 K to the temperature in 5 m depth. The depth of the 23 °C and 25 °C isotherms are denoted by 23 and 25.

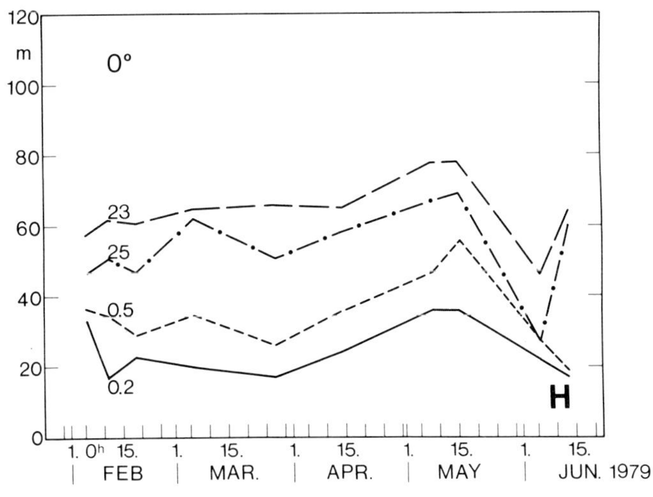

Fig. 5. Time variation of the mixed layer depth at the equator defined and denoted as in Fig.4.

4. THE MEAN CONDITIONS

The mean conditions, defined over our observation period, are somewhat artificial, because they include a warm period from January to mid-February, this is followed by a cooler period lasting until the beginning of March, then it stays warm until the onset of the summer cooling at the end of May (Fig. 2). Even if this data cannot produce a clear seasonal or annual mean, the effect of averaging is to reduce the effect of short period noise.

In figure 6a the extent of the different layers is shown. The surface mixed layer (M) decreases slightly from north to south. The thermocline (T) is clearly spread by about 30 m near the equator. The thermostad decreases by about 50 m with a minimum at 0°30' S. The squeezing of the thermostad is not only due to the spreading of the thermocline, but also to a clear doming from below (Fig. 3). The same effect is observed for the main thermocline (MT).

If one looks at the heat content within the individual layers, we find a rather similar feature (Fig.6b). The surface mixed layer is characterized by a decreasing heat content from the north to the south. The maximum in the thermocline is approximately balanced by the minimum in the thermostad. The observation of the good correlation between the heat content and the layer thickness is confirmed by the calculation of regressions. It shows correlation coefficients higher than 0.98 for all layers. It follows that the change in the layer thickness is the dominant process producing the change of the heat content within a layer.

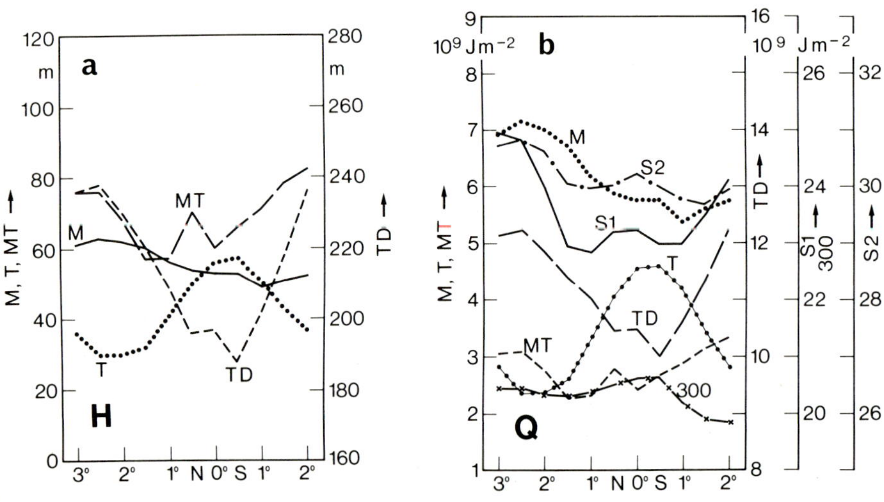

Fig. 6. (a) The vertical extent (H) and (b) the heat content (Q) of the various vertical layers calculated from an average over five months. The symbols denote the following layers:

M - Surface mixed layer
T - Thermocline
TD - Thermostad
MT - Main thermocline
S1 - Sum over the upper 4 layers
S2 - Sum over the complete water column of 600 m
300 - Sum over the upper 300 m

When one calculates the sum over the four layers (Q_{S1}), one finds a clear minimum between 1°30' N and 1° S. This sum corresponds to the heat content of the "Warmwassersphäre" (Defant, 1928) which is supposed to be the direct heat reservoir of climate (e.g. Krauss, 1980).

As the heat content is mainly determined by the individual layer thickness, the heat content between fixed depths, e.g. Q_{S2} between 0 m and 600 m, and Q_{300} between 0 m and 300 m, show much smaller variability. Q_{S2} follows the north-south decrease of the mixed layer, but shows a slight maximum at the equator. This is due to the dominating downward spreading of the thermocline.

Q_{300} shows the same effects but with less intensity. It is calculated to compare the present results with those of Merle (1980), who has extended his calculations down to this level.

5. THE TIME VARIATION OF THE LAYERS

One can define the heat content (Q) of a given layer as the product of the vertical mean temperature (t) within the layer and layer thickness (H). In consequence the time variation of the heat content can be due to a change of mean temperature and/or a change in layer thickness. From the meridional distribution (Figs. 6a and b) one sees that layer thickness is the dominating effect. This is not surprising if one considers the equation:

$$dQ = \rho c_p \ (Hdt + tdH)$$

From typical values in the various layer I calculated the increase of the mean temperature (dT) which is equivalent to a 1 m change in the layer thickness (Table 1).

	t	H	dT
Surface mixed layer	27 °C	60 m	0.45 K
Thermocline	19.5 °C	50 m	0.39 K
Thermostad	12.5 °C	225 m	0.06 K

Table 1. Typical numbers for the various layers; see text.

If one considers that the layer thicknesses are varying by tens of meters, a corresponding variation of the mean temperature would be incompatible with a definition of layers by fixed temperatures. Because the layers are chosen as regions where the vertical gradient is varying only slightly the mean temperature of the layer should stay constant. If not, a deviation from linearity has to be assumed. If the mean temperature decreases, the gradient in the upper part of the layer must become higher than the one in the lower portion.

The extent and the mean temperature of the three upper layers at 3 °N and at the equator for the ten sections is shown in figure 7. The comparison of the temperature and depth ranges shows that the normal contribution of the temperature change is about ten per cent. In the surface mixed layer and the thermostad exceeding rates are found. This is due to the fact, that the surface mixed layer has no upper temperature limit and the thermostad shows the highest variance in vertical gradients.

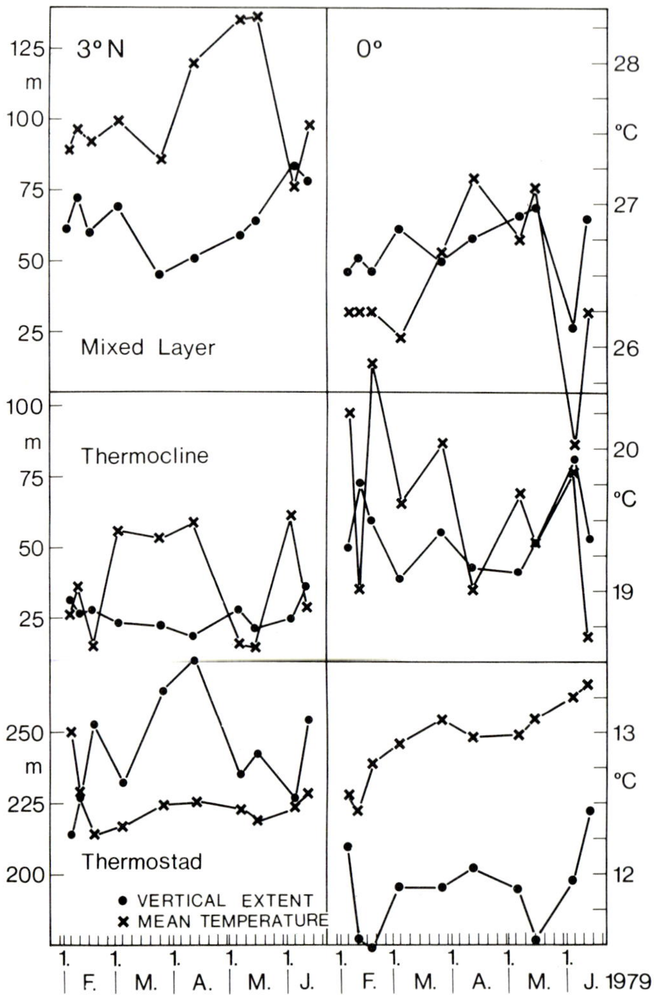

Fig. 7. Vertical extent and vertical mean temperature of three layers at 3° N and at the equator.

In the mixed layer, increasing layer depth correlates with increasing temperature at 3° N and at the equator. This can be understood by the fact, that during cold periods the uppermost part of the thermocline is included in the mixed layer defined by a fixed temperature.

In the thermocline at 3° N one finds in most cases an increasing layer thickness combined with decreasing temperature. Due to the equatorial thermocline spreading the variability increases towards the equator. During sections three and nine the upward spreading produces an increase of temperature. In the long period variation low temperatures correlate with a thin thermocline layer.

In the thermostad no general correlation is found. In February and May its thickness increases with decreasing temperature. Positive correlation is due to a decrease of the upper gradient in the thermostad, together with a decrease of the layer thickness. The observation, that this gradient decreases together with the thermostad thickness can be explained when one takes into account, that the thermostad is supported by mixing from above (e.g. Halpern, 1980). Decreasing mixing would increase the upper gradient and decrease the supply of well mixed water. In consequence the thickness of the thermostad would decrease. It is evident that this observation does not exclude other processes which can account for negative correlation and will be discussed later.

6. THE TIME VARIATION OF THE HEAT CONTENT

The meridional distribution of the time averaged heat content (Fig. 6b) of the upper 600 m corresponds to that of the mixed layer. Therefore one expects a correlation of the heat content with the mixed layer temperature (Fig. 2). Surprisingly the correlation is negative. The warmest temperatures from March to May correspond to the smallest heat content (Fig. 8). This is visible in the upper 300 m; more evident in the upper 600 m; and strongest in the "Warmwassersphäre".

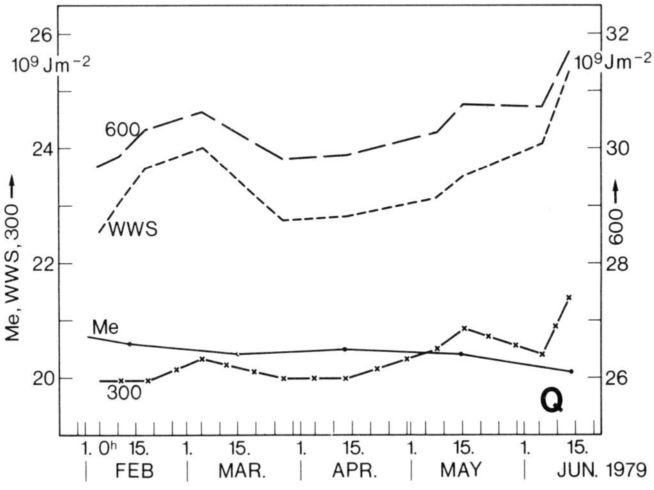

Fig. 8. The time variation of the meridional average from 3° N to 2° S of the heat content in the upper 300 m, 600 m and in the "Warmwassersphäre" calculated from the FGGE data (denoted by 300, 600 and WWS). The heat content in the upper 300 m, calculated by Merle from historical data is denoted with Me.

The trend within the upper 300 m can be compared with the results from Merle (1980). The strongly reduced amplitude seen in Merle's heat content compared to that found in this study is due to the space and time averaging in the former Nevertheless, the gross features correspond. There is a delay of several months between Merle's and the present data. In Merle's data, the springtime maximum occurs in April and the minimum is observed in July.

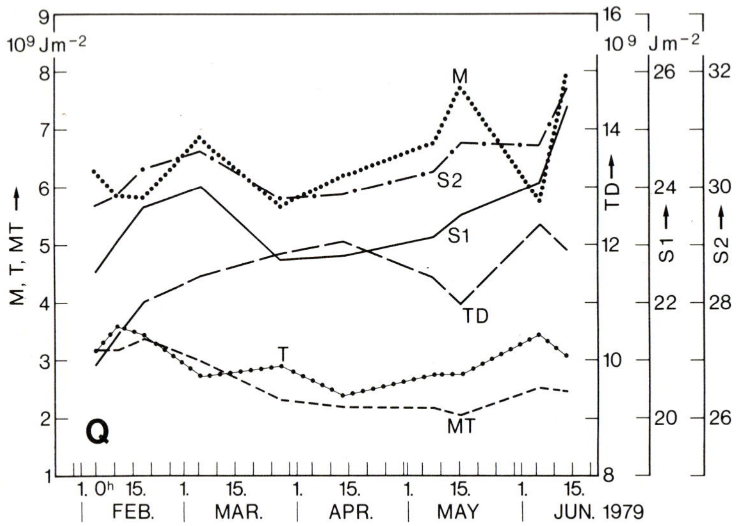

Fig. 9. The time variation of the meridional average from 3° N to 2° S in the layers shown in Fig. 6.

The contribution of the individual layers to the sum of the heat content is shown for the meridional average in figure 9. Whereas thermocline and main thermocline are varying very smoothly as do Q_{S1} and Q_{S2}, the mixed layer and thermostad interact intensively in a shorter time scale. This impression is biased because there are more extra equatorial stations than equatorial ones, averaging out equatorial thermocline spreading. In figures 10 and 11 for each layer, five latitudes between 3° N and 2° S are presented.

The heat content of the upper 600 m (Q_{S2}, Fig.10a) shows a smaller meridional variability as the individual layers. It should be noticed that the maximum in March is due to an increase at the southern stations whereas the maximum in June results from an increase in the north. The surface mixed layer (Fig. 10b, Q_M) shows a similar meridional dependence. The springtime maximum is preceded by even higher values in the north. The increase towards the maximum is nearly a month later than that of Q_{S2}. A striking feature occurs during the summer cooling, where the heat content continuously increases in the north but decreases dramatically south of 1° N with a minimum at the equator. The thermocline (Fig.11a, QT) shows highest variability at the equator reflecting equatorial thermocline spreading. The maxima occur in February and June, simultaneously with an increase of Q_{S2}. The mostly negative correlation between 1° N and 1° S can be explained by a meridional fluctuation of the thermocline maximum across the equator. In the thermostad (Fig. 11b, Q_{TD}) one finds the strongest variations furthest from the equator. A 180°-phase relation between the northern and the southern stations indicate a tilting of the lower boundary across the equator. The strongest increase of the meridional average of Q_{TD} in February and June coincide with maxima of Q_{TD} at the

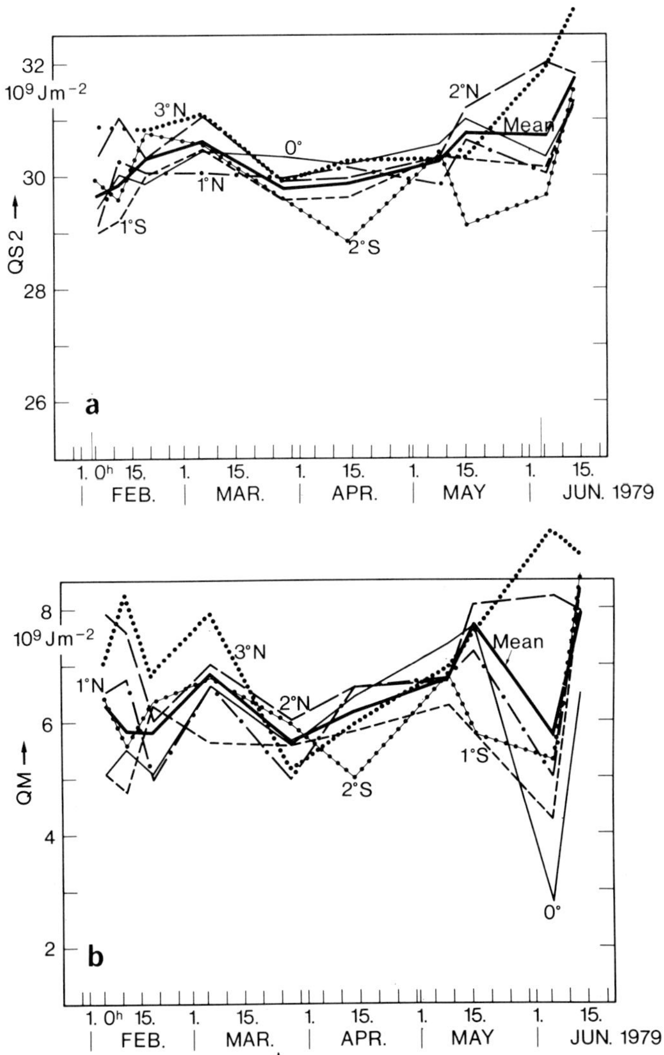

Fig. 10. The time variation of the meridional average from 3° N to 2° S of the heat content (a) in the layer from the surface to 600 m, (b) in the surface mixed layer.

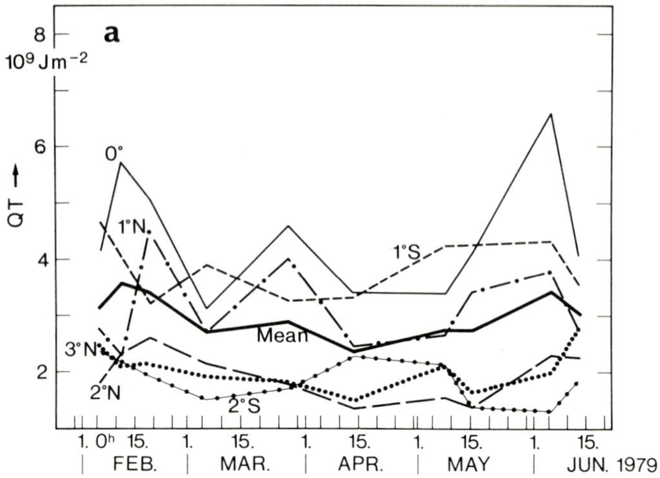

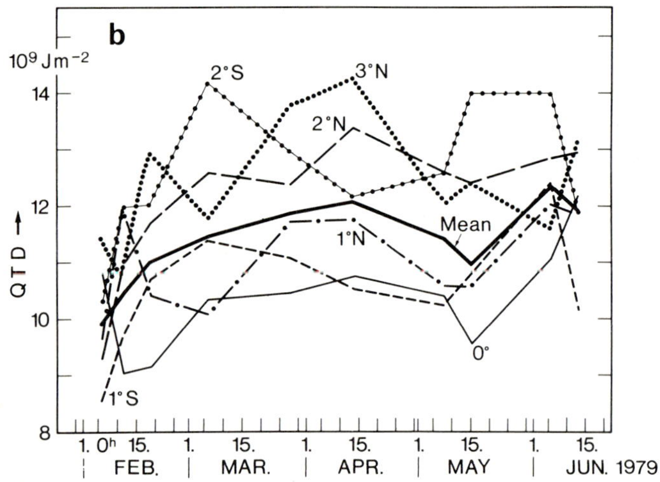

Fig. 11. The time variation of the meridional average from 3° N to 2° S of the heat content (a) in the thermocline, (b) in the thermostad.

equator. The increase of Q_{TD} proceeds from the south to the north, it plays the significant role in the increase of Q_{S2} because Q_M and Q_T cancel each other.

7. DISCUSSION AND CONCLUSIONS

The increase of the variations of the heat content with the layer depth (Fig. 8) and the correlation of periods of strong increase within the thermostad and the upper 600 m, indicate that the thermostad plays a significant role in the heat budget of the equatorial Atlantic. As the thermostad is not in direct contact with the heat sources, variable advection has to be taken into account to understand this observation.

The mean current system from January to March within the area of observation is shown by the data of the moored current meters (Fig. 12). The westward South Equatorial Current (SEC) is observed in the mixed layer. It supplies warm water from the border of the Gulf of Guinea. The eastward Equatorial Undercurrent supplies high saline water from the South Atlantic subtropical salinity maximum. It reaches down to the thermostad, where westerly flow is found in the Equatorial Intermediate Current (EIC), extending down to the main thermocline. This current is related to a west-east slope of the isotherms (e.g. Merle, 1980). The resulting pressure gradient creates a meridional convergence when the Coriolis force increases off the equator. This convergence is visible in the current measurements (Fig. 12) and as a doming in the temperature (Fig. 3). To the north and

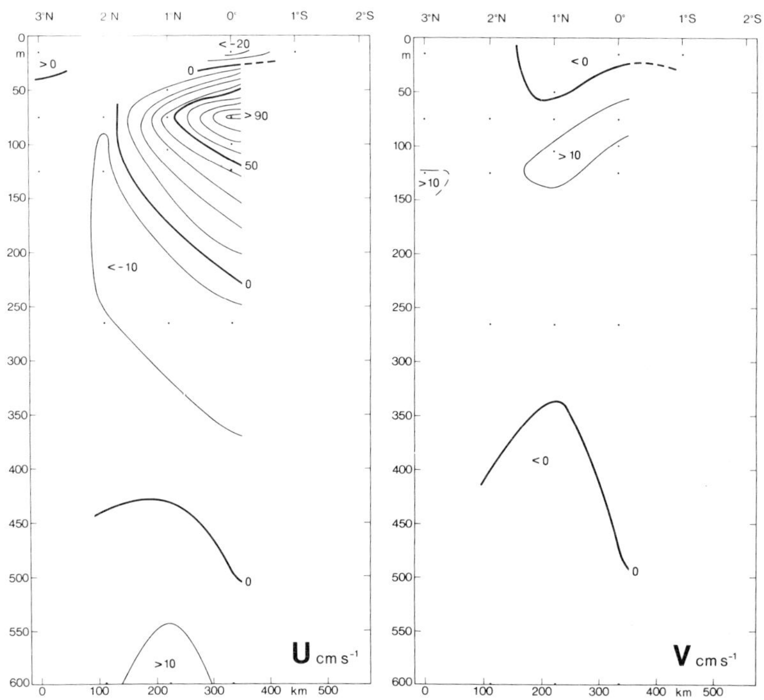

Fig. 12. The mean currents from moored current measurements between 31 January and 30 March 1979, u - zonal component, positive to the east, v - meridional component, positive to the north.

south of our observation area the main thermocline slopes upward giving rise to the North and South Equatorial Undercurrents (NEUC, SEUC, e.g. Voituriez, 1981). Both currents transport South Atlantic Central Water (SACW). The origin of the SACW is assumed to be the subtropical front in the South Atlantic (Defant, 1936). An increase of the transport by the undercurrents would widen the thermostad away from the equator, and therefore increase the heat content of the water column. Consequently, the doming will be enhanced. A further mechanism to produce thermostad water characteristics is the vertical mixing between high saline water of the EUC and the deep water below as pointed out by Halpern (1980). This process increases the thermostad at the equator together with the thermocline spreading. If this process is active, a decrease of the upper layer content should be observed.

The identification of these two processes is possible by looking for the most intense thermocline spreading which is found in sections three and nine. During both cases the heat content has increased in the thermostad and decreased in the mixed layer. Because the mixed layer shows the highest heat content in the north correlated with currents to the west, one can conclude that the increase in the thermostad is due to advection of heat by the SEC, and consecutive downward transport.

Slight thermocline spreading associated with intense doming of the lower boundary of the thermostad appears during sections four and six. This has to be interpreted as influence of the undercurrents.

The variation of the meteorological situation during the observation period is presented in figure 13. The periods of strong southeasterly winds during

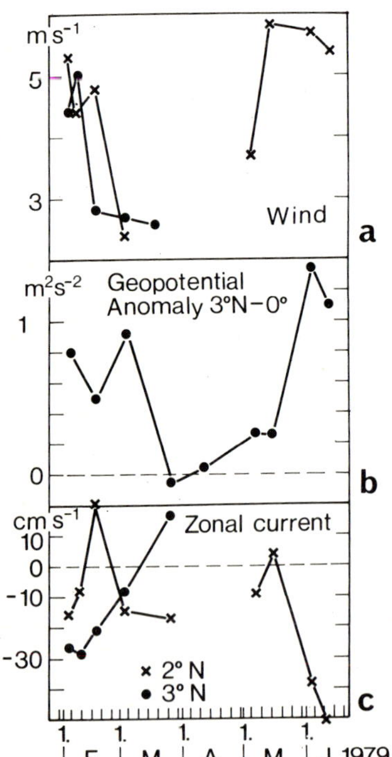

Fig. 13. (a) Daily means of the wind vector velocity measured on buoys at 2° N and 3° N. (b) The difference between the geopotential anomaly at 5 dbar relative to 500 dbar at 3° N and the equator. (c) The 4-day means of the zonal current at 75 m (positive to the east) from moored current meter measurements.

February and June are interrupted by a period of light winds when the Intertropical Convergence Zone (ITCZ) has moved southward (Köhne and Speth, 1982). The geopotential anomaly and the zonal currents vary concurrently with the wind. The periods of heat supply from above are correlated to the presence of the SEC. When the SEC weakens, thermostad heat content nevertheless increases by supply of SACW through the undercurrents.

It must be noticed that both processes can act simultaneously. This is the case in February, when the heat gain in the thermostad is higher than the heat loss from the layers above. In this case the undercurrents overcompensate the heat loss due to relaxing of the SEC and provide a net increase of the whole water column.

ACKNOWLEDGEMENTS

This data set could only be obtained through the direct participation of C. Brockmann and J. Meincke. A. Sy processed the complete STD-data set. I am very grateful to Mrs. Petersen for preparing the drawings and Mrs. Schuster for typing the manuscript.

REFERENCES

Cochrane, J.D., Kelly, F.J., Jr. and Olling, C.R. 1979. Subthermocline Countercurrents in the Western Equatorial Atlantic Ocean. J. Phys. Oceanogr., 9: 724-738.

Defant, A. 1928. Die systematische Erforschung des Weltmeeres. Zeitschr. Ges. Erdkunde: 459-505.

Defant, A. 1936. Die Troposphäre. Deutsche Atlantische Exped. "Meteor" 1925-1927, Wiss. Erg., Bd. VI, Teil 1 (3): 289-411.

Fahrbach, E. and Bauerfeind, E. 1982. The variability of the equatorial thermocline spreading as an indication of equatorial upwelling in the Atlantic Ocean. Oceanogr. trop., 17: 119-130.

Halpern, D. 1980. A Pacific equatorial temperature section from 172° E to 110° W during winter and spring 1979. Deep Sea Res., 27A: 931-940.

Hastenrath, S. and Lamb, P. 1978. Heat Budget Atlas of the Tropical Atlantic and Eastern Pacific Oceans. University of Wisconsin Press, Madison: 112 p.

Köhne, A. and Speth, P. 1982. Variation of the ITCZ in the Atlantic during FGGE. Trop. Ocean-Atmosphere Newsletter 9, p. 7.

Krauss, W. 1980. Golfstrom und Nordostatlantische Wärmeanomalie. Umschau 80, Heft 6: 167-174.

Merle, J. 1980. Seasonal Heat Budget in the Equatorial Atlantic Ocean. J. Phys. Oceanogr., 10 (3): 464-469.

Molinari, R.L., Katz, E., Fahrbach, E., Lass, H.-U. and Voituriez, B. 1983. Near surface temperature observations obtained in the equatorial Atlantic Ocean during FGGE (1979). Subm. to Proceedings of the May 1982 Liège Colloquium on Ocean Hydrodynamics.

Oort, A. and Vonder Haar, T.H. 1976. On the observed annual cycle in the ocean-atmosphere heat balance over the Northern Hemisphere. J. Phys. Oceanogr., 6: 781-800.

Sy, A. and Meincke, J. 1981. A comparison of hydrographic features observed in the Equatorial Atlantic during FGGE using a conventional CTD and a towed system. FGGE-Equatorial Oceanography, SCOR WG 47, Venice Conference Proceedings, 55-60.

Voituriez, B. 1981. Les sous-courants équatoriaux nord et sud et la formation des dômes thermiques tropicaux. Oceanologica Acta, 4 (4): 497-506.

FREE DRIFTING BUOY MEASUREMENTS IN THE INDIAN OCEAN EQUATORIAL JET

Gilles REVERDIN[1], Michèle FIEUX[1], Joseph GONELLA[1] and Jim LUYTEN[2]

Abstract

Twenty one buoy trajectories obtained in the Indian Ocean from May 1979 to June 1982, are used to infer the current system within 10° of the Equator. The currents were mainly zonal and manifested a dominant semi-annual cycle. They were confined within four degrees of the Equator. This cycle is predominantly related to the Monsoon wind regime; it is associated with zonal redistribution of mass, induced by the zonal wind component or by basin scale pressure gradients. Interannual variations are important, but they did not mask the occurence of the jet. Part of the variability was related to zonally propagating events, as the one responsible of the reversal of the autumn jet. They were also waves in the meridional component of the currents at the equator in the period range extending from 5 to 20 days. Meanders in the western Arabian Sea North of the Equator during the south-west Monsoon are probably connected to the intense eddy activity near the coast in this area.

1. INTRODUCTION

During 1979-1981, 21 drifting buoys were deployed in the equatorial Indian Ocean for the study of the seasonal variability of the equatorial currents. The investigation was made possible by the participation of France (Laboratoire d'Océanographie Physique, with the logistic support of the Terres Australes et Antartiques Françaises), England (IOS), Australia (CSIRO) and the United States (WHOI).

Due to the disappearance of the Coriolis force at the equator as well as its variation with latitude, equatorial areas are distinctive in two ways. First, zonal momentum transferred from the atmosphere to the ocean concentrates on the equator to form a narrow accelerating jet. Second, the equator acts as a wave

1. LOP Museum / CNRS LA 175 / 43, rue Cuvier — F-75231 Paris Cédex, France.
2. WHOI Woods Hole Mass. 02543 U.S.A.

guide. Flow changes induced by the presence of coastlines,variability in wind patterns and instabilities of the current system tend to propagate along the equator (Moore et al., 1977). Various observations reveal that equatorial waves occur over a very large domain, from 2 - 3 days to a year (see, for example : Eriksen, 1980 ; Luyten et al., 1982 ; Weisberg et al., 1979).

In the Indian Ocean, low frequency waves and semi-annual formation of an equatorial jet are prominent features. These phenomena are related to the seasonally varying monsoon circulation. Measurements made from 1973 to 1975 in the vicinity of Gan Island show that the jet is approximately in phase with the local east-west component of the wind stress (Knox, 1976).

The buoys were deployed near the equator, mostly in the western Indian Ocean, between May 1979 and November 1981, during the SINODE cruises (Surface Indian Ocean Dynamics Experiment). Current variability was evaluated for different time scales : the entire three year data set provides a picture of seasonal and interannual variability, while individual trajectories indicate the distribution of kinetic energy of a Lagrangian tracer in the 1 - 30 day range.

In this paper, we will evaluate, in a first part, the accuracy of a drifting buoy to serve as a tracer of near surface motions.We will then investigate the seasonal cycle of the currents, within the data strip within 10° of the equator. The width of the equatorial jet is evaluated, as well as the mass transport associated with it. In a third part, Lagrangian statistics are established, to find how the currents correlate in space and time. In a fourth part,wave like events in the buoy shift velocities are described. Wave packets propagating to the west, and associated to the disappearance of the Eastward equatorial jet, are analysed in view of the theory of the spin up of an ocean from rest. Meanders on two trajectories North of the Equator in August, September are then compared to investigate the propagation of the signal. Finally, the temporal variability for periods less than a month is studied to determine if trapped equatorial waves are observable as those being observed on moored current data in the equatorial regions (Luyten and Roemnick, 1982 ; Weisberg et al., 1979).

2. ACCURACY OF CURRENT MEASUREMENTS

The SINODE drifting buoys were drogued at 15 meters in order to couple the surface portion of the buoy to the surface mixed layer motion. However, the buoy motion is complicated by the surface wave action and the drag of the wind on the buoy. For these buoys, both effects could theoretically produce a 5 cm/s error in the measurements of the current, for wind intensities of 10m/s at 10 meters above the sea surface (Bervas,1978). The spectrum of the velocity of the buoy caused by the wind drag should be directly proportional to the spectrum of the wind (Kirwan & al., 1978). In May and June 1979, current profiler measurements from the Marion Dufresne, located within 100 km of some of the buoys were taken. The difference between the Eulerian and the Lagrangian eastward component of the velocity was less than 5 cm/s. At this time, the wind was to the east at 5 m/s. This gives an upperbound of the wind effect on drogued buoys for typical conditions in the equatorial Indian Ocean.

Most of the buoys were tracked by the Argos system. An analysis of the tracking accuracy was done for a buoy anchored at the equator in the eastern Atlantic Ocean. Standard deviation about the mean buoy position was 300 meters distributed in first approximation like white noise. During the SINODE experiment buoy positions were collected 6-8 times/day. This implies an error in one day mean current less than 1 cm per second ; it decreases for longer time scales and becomes negligible after two days (fig.1). There can also be an aliasing of the tidal currents at the semidiurnal period due to the sparsity of the positions. To test this effect, we simulated sampling by Argos on current meter data at 150 meter depth moored during F.G.G.E. in the equatorial Indian Ocean. No significant rejection of energy at lower frequencies was found. It appears therefore, that errors in sampling and positioning are not masking oceanographic current variability for time scales longer than a day. In order to filter out these errors the trajectories were smoothed with a one day filtering process.

3. SEASONAL CYCLE

The main objective of the program was to study the seasonal cycle of the equatorial currents because the data are scarce and because the circulation is primarily zonal on the basin scale, the data are averaged temporally and zonally. We chose a temporal resolution of one month. Due to the meridional displacements of the buoys during that time, the spatial resolution was chosen as 2°30' in latitude.

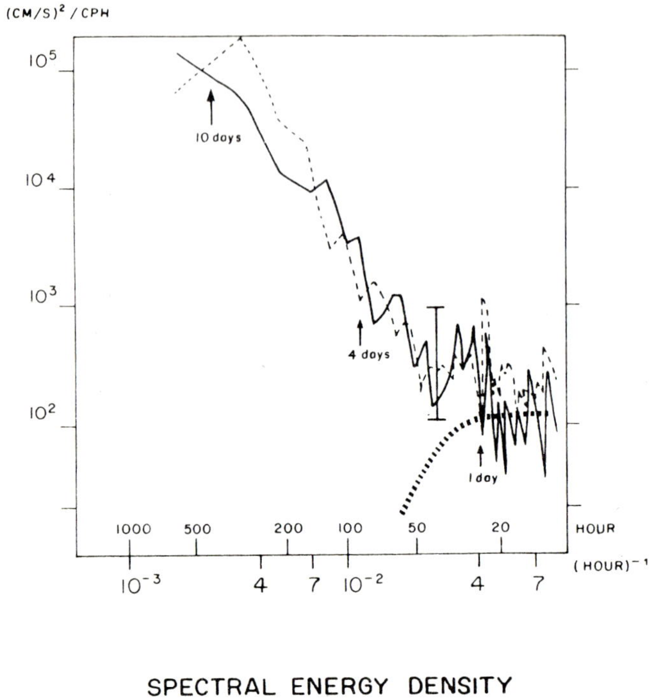

Fig. 1 : spectrum of the zonal and meridional component of buoy 2105 drift velocity. The 90 % confidence interval is indicated. In thick dashed line, noise spectrum due to the positioning by Argos (evaluated in the Gulf of Guinea).

Monthly means were constructed from the zonally averaged buoy velocities, combining the data from all three years. It is difficult to distinguish here between zonal inhomogeneities and interannual variations. However, for the basic features of the seasonal cycle, monthly mean buoy velocities are in agreement with historical data, including current meter and hydrographic observations (Table I, fig. 2).

The estimated variance due to zonal inhomogeneities is of the order of $(30\ cm/s)^2$. This variance is of the same order of magnitude as the variance obtained from the historical data, including part of interannual and zonal variations (Table I). This leads us to estimate, that interannual variations are probably not larger than $(30\ cm/s)^2$, or approximately 25 % of the seasonal cycle.

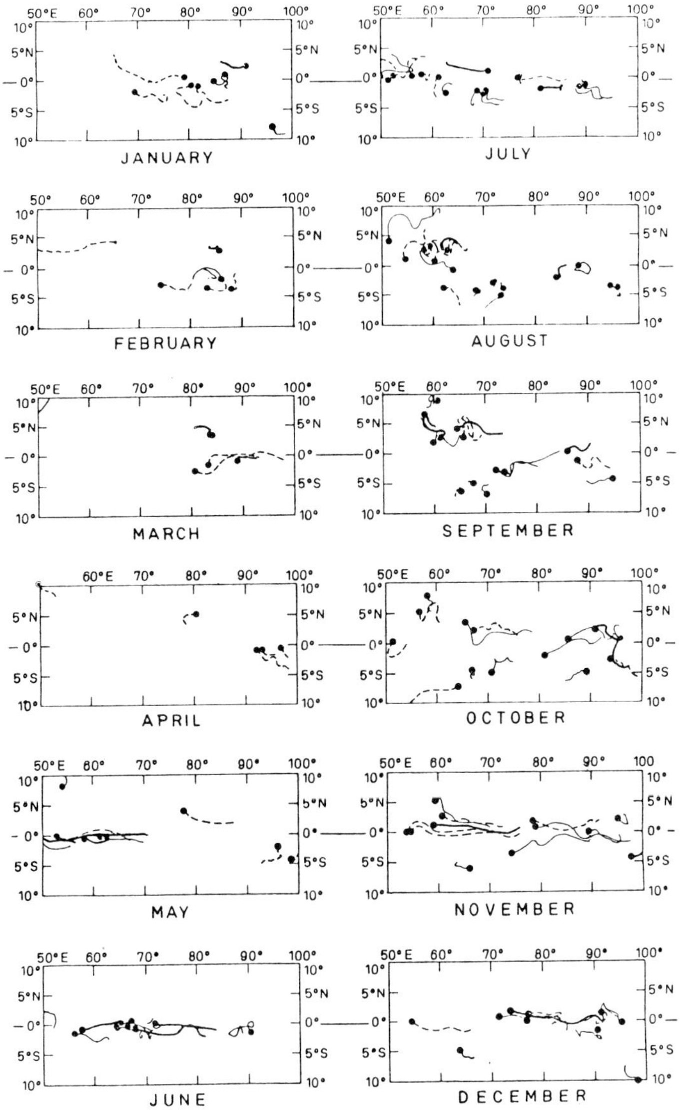

Fig. 2. Map of monthly buoy trajectories. The three years of data are indicated by different lines. Thin full line for May 1979 - April 1980; thick full line for May 1980 - April 1981; dashed line for May 1981 - June 1982. The dots correspond to the origins of the monthly drifts of the buoys.

TABLE 1

Equatorial current measurements, estimated from various papers (cm/s)

Year \ Month	J	F	M	A	M	J	J	A	S	O	N	D	Authors
1964			-60*	80*		40*							(1)
1973	-20	-80	-40										(2)
1973	20	-25	-5	100	80	50	30	-20	30	80	120	120	
1974	-50	-60	-50	10	80	50	40	0	10	60	80	80	(3)
1975	10	-40	0	40	10								
1975		-70	-40	-10	40								
1976			-30	0	50	30							(4)
1976				10	60								(5)
1977	-10	-22	25	60	55								(6)
Average	-10	-50	-25	36	53								
Variance	28	24	29	40	24								

N.B. The stars indicate that the measurements were too local in time or space.

Authors :

(1) Swallow (1967) - Sections on the Western Indian Ocean.
(2) Kort (1977) - Stations.
(3) Knox (1976) - Weekly near Gan 73° E .
(4) Leetmaa & Stommel (1980) - Sections.
(5) Neyman & al. (1978).
(6) Stommel & Regier (private communication) - Buoy drifts.

The main features of the monthly mean circulation (fig. 3) are :

- the reversal of the currents twice a year ;
- a very strong equatorial eastward jet in April-May and October-November ;
- a weak eastward flow in July-August ;
- and an antisymmetric circulation in January-February with eastward flow of the equator and westward flow north of it.

On the average, the current at the equator is directed toward the east. The jet is concentrated in the vicinity of the equator (at 5° N and 5° S, the currents are significantly weaker) and maximum monthly drift velocities are over 1m/s. The maximum daily currents observed on buoy trajectories are larger (1.4m/s). The current system is mainly zonal. The meridional velocities on the equator are possibly related to a Sverdrup balance (Mc Phaden, 1982). Farther from the equator, the wind fields tend to show that the meridional velocities are probably related to an Ekman transport.

To study further the seasonal cycle, we fitted by least squares, the meridional profiles of the zonal velocity for each month, by a Gaussian

$$F(y) = A \exp\left(- \frac{y^2}{2\,\ell^2}\right)$$

In this expression, y is the latitude in degrees and ℓ was chosen by least squares fit of the observations. The half-width is between 200 and 250 km.

There exist different ways of estimating theoretically the half-width of the jet. The scaling of the linear wind driven circulation, corresponds to a radius of deformation

$$\lambda = \left(\frac{N_0 h}{\beta}\right)^{1/2}$$

where h is the thermocline depth, N_0 a characteristic Brünt-Väisälä frequency of the upper ocean and β the equatorial value of the Rossby parameter (Philander et al., 1980). With typical values for the Indian Ocean of $h = 200$ m and N_0 in the range $10^{-2}\ s^{-1}$ to $2\ 10^{-2}\ s^{-1}$, we obtain for λ between 200 and 260 km. In terms of the equatorial wave theory, the equatorial radius of deformation depends on the vertical mode n, as

$$\lambda_n = \sqrt{\frac{c_n}{\beta}}$$

The wind forcing is probably projected mainly on the first and second modes. With values of the wave speeds c_n computed from profiles at 62° E from the Marion Dufresne in May and June 1979, we determine :

$n = 1$ $\quad c_1 = 2.50$ m/s $\quad \lambda_1 = 310$ km

$n = 2$ $\quad c_2 = 1.60$ m/s $\quad \lambda_2 = 250$ km

Observations made in the Indian Ocean in 1979 and 1980 indicate, using the vertical propagation of the semi annual cycle of the deep currents that the equatorial radius of deformation should have a value ~ 280 km (Luyten et al., 1982). Our value is slightly smaller. The difference is probably related to the fact that non linearities tend to narrow eastward jets significantly (Philander, 1979; Cane, 1979).

More than seventy per cent of the energy in the meridional profiles is contained in the fitted Gaussian. The amplitude of the Gaussian (fig. 4) gives an indication of the zonal transport of mass in the surface layer. The bars in fig.4 give an indication of the variance of the estimates. The variance is due to the interannual variability, the zonal inhomogeneity or the small time scale variability in the currents measured by the buoys, but also to the fact that the Gaussian does not fit exactly the real profiles. When the currents are symmetrical about the equator, the Gaussian should correctly represent the jet. This is evidently not the case in January and February,neither in July and August.As the winds are predominantly cross-equatorial at this time,the currents should be antisymmetrical(Philander & al., 1981[b]). Therefore, the apparent errors in January

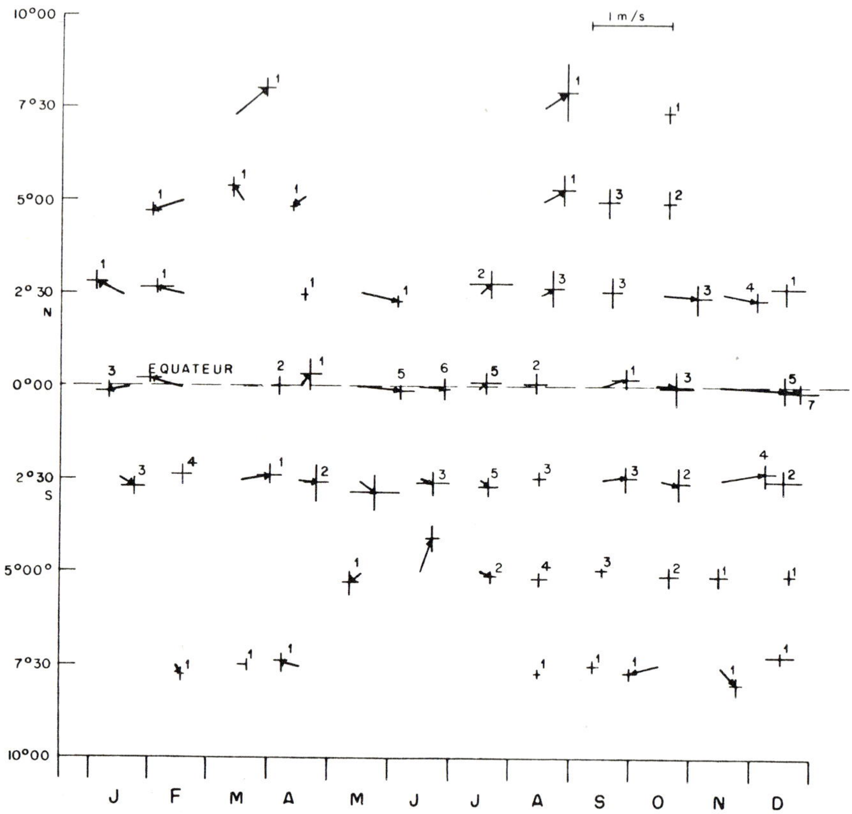

DRIFTING BUOYS VELOCITIES

INDIAN OCEAN-1979-1982

Fig. 3 : monthly mean currents obtained for each 2°30' latitudinal strip. The crosses indicate the variance of the daily current distribution about the average represented by the arrows, which start at the grid points.

and February represent the current system observed during these months. Nevertheless, the amplitude of the Gaussian should still provide a good indication of the zonal transport of mass in equatorial areas.

It is interesting to compare the currents with the climatological variations of the 20°C isotherm in the Indian Ocean west of 65°E within 5° of the equator (Quadfasel, 1982), as obtained from historical data. Assuming no entrainment through the thermocline, the equation relating the depth h of the 20°C isotherm to the transport of the jet is :

$$\frac{d(hS)}{dt} + Tj + Tb = 0$$

where S is the surface of the domain, Tj is the transport by the jet and Tb is the mass transport through the northern and southern boundaries. If Tb is negligible, Tj and h should be in quadrature. Such a phase lag between h and the jet magnitude is observable (fig.4). In order to evaluate quantitatively such a balance, the transport by the jet is computed by assuming a homogeneous depth of 80 meters affected by the eastward transport. We also substract the mean eastward transport which corresponds to a mean jet magnitude of 25 cm/s.

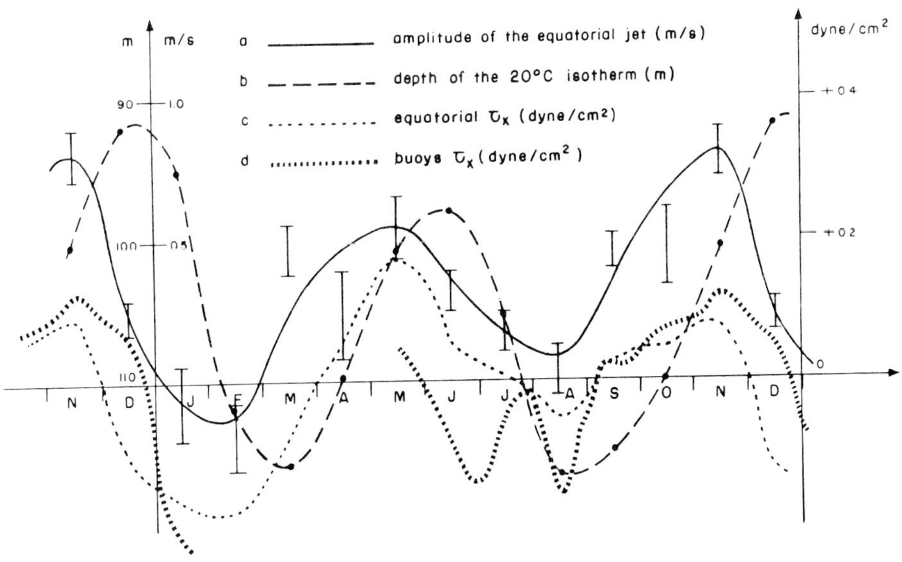

Fig. 4 : the equatorial seasonal cycle.
a the amplitude of the jet deduced from buoy drift data. The vertical bars give an estimate of the variance for each monthly estimate. The curve was fitted visually.
b the depth of 20°C isotherm averaged over a domain west of 65°E and between 5° North and 5° South, according to Quadfasel (1982).
c and d represent the zonal wind stress for the periods : first January 1979 - 11 th November 1979, first March 1981 - 7 th March 1982. The wind stress was averaged over a five degrees wide equatorial band (c), and along the buoy tracks in the same band (d). τ_x is computed from bulk aerodynamic formula :

$$\tau_x = \rho\, C_D\, |V|\, V_x \qquad \text{with} \qquad \rho C_D = 2\ 10^{-6}$$

The evaluated Tj and $-\frac{d(hS)}{dt}$ present some striking differences (table 2), but the orders of magnitude are comparable.

Table 2

Estimated seasonal transports of mass in boundary layer

Month		J	F	M	A	M	J	J	A	S	O	N	D
1	10^6 m^3/s	-10.5	-20	-3	6	11	2	-6.5	-9	2	15	22	-4
2	10^6 m^3/s	-10	-12	0	10	8	0	-13	-4	4	8	11	0

1) Jet eastward transport after substracting the mean yearly average.
2) Transport associated with the variation of mass in the western Indian Ocean.

The transport by the jet is 50% larger than values that can be explained exclusively by the changes in the mass content in the western Indian Ocean, between 5° N and 5° S (Table II). There are numerous reasons for this. In the first place there could be a westward transport in the thermocline above the 20° C isotherm, which is not accounted for with buoy drifts. This is probably the case from May to January, where the pressure gradient might be directed toward the west (Eriksen, 1979). In the second place, the variations of the 20° C isotherm in the west are not restricted within 5° of the equator (Quadfasel, 1982). There is in particular a very large depression of the thermocline in the Northwestern Arabian Sea, offshore of Somali, in the great gyre during the Southwest Monsoon (Swallow et al., 1983). In the third place, there might be some mixing, due to upwelling along the Somali coast, or entrainment through the thermocline during the summer months, when the winds blow strongly from the southwest over the western Arabian Sea.

The eastward wind stress, within 2° of the equator, is also given on fig. 4. We used the winds provided by the European Center of Medium-range Weather Forecasts (ECMWF) for 1979 (F.G.G.E. analyzed fields) and from March 1981 to March 1982. These two periods cover most of the buoy trajectories. We observe an in-phase relationship between the currents and the winds. This was discussed in details elsewhere (Knox, 1976 ; Cane, 1980 ; Mc Phaden, 1982) as being a proof that the currents in the middle of the Indian Ocean in the vicinity of Gan Island (0° 30'S, 73°E) respond locally to the winds. The location of Gan is well suited for this kind of argument, as the eastward wind stress is quite large there. Further to the west, the eastward component of the wind stress is not so well defined. In 1979, as in 1981, no well defined eastward wind stress occured during the autumn transition period. In 1979, the eastward wind stress was more than two times smaller than what was observed in 1981 and did not exceed

5.10^{-2} dyne/cm^2, when averaged from 50° E to 95° E. With such an intensity almost half a year would be needed to accelerate the jet to the observed values. However, there is no striking difference between maximum velocities of autumn jet during the two years (fig. 8). It is possible that the jet is caused by the accumulation of mass above the thermocline in the western Indian Ocean, during the Southwest Monsoon. When the meridional wind relaxes along the East African coast, there is probably a mass redistribution of the water accumulated mainly North of the equator (probably related to the curl of the wind stress in this area, as suggested by Cox , 1979). This might create an eastward surface jet at the equator, caused by the conservation of absolute vorticity.

4. SPATIAL AND TEMPORAL VARIABILITY

The statistical properties of the lagrangian velocity time series provided by the buoys will be established in this chapter. The advantage of lagrangian statistics is twofold . In the first place it gives indication of how properties conserve themselves along the flow. In the second place, it affords the opportunity of investigating the spatial scales, when numerous buoys are present at the same time. Eulerian measurements are complementary for the study of the correlation in time. An array of current meters can also give some indication of the correlation in space. They are, however, very expensive for that purpose.

We will restrict the aim of this chapter to the statistical study of quantities : the monthly mean drift velocities, and the fluctuations from this mean. Due to our ignorance of a precise zonal mean, spatial variability was assessed by studying the structure functions

$$S'_{uu}(0,\Delta x) = <[\overline{u}(x) - \overline{u}(x+\Delta x)]^2/<\overline{u}>^2>$$

$$S'_{vv}(0,\Delta x) = [v(x) - v(x+\Delta x)]^2$$

where the overbars indicate an average of the buoy velocity over monthly track segments, the brackets refer to an ensemble average over the buoys contained in the same 2°30' latitudinal band during one given month and Δx is the zonal separation. The zonal component of the spatial inhomogeneity is assumed to be proportional to the mean current (We eliminated the months with weak eastward or westward flows). The spatial variances can thus be estimated, by considering the limit of the function S'_{uu} and S'_{vv} for large separations. They are found (fig. 5) to be roughly within the range of 10- 12 cm/s for the meridional velocity component and within the range of 30 - 40 % of $<\overline{u}>$ for the zonal velocity component. These intervals are valid at the 90 % confidence interval referring to a student T test. The spatial inhomogeneities are not correlated for separations larger than 10° - 14° for $\overline{u}$ and 14° - 18° for $\overline{v}$ (fig. 5).

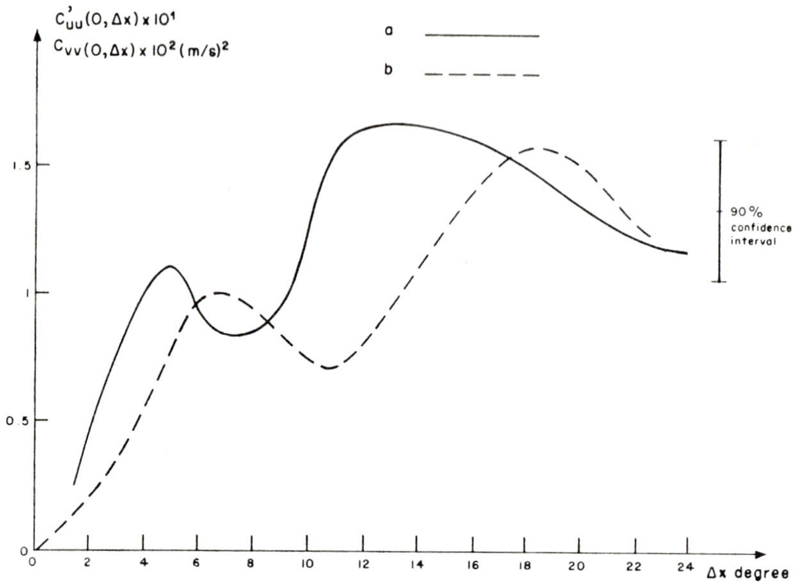

Fig. 5 : spatial structure functions for the monthly buoy drifts within 5 degrees of the equator.
a zonal component, normalized by the mean monthly velocity.
b meridional component.

It is also interesting to see how smaller time scale fluctuations are correlated in space; for this purpose we evaluated the correlation function

$$C_{u'u'}(0,\Delta x) = <[u'(t,x)\ u'(t,x+\Delta x)]/\sigma_u^2>$$

$$C_{v'v'}(0,\Delta x) = <[v'(t,x)\ v'(t,x+\Delta x)]/\sigma_v^2>$$

where the fluctuations $u'(t,x)$ and $v'(t,x)$ are defined as

$$u'(t,x) = u(t,x) - \overline{u}(t,x) \quad \text{and} \quad v'(t,x) = v(t,x) - \overline{v}(t,x)$$

There is a large decrease of the correlation for large Δx, when we eliminate the long time scales in u', by taking for u a moving average along the buoy (fig. 6). The small time scales decorrelate in space over shorter separations Δx than the average $\overline{u}$. Typically there is no correlation at Δx of the order of 6 to 8 degrees. The effects of correlation in the meridional direction should also be evaluated. Unfortunately, they cannot be seriously handled with lagrangian measurements. It is probable that the correlation scales should be smaller latitudinally than longitudinally, as it is found for deep current measurements (Eriksen, 1981).

In order to investigate the time scales characteristic of these fluctuations, a lagrangian correlation function was evaluated by computing along each monthly buoy track segment initiated at time t, the quantity $u_i'(t)\ u_i'(t+\Delta t)$ and by averaging it over all buoys assuming that these small time scales are statistically stationary and that they are homogeneous in space in a $5°$ strip centered along the equator. We estimate the functions

$$C_{u'u'}(\Delta t,0) = <[u_i'(t)\ u_i'(t+\Delta t)]/\sigma_{u'}^2>$$

$$C_{v'v'}(\Delta t,0) = <[v_i'(t)\ v_i'(t+\Delta t)]/\sigma_{v'}^2>$$

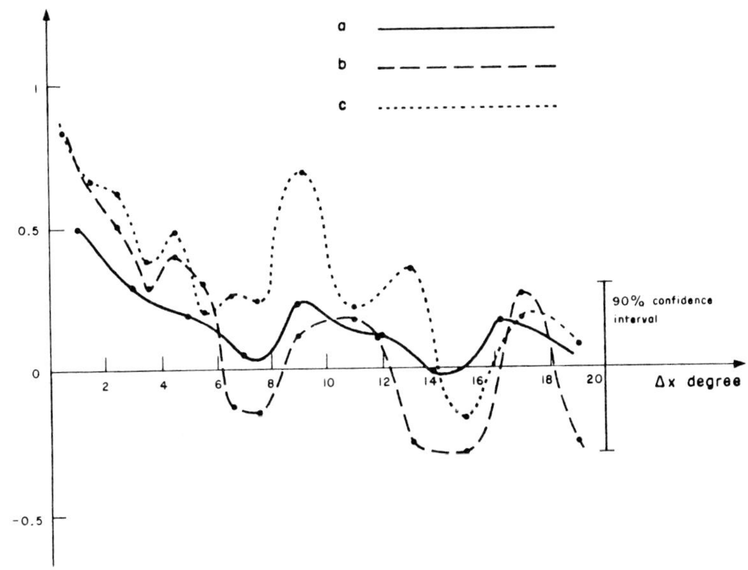

Fig. 6 : spatial correlation functions for the fluctuations
a zonal component (eliminating the large time scales, with a linear moving filter).
b meridional component.
c zonal component (without elimination of long time scales).

The variances $\sigma_{u'}$ and $\sigma_{v'}$ are nearly equal to 25 cm/s (the larger periods were eliminated by taking for u a moving average). We observe (Fig. 7) a very steep decrease of $C_{u'u'}$ and $C_{v'v'}$ for small Δt, with a sign reversal after 3 to 4 days. This is followed by some oscillations, slightly different from zero at the 90% confidence level, but not different from zero at the 95% confidence. This is typical of some large band width spectra (band-width $\sim$1/14 day^{-1}). We consider that it corresponds approximately to 4 independent degrees of freedom in a one month period.

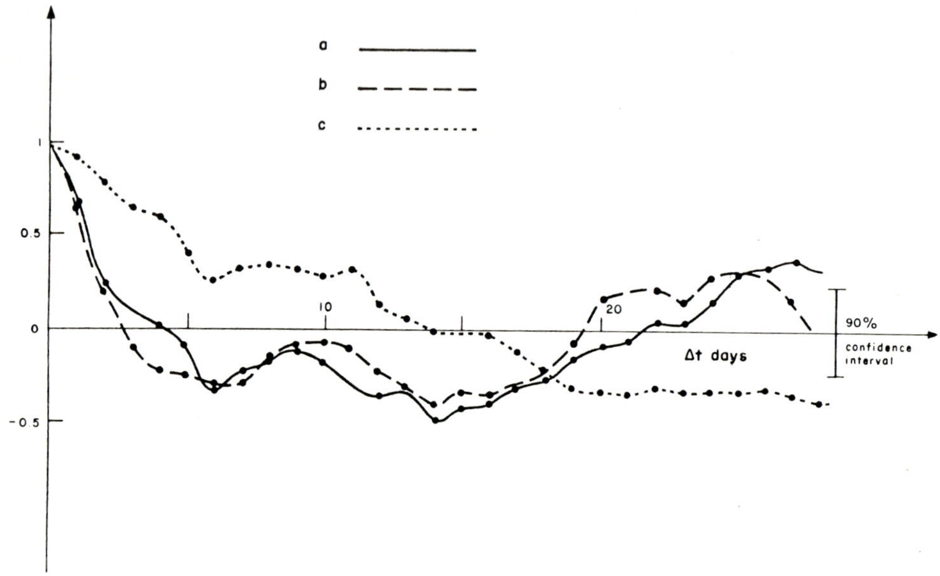

Fig. 7 : temporal correlation functions.
a zonal component (eliminating the large time scales, with a linear moving filter).
b meridional component.
c zonal component (without eliminating large time scales).

5. EQUATORIAL WAVES

Part of the observed zonal variability can be explained by the occurence of propagating events with periods less than a month. The intense eastward jet in the previous section, was present in November and disappeared at the end of November or December in the eastern sector of the Indian Ocean. The jet disappeared abruptly, as deduced by buoy trajectories (fig.8) in 1979 and in 1981. In 1980, nothing can be said, as the data were scarce. During both years, during which there was sufficient data, the disappearance of the jet propagated from the east to the west. Its speed was about 55 cm/s in 1979 and 40 cm/s in 1981.

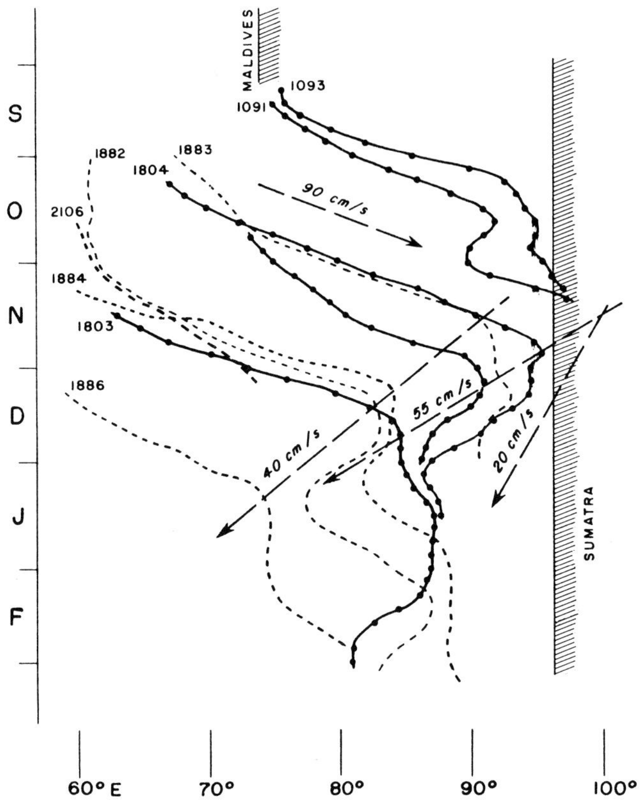

Fig. 8 : buoy trajectories between October and February, within 3° of the equator. The vertical scale corresponds to time (days in european numerals, months in roman numerals). The horizontal scale corresponds to longitude. Full lines for 1979, dashed lines for 1981.

The reversal of the zonal velocity, can be related to Rossby wave fronts (Gonella et al., 1981; Luyten and Roemnich, 1982). According to the linear theory of the spin up of the ocean (Philander al., 1980 ; Cane, 1979), the evolution of the flow from rest can be schematically described as follows. At first, an accelerating equatorial jet forms, associated with a deepening of the equatorial mixed layer on a time scale of the order of 10 days. We will assume, for simplicity, that there is only one excited vertical mode. A Kelvin wave front generated at the western boundary and a Rossby wave front generated at the eastern boundary suppress the acceleration in the surface layer. The Kelvin front travels three times faster than the fastest Rossby front. When the Kelvin wave has crossed three quarters of the basin, the two fronts will add their effects, resulting in a deceleration of the currents. This deceleration will stop after passage of additional Rossby fronts generated at the eastern boundary, as a result of the reflection of the Kelvin front. The second Rossby front follows the first one, with a delay of 20 to 40 days for the size of the Indian Ocean (for the first or second vertical mode respectively), and leaves very small currents behind.

The observed disappearance of the jet is probably related to this Rossby wave fronts generated at the eastern boundary and superposed on an unaccelerating flow. The propagation of the observed front has the same velocity as typical long Rossby waves corresponding to the second vertical mode. This description has to be seriously revised, when taking into account the non-linearities, as Rossby waves are particularly sensitive to these effects, due to their small group velocity (the surface jet speed is 90 cm/s) (Philander, 1979[b] ; Mc Phaden et al., 1979).

The changes of the propagation velocity of the front from one year to the other, can be partially attributed to different vertical structure of the currents (presence or absence of an undercurrent). Other effects could affect the phenomena. Changes in the wind stress for example, or instabilities of the currents, producing a meandering of the jet. As already stated for 1979, the wind stress magnitude could not explain the amplitude of the currents.

After abrupt disappearance of the jet, there is some flow reversal, which could be due partially to the non-linear character of the flow or overshooting. The November front in 1979 west of the Maldive Islands (73° East) was not clearly identified. Two explanations are suggested : the surface wind may have been out of phase with the Rossby front in those areas or part of the energy was dissipated or blocked by the Maldive Island chain (Du Penhoat, 1982 ; Yoon, 1981).

In August and September, several buoys meandered in the western Indian Ocean north of the equator in an eastward mean current. Interesting properties can be deduced from two buoy trajectories separated roughly by 70 km. The meanders propagated westward with a phase speed of 23 cm/s (fig. 9). As this phase speed was of the same order of magnitude as the mean currents (19 cm/s), a change of reference frame was necessary to determine the characteristic period of the phenomena in a fixed reference frame. The period of the meander was in the range 15 - 30 days with a wave length of 700 km.

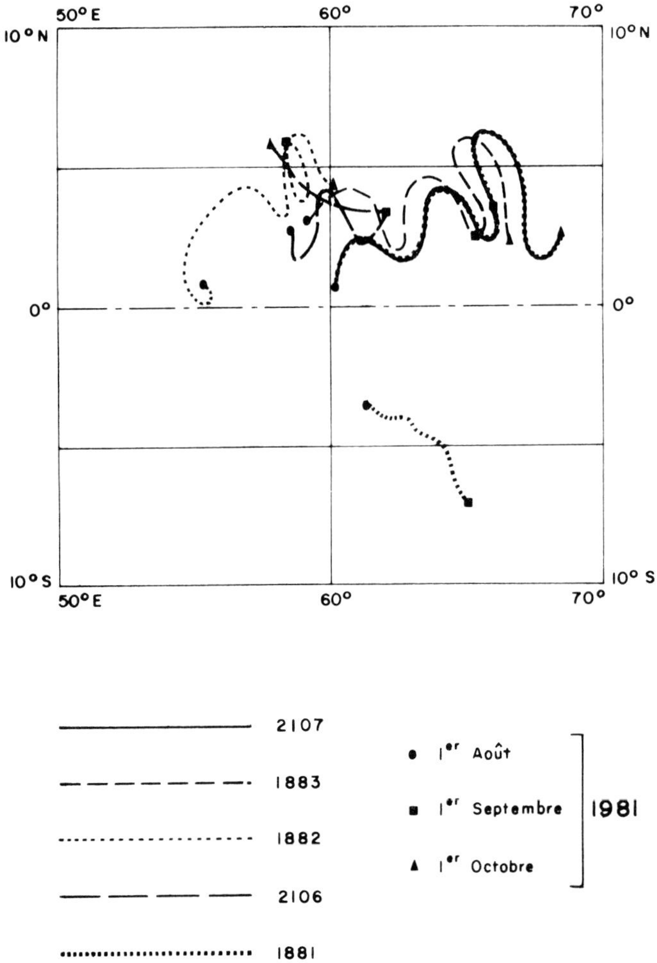

Fig. 9 : August to November 1981 buoy tracks in the western Indian Ocean.

They could be related to waves excited at the western boundary by instabilities of the current system north of the equator. They could also be excited by local changes of the northward component of the Monsoon winds (Delecluse, personal communication). The ECMWF analyzed data show that in 1981 the relaxation of the monsoon winds at the equator occured suddenly near the 15th of September, long after the appearance of the meandering. Further studies are required on this subject in order to infer the effect of wind changes in interaction with a western boundary. An analysis of the buoy trajectories in terms of the linear theory of equatorial waves is probably not adequate due to the importance of non-linearity in this area, located north of the equator.

Finally, evidence of 7 - 20 day waves was found at the equator in the meridional component of buoy velocities (fig. 10, fig. 11). Considering the spectra in this period range (Table III), it appears that the waves are present mainly in November and in December, as well as in April and in May. Their peak amplitude is larger than 20 cm/s. The energy decays rapidly with latitude as shown by the ratio of meridional kinetic energy over zonal kinetic energy (Table III). The signal seems to be correlated over large distances along the equator (1000 km). There is some evidence on the data of westward phase propagation at high speed (about 1,50 m/s). The central period of this energy band, is approximately 12 days in the lagrangian frame of reference. There should not be very large nonlinear interaction between such waves and the jet because of the large phase speed (1,50 m/s), compared to the jet speed (0,90 m/s).

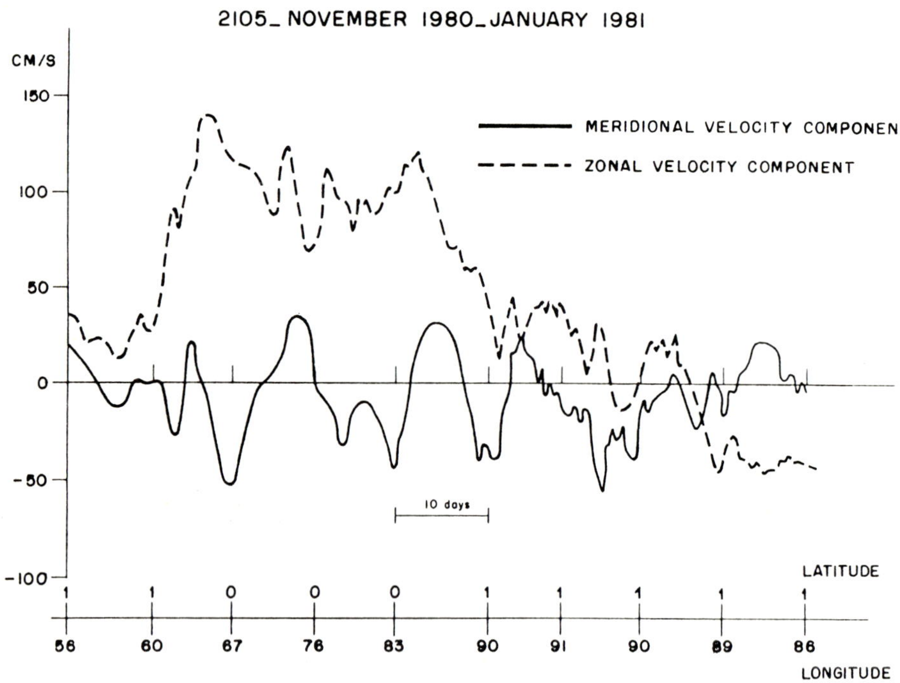

Fig. 10. November 1980 to January 1981 meridional velocity (dashed line) and horizontal velocity (full line) of buoy 2105. Longitude and latitude are indicated every 10 days.

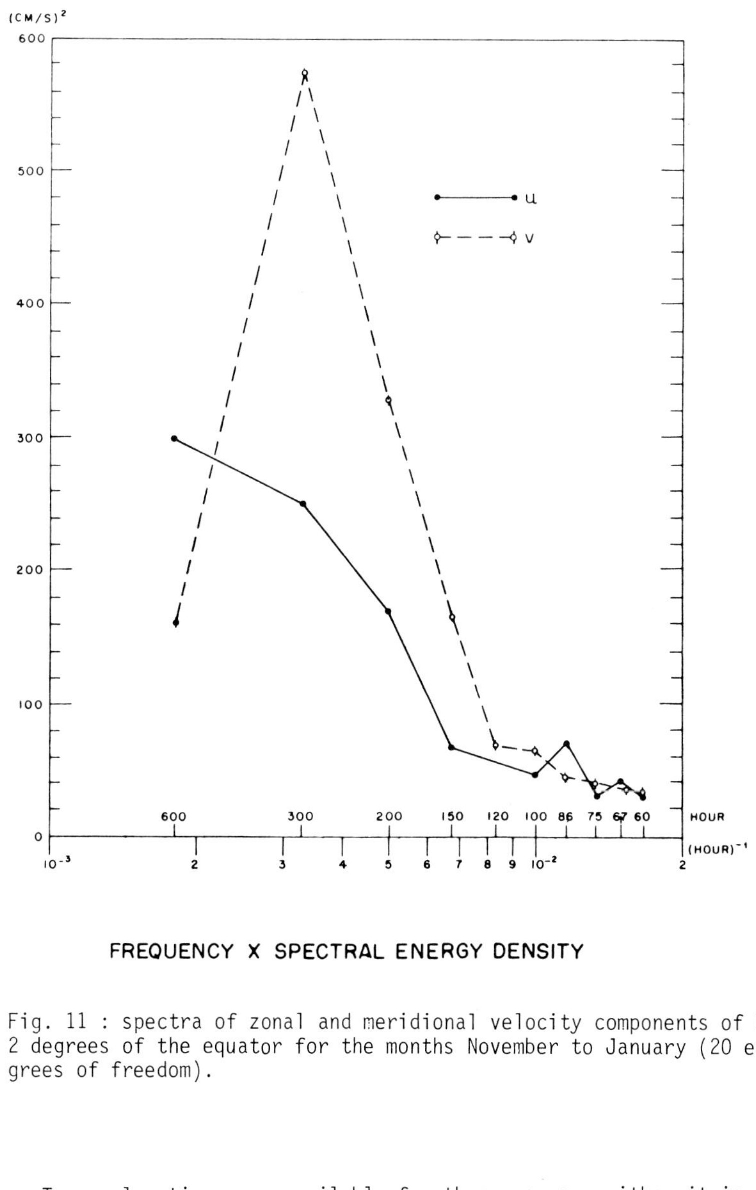

Fig. 11 : spectra of zonal and meridional velocity components of buoys within 2 degrees of the equator for the months November to January (20 equivalent degrees of freedom).

Two explanations are available for these waves : either it is a fluctuating equatorial current induced by the wind which is possible, because the inertial frictional time scale of spin-up of the upper ocean is approximately 10 days. Or they are oceanic equatorial waves. To test the first hypothesis, we looked at the wind data (ECMWF analysed fields). There is more energy at these periods in the zonal component than in the meridional component of the wind stress. Furthermore the energy is not larger during the transition months. As there are few data assimilated for the analysis in the equatorial Indian ocean, the spectra are probably not very reliable; however, it seems unlikely that the local influence of the winds could explain completely the upper ocean variability.

TABLE 3

Energy of the currents for periods between 7 and 20 days in (cm/s)

a) Seasonal cycle for buoys within 2 degrees of the equator.
b) Meridional dependence, when comparison is possible.

For both tables : 1) meridional kinetic energy; 2) zonal kinetic energy; 3) ratio of meridional over zonal kinetic energy; 4) number of buoys.

a)

	J	F	M	A	M	J	J	A	S	O	N	D
1	150	270	280	750	310	210	440			650	960	680
2	190	210	270	320	170	150	520			1180	150	420
3				1.28							1.50	
4	4	2	3	3	2	3	3			3	4	7

b)

Latitude y (degrees)	$0 < y < 2$	$2 < y < 4$	$4 < y < 8$
1	590	430	240
2	460	390	220
3	1.27	1.12	1.10
4	14	20	16

These waves affect primarily the meridional velocity component. They could correspond to mixed Rossby gravity waves (Moore et al., 1977). Their energy would propagate eastward and could be generated by impulsive meridional winds (Delecluse, personal communication). They could also be due to instabilities of the equatorial jet, as it coincides in time with it. This last point is reinforced by the fact, that despite very different wind conditions in the autumn of 1979 and 1981, the jet had the same intensity (fig. 8) and the waves are more energetic in 1981 than in 1979. It could also explain, why it is not so frequent in April-May,as the jet is not so intense. However,analytical two-layers stability analysis tended to prove that eastward equatorial jet are still stable for larger velocities (Philander, 1976).

6. CONCLUSION

Drifting buoys tracked by the Argos system have been an accurate tool for studying the current dynamics of the upper ocean for periods longer than one day. In the Indian Ocean, with 21 buoys launched over a three years period, we evaluated the seasonal cycle of the zonally averaged currents. Two maxima of the eastward equatorial jet were observed in April-May and October-November, separated by weaker currents. Despite the dispersion of the data over three years and different longitudes, the seasonal cycle of the jet could be estimated to have a variance of 20 cm/s, and the variations of the current with longitude corresponded to a variance of 30% of the mean current. The jet was confined within 4° of the equator. The half-width was smaller than what could be predicted from linear theory. Events on smaller time scales, less than a month, are correlated on a space scale less than 10 degrees of longitude.

Some of the current variability could be related to propagating events. The reversal of the autumn equatorial jet was found to be propagating to the west at a speed of 40 cm/s in 1981 and 55 cm/s in 1979. There is some question on the interpretation of this as an occurence of a Rossby wave front, as the jet should influence the propagation of the wave. Energy for periods between 5 and 20 days was concentrated in the meridional velocity component. This appears typical for mixed Rossby gravity waves. A plausible interpretation is, that the waves could be generated by instabilities of the equatorial surface jet.

Meanders in August and September were followed by two buoys. They were probably part of the large eddy activity concentrated north of the equator during the summer Monsoon in the western Indian Ocean. They were propagating westward.

The occurence of wave-like events, which are not directly correlated with the wind, restrict the domain of frequencies where the direct local influence of the wind on the currents (related to the boundary layer structure) can be seen . It is probably still possible to derive from the wind data some relationships between the formation of the equatorial jet and events in the zonal wind stress lasting from 5 days to one month. For longer periods, the buoys are probably not well suited in equatorial regions, as they do not stay in the same latitudinal strip for a very long time. Furthermore, waves excited at the boundaries would have enough time to propagate some remote effects, which could perturb the buoy trajectories.

Acknowledgements

The twenty one buoy trajectories would not have been obtained without a wide cooperation between all the members of the INDEX group IOS (J Swallow) ; CSIRO (G Creswell) ; WHOI (J Luyten, G Needell) ; LOP Museum Paris and the support of NOAA (USA) and TAAF (France). J.F. Murail assisted in the data analysis. The drawings were done by Mrs Christoforides.The typewriting was made by J Maheux. We are greatly indebted to R. Molinari for his useful comments on the manuscript.

REFERENCES

BERVAS J.Y., 1978 - Influence des forces de trainée sur la dérive d'une bouée équipée d'une ancre flottante. - Rapport interne TDI, COB CNEXO

CANE M., 1979 - The response of an equatorial ocean to simple wind stress patterns - II : Numerical results. - Journal of Marine Research, Vol. 37, pp. 253-299.

CANE M., 1980 - On the dynamics of equatorial currents, with application to the Indian Ocean. - Deep Sea Res., Vol. 27, pp. 525-544.

COX M., 1979 - A numerical study of Somali current eddies - J. Phys. Oc., Vol. 9, pp. 311-326.

DU PENHOAT Y., M.A. CANE and R.J. PATTON, 1982 - Reflections of low-frequency equatorial waves on partial boundaries - in the present volume.

ERIKSEN C., 1981 - Deep currents and their interpretation as equatorial waves in the western Pacific Ocean - J. Phys. Oc., vol. 11, pp. 48-70.

GONELLA J., M. FIEUX and G. PHILANDER, 1981 - Mise en évidence d'ondes de Rossby équatoriales dans l'Océan Indien à partir de bouées dérivantes - C.R. Acad. Sc. Paris, T. 292, pp. 1397-1399.

KIRWAN A.D., G. Nc NALLY and PAZAN S., 1978 - Wind drag and relative separations of undrogued drifters - J. Phys. Oc., Vol. 8, pp. 1146-1150.

KNOX R., 1976 - On a long serie of measurements of Indian Ocean equatorial currents near Addu Attol - Deep Sea Res., Vol. 23, pp. 211-221.

KORT V.G., 1977 - Equatorial current in the Indian Ocean during the Northeast monsoon- Oceanology, Vol. 17, n° 2, pp. 115-120.

LEETMAA A. and H. STOMMEL, 1980 - Equatorial current observations in the Western Indian Ocean in 1975 and 1976 - J. Phys. Oc., Vol. 10, pp. 258-269.

LUYTEN J. and ROEMNICH D., 1982 - Equatorial currents at semi-annual period in the Indian Ocean - J Phys. Oc., Vol. 12, pp.

MC PHADEN M., 1982 - Variability in the central equatorial Indian Ocean - Part I : Ocean Dynamics - Journal of Marine Research, Vol. 40, pp. 157-176.

MC PHADEN M. and R. KNOX, 1979 - Equatorial Kelvin and inertio-gravity waves in zonal shear flow - J. Phys. Oc., Vol. 9, pp. 263-277.

MOORE D. and G. PHILANDER, 1977 - Modeling of the tropical oceanic circulation - The Sea, Vol. VI, Wiley Interscience, Ny, PP. 319-361.

NEYMAN V.G., V.A. BUBNOV and V.D. YEGORIKHIN, 1978, -Investigation of equatorial currents in the western part of the Indian Ocean.

PHILANDER G., 1979 - Nonlinear coastal and equatorial jets - J. Phys. Oc., Vol. 9, pp. 739-747.

PHILANDER G., 1979[b] - Equatorial waves in the presence of the equatorial undercurrent - J. Phys. Oc., Vol. 9, pp. 254-262.

PHILANDER G. and P. DELECLUSE, 1982 - Coastal currents in low latitude - submitted to Deep Sea Research.

PHILANDER G. and R. PACANOWSKI, 1980 - The generation of equatorial currents - J. Geoph. Res., Vol. 85, n° C2, pp. 1123-1136.

PHILANDER G. and R. PACANOWSKI, 1981 - Response of equatorial ocean to periodic forcing - J. Geoph. Res., Vol. 86, n° C3, pp. 1903-1916.

PHILANDER G. and R. PACANOWSKI, 1981[b] - The oceanic response to cross-equatorial winds with application to coastal upwelling in low latitudes -Tellus, Vol. 33, pp. 201-210.

QUADFASEL D.R., 1982 - Low frequency variability of the 20°C isotherm topography in the western Indian Ocean - J. Geoph. Res., Vol. 87, n°C3, pp. 1990-1996.

SWALLOW J.C., 1967 - Equatorial undercurrent in the western Indian Ocean in 1964 - Studies in tropical Oceanography, Miami, n° 5, pp. 15-36.

SWALLOW J.C., R.L. MOLINARI, J.G. BRUCE and O. BROWN - Development of near surface flow pattern and water mass distribution in the Somali basin in response to the Southwest Monsoon of 1979 - submitted to J. Phys. Oc.

WEISBERG R.H., A. HORIGAN and C.COLIN, 1979 - Equatorially trapped Rossby-gravity wave propagation in the gulf of Guinea. J. Mar. Res., vol 37, pp.67-86.

YOON J.H., 1981 - Effects of Islands on equatorial waves - J.Geoph. Res., Vol. 86, n° C 11, pp. 10913-10920.

INVESTIGATION OF SMALL-SCALE STRUCTURE OF HYDROPHYSICAL FIELDS IN THE EQUATORIAL REGION OF THE INDIAN OCEAN.

N.N. KORCHASHKIN, I.D. LOZOVATSKY

The results of the investigation of the small-scale structure of velocity, electroconductivity and temperature fields in the ocean are considered using the data from measurements in the region of the equatorial currents and in the area of the Mascaren ridge in the Indian ocean. Thus, according to the data from vertical soundings at the equator in the depth range 0 - 1000 m, layers of intensive current velocity fluctuations, with thicknesses a few tens of meters, were revealed, located in the lower part of the upper quasiuniform layer (50 - 80 m) and under the sharp thermocline, at a depth of 160 - 230 m, which is the upper boundary of the equatorial subsurface Tareev current. In the upper part of the thermocline (80 - 160 m) where the vertical density gradient is sharply increased, the level of small-scale velocity fluctuations is much smaller. Spectral analysis of velocity and temperature fluctuations shows that in the lower part of the upper quasiuniform layer and in the weakly stratified layers of the thermocline, the velocity spectra S_u and the temperature spectra S_t display the inertial-convective range of scales where the Kolmogorov-Obuchov "- 5/3 law" holds. Immediately after the inertial range, there is a viscous range in the spectra S_u with sharp decrease of the spectral density level and the range of power law "- 1" in the spectra S_t predicted by Batchelors's theory — which then goes over into the viscous-conductive range. Thus, according to the data from the field measurements, a good agreement between sea turbulence spectra both of velocity and temperature and the results of laboratory measurements and known relations was obtained. The characteristic estimates of dissipation of velocity ε and temperature ε_t irregularities, obtained using the results of these measurements, are of the order $10^{-3} - 10^{-1}$ cm^2/s^3 for ε and $10^{-6} - 10^{-5}$ $°C^2/s$ for ε_t.

The law of dependence of the dimensionless relationship $\varepsilon/b\,V_z^2$ on a Richardson number $Ri = N^2/V_z^2$ is illustrated (fig. 1) in order to explore an interrelation of turbulence characteristics. Here b is the variance of velocity fluctuations, V_z a mean vertical gradient of the mean flow, N the Väisälä-Brunt frequency. As one can see in fig. 1, the experimental points can be approximated by a straight line whose angular coefficient is nearly 1/2. Thus, when we go from the domain of the upper equatorial current to the opposing

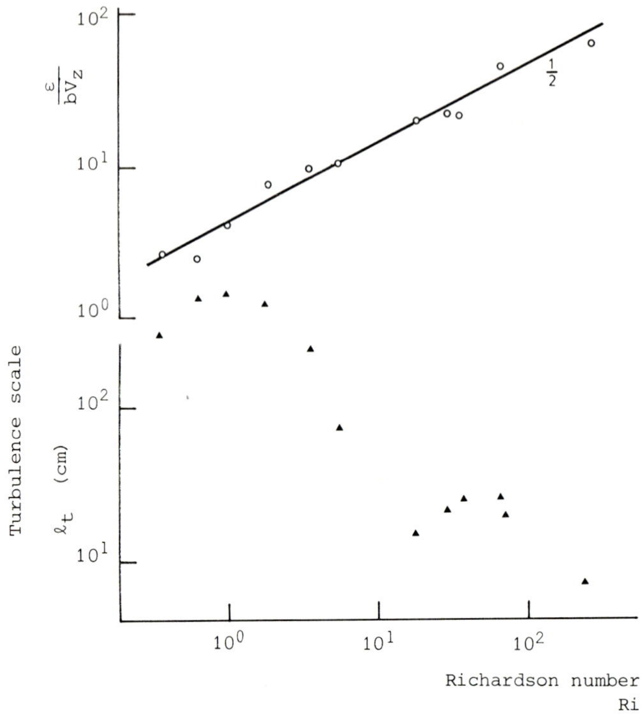

Fig. 1. The dependence of the dimensional relation ε/bV_z and the turbulence scale ℓ_t on the Richardson number.

Tareev current, the dependences between the turbulence characteristics b and ε and the parameters N and V_z have the following form :

$$\frac{\varepsilon}{bV_z} = c\,Ri^{1/2} \tag{1}$$

where $c \sim 4$. It follows in fact from formula (1) that

$$b = \frac{\varepsilon}{4N} \tag{2}$$

We can express ε with the help of b and the turbulence scale :

$$\varepsilon = \frac{A\,b^{3/2}}{\ell_t} \tag{3}$$

where A is a universal dimensionless constant. Now from (2) we obtain that

$$\ell_t = \frac{A}{4}\,\frac{b^{1/2}}{N} = \frac{A}{8}\,\frac{\varepsilon^{1/2}}{N^{3/2}} = \frac{A}{8}\,L_0$$

where $L_0 = \varepsilon^{1/2} N^{-3/2}$ is the Ozmidov buoyancy scale. A dependence of the turbulence scale on the Richardson number is represented in fig. 1. As one could expect, a general tendency to a decreasing ℓ_t with increasing Ri can be seen. Nevertheless, the clearest dependence $\ell_t(Ri)$ takes place in a bounded region of Ri values between 1 and 10. When $Ri < 1$, i.e. at weak hydrodynamical stability, the turbulence scale reaches 4 - 6 m, and when $Ri > 10$ it is not larger than 10 - 15 cm. In view of the condition in the formula (4) $A/8 \ll 1$, the relation obtained between ℓ_t and L_0 shows that the characteristic scale of the velocity field variation is less than that of the density field. It means that the fine structure in the profile of $V_z(z)$ must be expressed more strongly than in the profile of $N^2(z)$ and must have a smaller vertical scale.

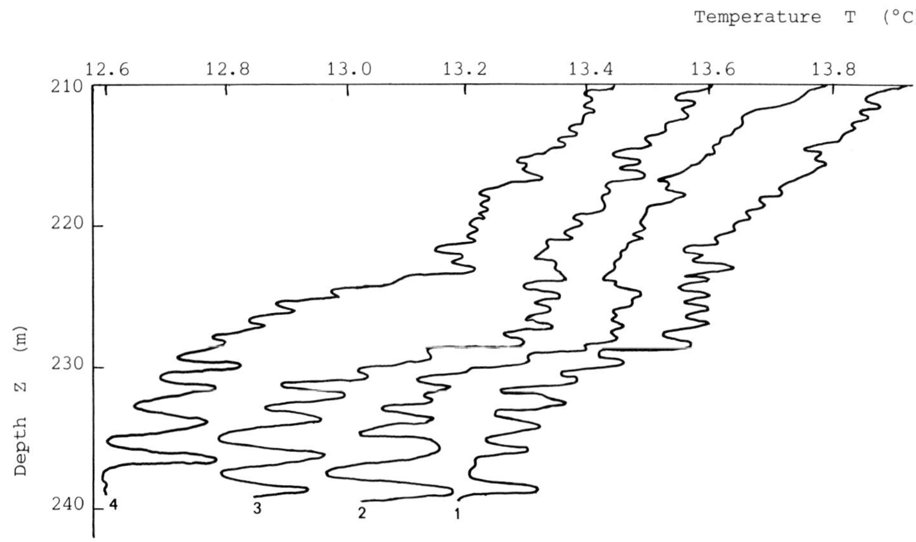

Fig. 2. The series of repeated soundings with extremely fine-structure layering of temperature profile.

The series of repeated soundings in the region of the slope of the Mascaren ridge permited us to see successive stages of development of extremely fine structural layering of the temperature profile in the depth range 215 - 250 m (fig. 2). The transformation of this layer, located in the frontal zone under the temperature inversion, stabilized hydrostaticaly due to salinity, leads to the formation of a very thin layer of water with thickness 50 - 100 cm; the sign of the temperature gradient alternates from one layer to another. The level of velocity fluctuations within the compass of the layer considered was markedly lower than at its upper and lower boundaries, and the horizontal temperature gradient was about $3\ 10^{-6}$ °C/cm. Our estimates show that we have observed the process of development of the fine-structure due to lateral convection.

Long-term towing of the turbulimeter in the thermocline in the region of the propagation of intensive internal waves of large amplitude, which were discernible on the ocean surface as alternating strips of choppy surface waves or rips, permitted to find the relationship between the characteristics of the sea surface state, parameters of internal waves and changing of the level of small-scale velocity and electroconductivity fluctuations in the water body. It appeared that when a large internal solitary wave or wave train passes, the intensity of fluctuations of electroconductivity is sharply increased and the level of velocity fluctuations is changed relatively little. In many cases the regions of high values of electroconductivity fluctuations are in accordance with the moments of passing of strips of rips above the turbulimeter. Thus, in most cases, the passing of internal waves do not change the density of turbulent energy in space. When the horizontal temperature gradient in the thermocline changes sharply due to propagation of internal waves, the intensity of small-scale electroconductivity fluctuations, which is proportional to the value of this gradient, is also sharply changed.

VELOCITY FIELD FINE STRUCTURE AND SHEAR INSTABILITY OF CURRENTS IN EQUATORIAL REGIONS OF THE OCEANS.

R.V. OZMIDOV, M.L. PYZHEVICH

Numberous simultaneous measurements of the fine vertical structure of the density and velocity fields in the Indian and Pacific oceans have been performed in the course of 22nd cruise of R/V *Dmitry Mendeleev* in 1979 (Ozmidov, 1979). The sounding, giving information on the fine vertical structure of salinity and temperature fields, has been used as well as an acoustic measurement system for recording the modulus and direction of the current velocity vector. The measurements were performed with vertical sounding from a drifting vessel. The speed of sounding ranged from 0.5 m/s to 0.8 m/s . The distance between the winches from which the measurements were taken was equal to 70 m .

Data from repeated soundings in a number of regions in the Pacific show a differing space-time variability of current velocity field fine structure.

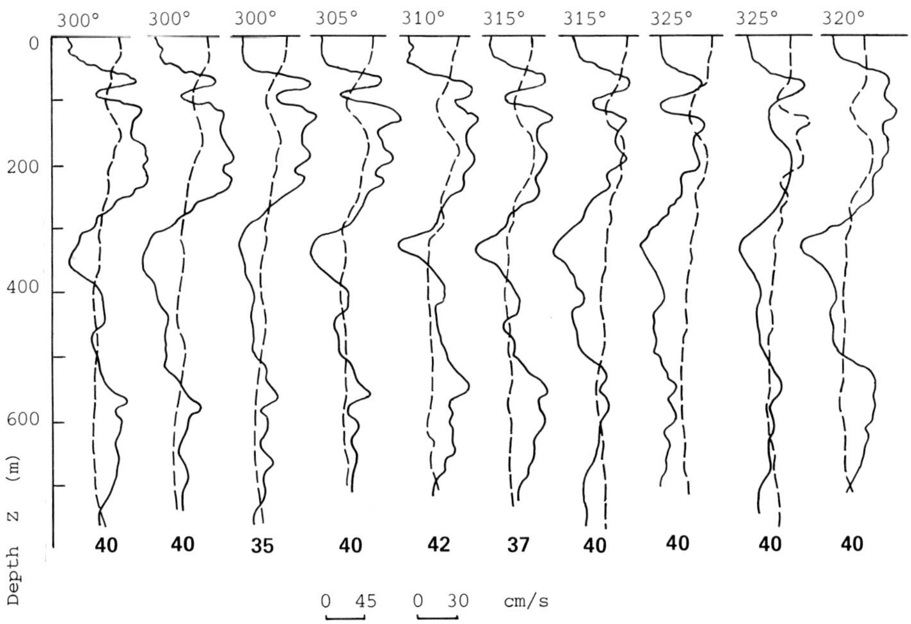

Fig. 1. Profiles of absolute values of flow velocity (dashed line) and its direction (solid line), according to the data of numerous soundings at station 1785.

Fig. 1 shows the data of 10 successive measurements of absolute values of flow velocity and direction, carried out at time intervals 40 minutes in the North-West part of the Pacific ocean at station 1785 (Ozmidov, 1979). Note the relatively strong time (more precisely, space-time) variability of the flow direction. But the profile of the absolute value of flow velocity is more persistent here from sounding to sounding. This bears evidence of oscillations of the flow velocity vector relative to some local axis. The measurements carried out at the equator at station 1803, however, are more persistent both in the flow direction and absolute velocity values; this is apparently characteristic of the systems of equatorial currents (fig. 2).

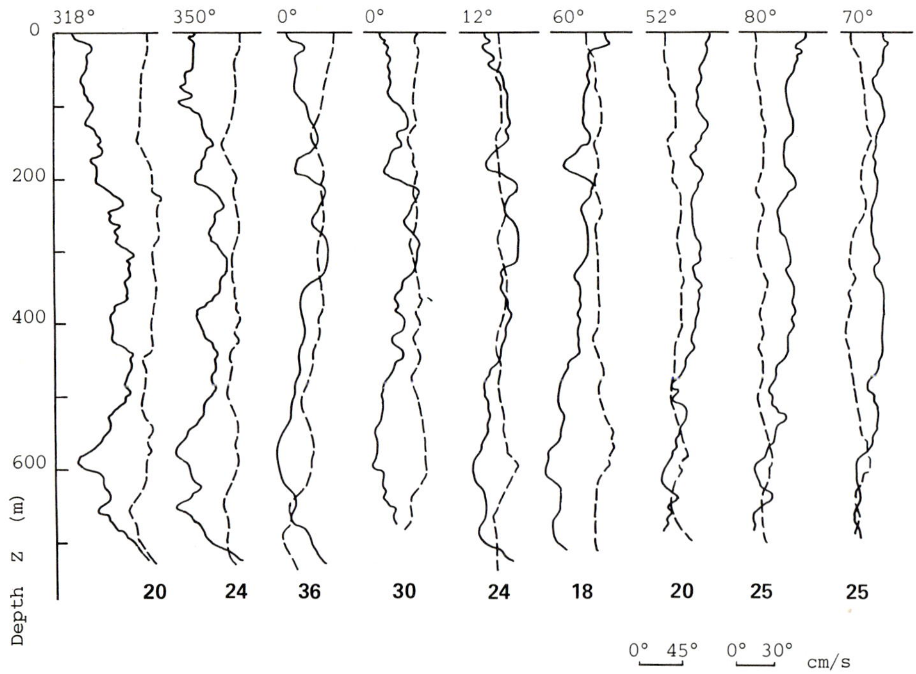

Fig. 2. The same as in fig. 1 for station 1808.

The structure of vertical profiles of fields in the ocean is characterized by unordered alteration, with depth, of layers with small and large values of vertical gradients; the amplitudes of these non-uniformities vary little with depth, as distinct from the non-uniformities of vertical structure of the temperature and salinity fields which usually have tendency to decrease with depth.

Table 1 gives, as a typical example, maximal, minimal and mean values of velocity components, their vertical derivatives and the square of absolute velocity values, computed using sounding n° 5 at station 1785. The table also

TABLE 1

Signal moment	U (cm/s)	V (cm/s)	$\frac{dU}{dZ}$ (s^{-1})	$\frac{dV}{dZ}$ (s^{-1})	$\frac{dV}{dZ}$ (s^{-2})
Minimum	$-2.2\ 10^{1}$	$2.2\ 10^{1}$	$-5.3\ 10^{-3}$	$-4.2\ 10^{-3}$	$9.1\ 10^{-9}$
Maximum	8.2	$3.3\ 10^{1}$	$-9.5\ 10^{-3}$	$-3.9\ 10^{-3}$	$9.1\ 10^{-5}$
Mean value	-5.4	$2.9\ 10^{1}$	$-2.1\ 10^{-5}$	$-1.4\ 10^{-4}$	$1.1\ 10^{-5}$
Variance	8.2	2.5	$2.7\ 10^{-3}$	$-1.9\ 10^{-3}$	$1.3\ 10^{-5}$
Asymmetry	$-1.2\ 10^{2}$	-7.3	$1.4\ 10^{-2}$	$-2.5\ 10^{-3}$	$7.6\ 10^{-3}$
Excess	$9.8\ 10^{-3}$	$1.0\ 10^{-2}$	$2.1\ 10^{-2}$	$3.0\ 10^{-3}$	$5.3\ 10^{-3}$

gives the values of variance, asymmetry and excess of the distributions of all these quantities.

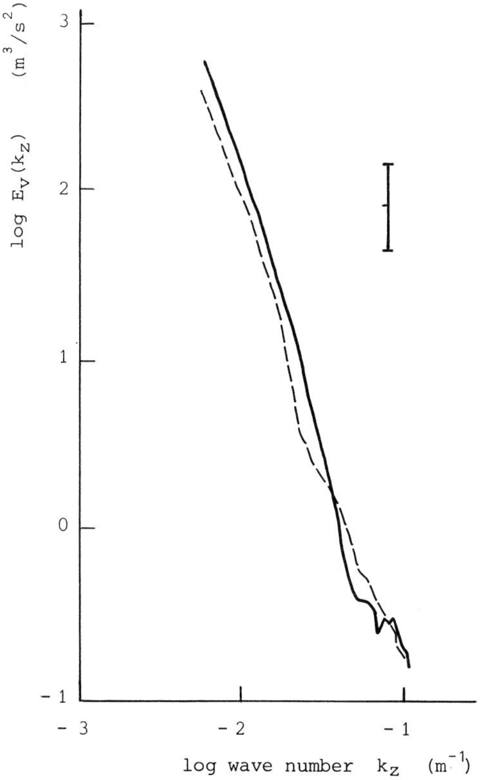

Fig. 3. Spectral densities of vertical non-uniformities of zonal (a) and meridional (b) flow velocity components accordingly to the data of the sounding n° 5 at the station 1785. The vertical line represents the 95 % confidence interval.

The interesting information on non-uniformities of the velocity field gives spectral-density functions. As an example, fig. 3 shows the spectral density functions of vertical non-uniformities of zonal (a) and meridional (b) velocity components, according to the data from sounding n° 5 at station 1785, in the depth range from 20 to 664 m. The calculations were performed using a Fourier transform of the correlation function with 26 degrees of freedom. The range of vertical scales of the velocity field non-uniformities investigated ranges from 8 to 140 m. The spectral curves fall uniformly with wave number k_z, with mean slope close to -3 (in logarithmic scale), and for the zonal velocity component this slope tends to diminish with increasing k_z. For $k_z = 0.03\ m^{-1}$ (which corresponds to the non-uniformity scale ~33 m), there is an inflection of the spectral curve. This scale, presumably, can be a boundary line between

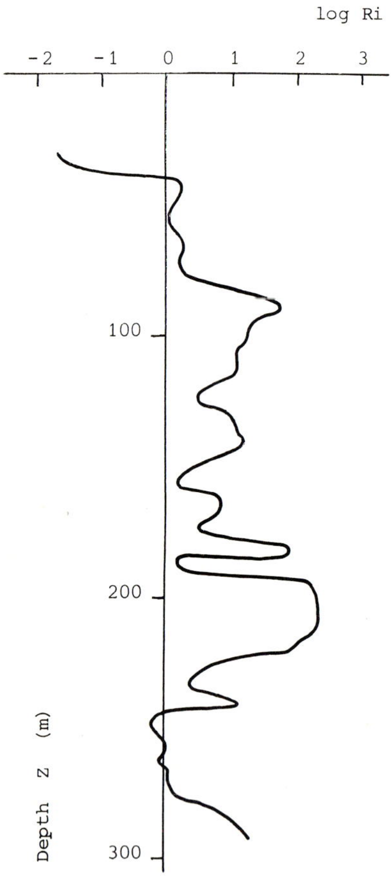

Fig. 4. An example of vertical profile of the logarithm of the Richardson number with depth resolution 10 m approximately.

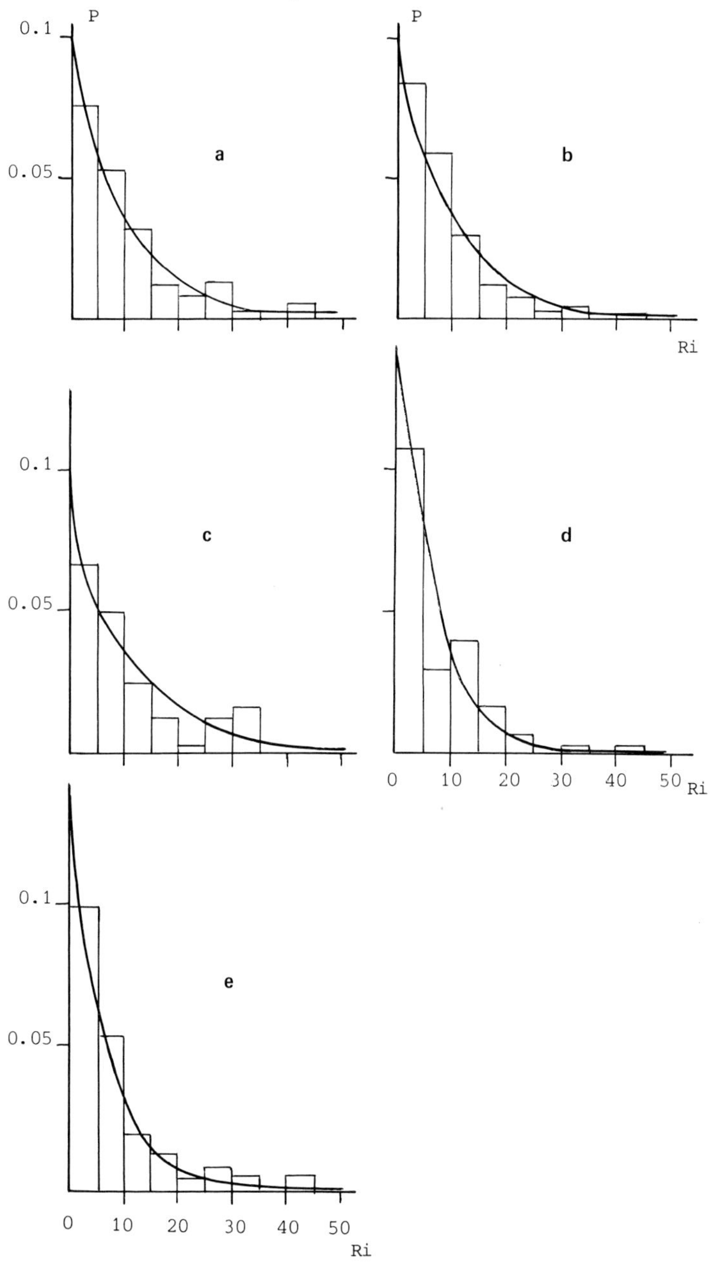

Fig. 5. Histograms of distribution of Ri and corresponding fitting curves of the exponential law with different values of λ for five series of measurements.

non-uniformities with different mechanisms of formation. Using the data of simultaneous measurements of the fine structure of velocity and density fields, the values of the local Richardson number Ri were calculated. The vertical distributions of Ri show that, in a number of cases, in some thin layers, the conditions are created for the onset of shear instability and generation of turbulence (fig. 4). In particular, it came out that for a region of the Indian Ocean (near Sumatra) the values of log Ri ranged from - 1.7 to 2.5 and in the 30 - 40 m layer the value of Ri was smaller than the critical value 0.25 (log Ri < - 0.6). In deeper layers, the bulk of the Ri - values ranged from 1 to 15.

It is shown that the empirical distributions of Ri can be approximated by the exponential-law curves

$$\Phi(Ri) = \lambda e^{-\lambda Ri}, \quad Ri > 0$$

where λ is the parameter of the distribution.

It is interesting that in some regions of the ocean there may be, presumably, a rather stable distribution of Ri with approximately constant values of λ. Presumably these distributions can be considered as a sufficiently representative characterization of the fine structure of hydrophysical fields in a given region of the ocean.

REFERENCES

Ozmidov, R.V., 1979. 22nd cruise of R/V *Dmitry Mendeleev*. Oceanology, vol. 19, n° 5, 948-954.

TURBULENCE IN THE CORE OF THE EQUATORIAL UNDERCURRENT

CARL H. GIBSON

Department of Applied Mechanics and Engineering Sciences and Scripps Institution of Oceanography, University of California at San Diego, La Jolla, California 92093

ABSTRACT

Estimates of space time average velocity and temperature dissipation rates $\overline{\varepsilon}$ and $\overline{\chi}$ from microstructure measurements are particularly difficult in strongly stratified ocean layers such as the equatorial undercurrent high velocity core and the seasonal thermocline because the turbulence in such layers is extremely intermittent in time and patchy in space. In the seasonal thermocline at Ocean Station P, 50°N 145°W, χ was found to be lognormally distributed with variance $\sigma^2_{\ln\chi} = 5.3$, larger than values in the mixed layer above and in less stratified layers below. A similar pattern appears to exist in the equatorial undercurrent, with maximum $\sigma^2_{\ln\chi}$ occurring near the depth of maximum stratification in the high velocity core. Large data samples must be averaged to reliably estimate $\overline{\chi}$. The most probable χ value for a short record, χ_{mode}, is less than χ_{mean} by a factor of $\exp(-1.5\ \sigma^2_{\ln\chi})$ if χ is lognormal. Therefore, in the seasonal thermocline data mentioned, individual dropsonde χ profiles are likely to underestimate χ_{mean} by a factor of 3.5×10^{-4}, and this was observed. Combining all available χ measurements in the core layer of the equatorial undercurrent gives a $\sigma^2_{\ln\chi}$ value of about 7 and χ_{mode}/χ_{mean} of 3×10^{-5}. Clearly short χ records are subject to extremely large sampling errors: several orders of magnitude. Dropsonde measurements of χ in the undercurrent core by Gregg (1976) are less than towed body measurements of Williams and Gibson (1974) by 2-4 orders of magnitude and dropsonde measurements of ε by Crawford (1976, 1982) and Osborn (1980) are less than towed body measurements of Williams and Gibson (1974) and Belyaev et al. (1975) by similar large factors. This discrepancy is attributed to large undersampling errors in the dropsonde χ and ε values which are based on much smaller sample sizes than the towed body values. It is important to know the true average $\overline{\varepsilon}$ and $\overline{\chi}$ dissipation rates at core depths. If the dropsonde values of 10^{-5} cm^2/s^3 and 10^{-8} °C^2/s are representative then the undercurrent core is effectively nonturbulent and provides an insulating barrier to vertical heat transfer at shallow depths. If the towed body values of 10^{-1} cm^2/s^3 and 10^{-5} °C^2/s are closer to the true averages, then the equatorial undercurrent possibly plays a dominant rather than an unimportant role in tropical heat and mass transfer, and may therefore be an important factor in the determination of planetary climate.

1. INTRODUCTION

The intensity of turbulence, mixing and vertical diffusion as a function of depth in the equatorial undercurrent is important to many oceanographic problems. For example, the heat transfer and heat storage capacity of equatorial waters is important to numerical models of climate, but these properties of the flow depend critically on how rapidly and to what depths turbulence can diffuse solar heating down from the surface. Because the current speeds and turbulence levels are much larger than in surrounding waters, the equatorial undercurrent may constitute the major mechanism of heat transfer and mass transfer over a wide band of equatorial latitudes, even though the current itself is confined to a narrow range within a degree or two of the equator. For simplicity we shall refer to the undercurrents of the Atlantic and Pacific as a single flow.

Several factors suggest that turbulence levels in the undercurrent are substantially above the oceanic average. The most striking evidence is the broadening of iso-oxygen concentration surfaces and iso-salinity surfaces about the undercurrent velocity core. In the central Pacific, where the core layer is about 110 m deep, isotherms and iso-oxygen surfaces found at 120 m depths at 3°N and 3°S are depressed to depths of 300-400 m while the surfaces at 100 m are pushed up to only 40 m or so on the equator. It has been suggested that the broadened scalar property patterns may be attributed to geostrophic effects or advection from the sides. However, the exceptionally high biological productivity of the region, comparable to that of coastal zones, suggests excellent ventilation due to undercurrent turbulence is present at all depths in the undercurrent. This includes the high velocity core since, to maintain such a high productivity, nutrients must somehow be transferred up to the plants at the surface from below the core and oxygen must be transported from the surface down through the core to supply the increased biological oxygen demand at all levels below. Turbulent mixing would seem to be the most likely transport mechanism. Without turbulence, side advection could broaden the scalar property patterns, as observed, only if it resulted in upwelling above the core and downwelling below. However, Halpern has preliminary evidence (presented at the XIV Liége colloquium) that in the Pacific at 0° 110°W the upwelling divergence occurs at 160 m depth, which is well below the core, and that the vertical upwelling may actually be a maximum near the core depth of 75 m rather than zero. Figure 1 shows the north-south contours of salinity, density and temperature across the equator at 155°W in July, 1972, from Gregg (1976). A high salinity tongue extends

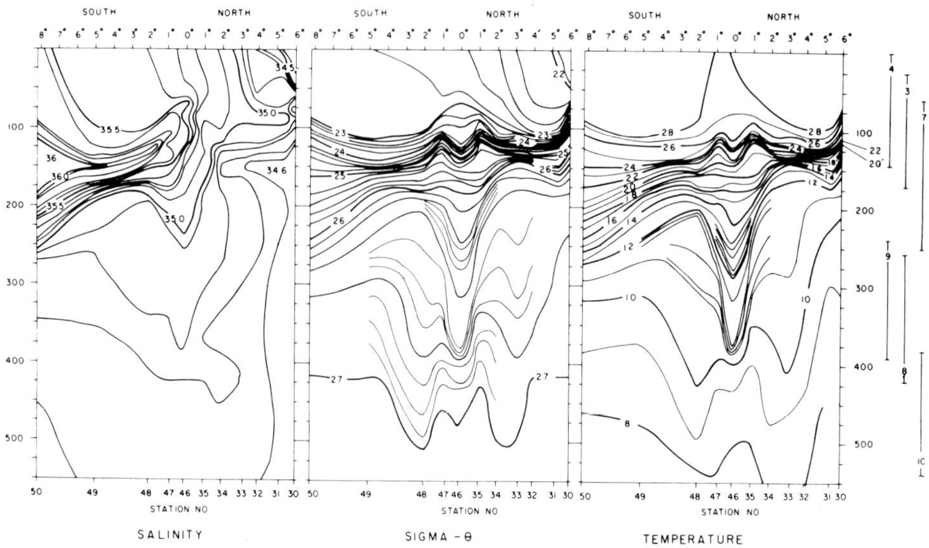

Fig. 1. North-south contours of salinity, density and temperature at 155^{0}W, July 13-22, from Gregg (1976). Note the convergence of low and high salinity tongues at 150 m depth indicating vertical upwelling may exist at the core depth of 110 m. The tongues broaden at ± 1.5^0 when they encounter the undercurrent and are apparently mixed by turbulence at all depths, including the velocity core. Depth ranges of the six MSR dropsonde data records are shown at the right.

toward the equator from the south and a low salinity tongue extends toward the equator from the north at depths of 140-170 m which is below the undercurrent core depth of 110 m estimated by Gregg (1976). Both tongues terminate and broaden abruptly at the northern and southern undercurrent boundaries at ± 1.5^0. This pattern also suggests that vertical upwelling at the equator begins at a depth below the high velocity core, and that an upward vertical velocity exists at core depths. This flow pattern has been modeled by Wyrtki (1982).

The cause of the upwelling is usually attributed to poleward divergence due to Ekman wind drift, although perhaps this divergence is enhanced (or dominated) by an estuarine-like pumping due to the large surface buoyancy flux at the equator, both from rainfall and solar heating, in combination with the enhanced mixing caused by the undercurrent. Such an estuarine pumping has been suggested by Tully and Barber (1960) to account for the vertical salinity distribution northward of the sub-Arctic boundary in the Pacific ocean. They estimate a vertical upwelling velocity of 20 ± 10 m/year. Gammon et al. (1982) estimate a comparable vertical upwelling rate; that is, 12-14 m/year in the same region from the vertical freon distribution. The magnitudes of the upwelling and diverging velocities increase with the turbulent mixing rate. At some level of turbulent mixing such estuarine pumping must be considered as a driving force for the undercurrent itself, but this is beyond the scope of the present paper.

Without turbulent diffusion, upwelling from below the core should cause the iso-concentration surfaces to dome and crowd together at the velocity core rather than broaden and deepen as observed. Geostrophic effects in such a complex meandering current system are difficult to assess.

Helm et al. (1980) have shown a well-defined separation between the high velocity core and the high salinity core in the Atlantic equatorial undercurrent, with the salinity core 15-40 m above the velocity core and about 10 nautical miles south when the velocity core was north of the equator. The separation is attributed to an asymmetry in the salt fluxes from the salinity core, with stronger turbulent mixing in the high velocity core than in the high shear layer above. In the present paper we also infer a maximum mixing rate in the velocity core, rather than a minimum as proposed by Osborn (1980) and others.

The first attempts at direct measurements of turbulence and turbulent mixing in the equatorial undercurrent were made using towed bodies. Williams and Gibson (1974) measured high frequency temperature fluctuations in the Pacific equatorial undercurrent by towing khz response sensors in the velocity core at 0° and 1°N in April 1971, and Belyaev et al. (1975a,b) measured high frequency velocity fluctuations by towing heated conductivity sensors at several depths in the Atlantic equatorial undercurrent during January-February of 1971. Subsequent microstructure studies in the undercurrent have been carried out with dropsonde mounted sensors. Gregg (1976) reports six temperature microstructure profiles through the Pacific equatorial undercurrent at 0° 155°W in June, 1972, with depth ranges shown in Figure 1. Crawford (1976, 1982) has studied velocity microstructure in both the Atlantic and Pacific equatorial undercurrents in 1974 and 1979, respectively. Osborn (1980) and Osborn and Bilodeau (1980) discuss measurements of both velocity and temperature microstructure in the Atlantic equatorial undercurrent at 24-33°W from the 1974 GATE studies.

The results of the towed body and dropsonde attempts to measure undercurrent turbulence and turbulent mixing parameters are quite different at all depths, but the most striking differences occur at the depth of the high velocity core. The purpose of the present paper is to review these differences and to consider possible explanations.

The explanation put forth by Gregg (1976) and Crawford (1982) is that towed body turbulence sensors are subject to high noise levels due to vibrations. However, the turbulence and temperature fluctuation levels detected by towed bodies in the equatorial undercurrent are several orders of magnitude higher than noise levels of the instruments indicated by towing in other regions outside the undercurrent. Belyaev et al. (1975a) report spectra with ε values as low as 10^{-4} cm^2/s^3 compared to 10^{-1} cm^2/s^3 measured in the undercurrent. Williams and Gibson (1974) spectra (shown in Figure 4) have noise peaks at harmonics and subharmonics of the 60 Hz line frequency, but do not indicate any vibrational contamination. Gregg (1976) suggested vertical vibration of the sensors in the ambient vertical temperature gradient could account for the measured signal, but Schedvin (1979) mounted accelerometers on the same towed body and found that vibration levels in the dissipation bandwidth are lower than those required by Gregg's model by 7 orders of magnitude.

Another possible explanation of the discrepancy is that the very high degree of turbulence intermittency at the undercurrent core depth may bias the dropsonde estimates of mean turbulence and temperature dissipation rates $\overline{\varepsilon}$ and $\overline{\chi}$, as discussed by Gibson (1981). These mean quantities are dominated by a few, rare, intense patches in an intermittent turbulence layer. Short data samples, such as only a few dropsonde profiles through the intermittent layer, will very probably not intersect the dominant patches and will therefore underestimate $\overline{\varepsilon}$ and $\overline{\chi}$ by large factors if the intermittency is large. Because ε and χ appear to be lognormal in stratified ocean layers, we may estimate the bias, or "undersampling error", by using the properties of the lognormal probability distribution. If only one small record is available, its ε and χ represent "mode" values of the random variables, by definition. However, if χ is lognormal, then $\chi_{mode} << \chi_{mean} = \chi_{mode} \exp(1.5\,\sigma^2_{\ln\chi})$. If several independent data records are available, then the variance of $\sigma^2_{\ln\chi}$ may be estimated and the "undersampling error factor" $\exp(1.5\,\sigma^2_{\ln\chi})$ computed. The best estimate of χ_{mean} is then given by $\chi_{mean} = \exp\mu \exp(\sigma^2_{\ln\chi}/2)$ where μ is $\overline{\ln\chi}$.

In the following we discuss the definitions of ε and χ and compare measured values and the distribution of χ in the seasonal thermocline with those in the undercurrent core. The latter values are used to estimate $\sigma^2_{\ln\chi}$, μ and χ_{mean} as a function of relative position in the undercurrent. Osborn (1980) divides the undercurrent into three depth zones: the core, the shear zone above the core, and the shear zone below the core. We will treat the undercurrents of the Atlantic and Pacific as a single flow and χ as a lognormal random variable whose statistical properties do not vary from ocean to ocean but depend only on the depth zones as classified by Osborn. Such a treatment is very crude, reflecting our primitive state of knowledge of the undercurrent and the fragmentary nature of the microstructure data. In the final section the vertical diffusivity and heat transport rates indicated by the lognormal model are compared to smaller values of Osborn (1980) and Gregg (1976) which do not account for undersampling errors caused by turbulence intermittency. Very large differences are found, particularly in the high velocity core layer where the probable undersampling error is over four orders of magnitude.

Evidence is accumulating that isolated dropsonde profiles of ε and χ may be quite unreliable as estimates of $\overline{\varepsilon}$ and $\overline{\chi}$ in most ocean layers due to such undersampling errors. Gregg (1977) estimates a vertical diffusivity of only 10^{-2} cm^2/s in the main thermocline of the north Pacific compared to 1.2-1.3 cm^2/s estimated by Gammon et al. (1982) from freon profiles. Gammon et al. (1982) point out that in the region of their measurements side advection will tend to reduce their vertical diffusivity estimate, so 1.2-1.3 cm^2/s actually represents a lower bound of the true value. Gibson (1982) infers that, based on the large vertical overturning scales measured by Gregg (1980), the Gregg (1977) microstructure is fossil turbulence in an advanced state of decay. Consequently, the space time average $\overline{\chi}$ and vertical diffusivity K will be underestimated by large factors if the Gregg (1977) values of χ and K are taken as representative. In section 4 an intercomparison test between towed body and dropsonde χ measurements is reported. A very large difference was found between individual χ profiles and average χ profiles, particularly at depths where the turbulence is most intermittent. This is consistent with the view presented in the present paper that the discrepancy between towed body and dropsonde ε and χ measurements in the undercurrent high velocity core are due to the scatter induced by high turbulence intermittency in this layer rather than due to any measurement noise in the towed body data.

2. COMPARISON OF TOWED BODY AND DROPSONDE DISSIPATION RATES ε AND χ

The most important parameters of turbulence and turbulent mixing of temperature are the viscous dissipation rate ε and the temperature dissipation rate χ. Velocity fluctuations are stirred down to small scales and damped by viscosity at a rate

$$\varepsilon = \nu(u_{i,j} + u_{j,i})^2/2 \quad ; \quad i,j = 1,2,3 \quad . \tag{1}$$

Temperature fluctuations are also reduced to small scales by turbulent stirring and are dissipated by molecular diffusion at a rate

$$\chi = 2DT,_i^{\,2} \quad ; \quad i = 1,2,3 \quad . \tag{2}$$

where ν and D are molecular viscosity and diffusivity coefficients, respectively. Repeated indices indicate summation from 1 to 3, and commas indicate gradients in the direction denoted by the indices i or j. Both ε and χ are entropy production terms which characterize the cascade of turbulent kinetic energy and temperature variance over the full range of length scales from the scale of production, at meters or kilometers, to the scales of dissipation, at centimeters or millimeters.

The scatter of ε and χ values measured by tow bodies and dropsondes is shown in Figure 2, along with the profiles of eastward velocity, salinity and temperature found at 0^0 150^0W in the Pacific equatorial undercurrent in April, 1971,

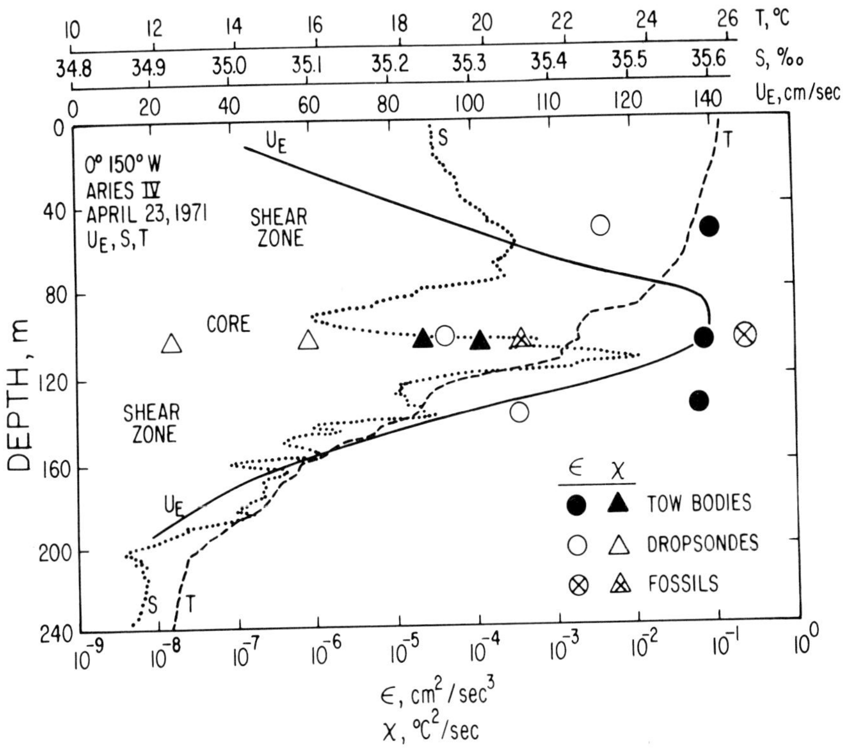

Fig. 2. Scatter of ε and χ in all equatorial undercurrents. Towed body ε values (dark circles) are larger than dropsonde values (open circles), especially at core depths. Towed body χ values (dark triangles) in the core are much larger than dropsonde values (open triangles). Dropsonde fossil temperature turbulence patches in the core layer shown in Fig. 3 imply previous ε_0 and χ_0 values at fossilization (crossed circle and triangle) close to the values measured from two bodies. See text for the sources of the data.

during the ARIES IV cruise of Williams and Gibson (1974). Values are grouped by shear zone and represent both Atlantic and Pacific measurements. Tow body ε values of Belyaev et al. (1975a) in the Atlantic were constant in all zones at about 0.1 cm^2/s^3 and agree with the Williams and Gibson (1974) values at 1^oN and 0° in the Pacific at 150^oW. Dropsonde ε values are from Osborn (1980) and Crawford (1976) in the Atlantic, which give 10^{-3} cm^2/s^3 above the core, 10^{-5} in the core and 10^{-4} below the core; lower than the tow body values by factors of 10^{-1} - 10^{-3}. Dropsonde χ values for the core layer range from 10^{-8} $^oC^2/s$ from Gregg (1976) to 10^{-6} $^oC^2/s$ from the Osborn and Bilodeau (1980) measurements in the Atlantic. Tow body χ values of 10^{-5} $^oC^2/s$ at 1^oN and 10^{-4} $^oC^2/s$ at 0° 150^oW were reported by Williams and Gibson (1974), larger than the dropsonde χ values by factors of 10^2 - 10^4.

Also shown in Fig. 2 are "fossil" dropsonde ε_o and χ_o values inferred from temperature microstructure patches detected by Gregg (1976) using the fossil turbulence model equations of Gibson (1982). These microstructure patches are shown in Figure 3a,b and were identified by Gregg (1976) as "puzzling: although the Cox number is low, the linearity of the profiles is unusual and suggests strong vertical mixing, either present or past." The pattern is typical of fossil temperature turbulence patches, as discussed by Gibson (1982), in which the turbulence has been damped by the stratification leaving a "fossil turbulence" remnant in the temperature field. The fossil preserves the vertical overturning scale L_T of the previous turbulence activity, where ε and χ values at fossilization are given by

$$\varepsilon_o = 2.8\, L_T^2\, N^3 \tag{3}$$

and

$$\chi_o = 0.43\, L_T^2\, (\partial\overline{T}/\partial z)^2\, N \tag{4}$$

both expressions from Gibson (1982). The patches are 1.7-2.3 m in vertical extent and lie between the salinity minimum at 90 m and the salinity maximum at 119 m. The two patches at 100 m contain numerous $\partial T/\partial z$ zero crossings, which indicates internal density inversions and a probable turbulence source since the ambient salinity profile is stably stratified. The ambient N value is about 1.4×10^{-2} rad/s and $\partial\overline{T}/\partial z$ is about 10^{-3} $^oC/m$. From these values and (4) ε_o is about 0.1-0.4 cm^2/s^3 and χ_o is $(1\text{-}3) \times 10^{-4}$ $^oC^2/s$, close to the towed body values. The patches lie in the high velocity core depth range and suggest much higher χ values were present at previous times than were observed by Gregg (1976).

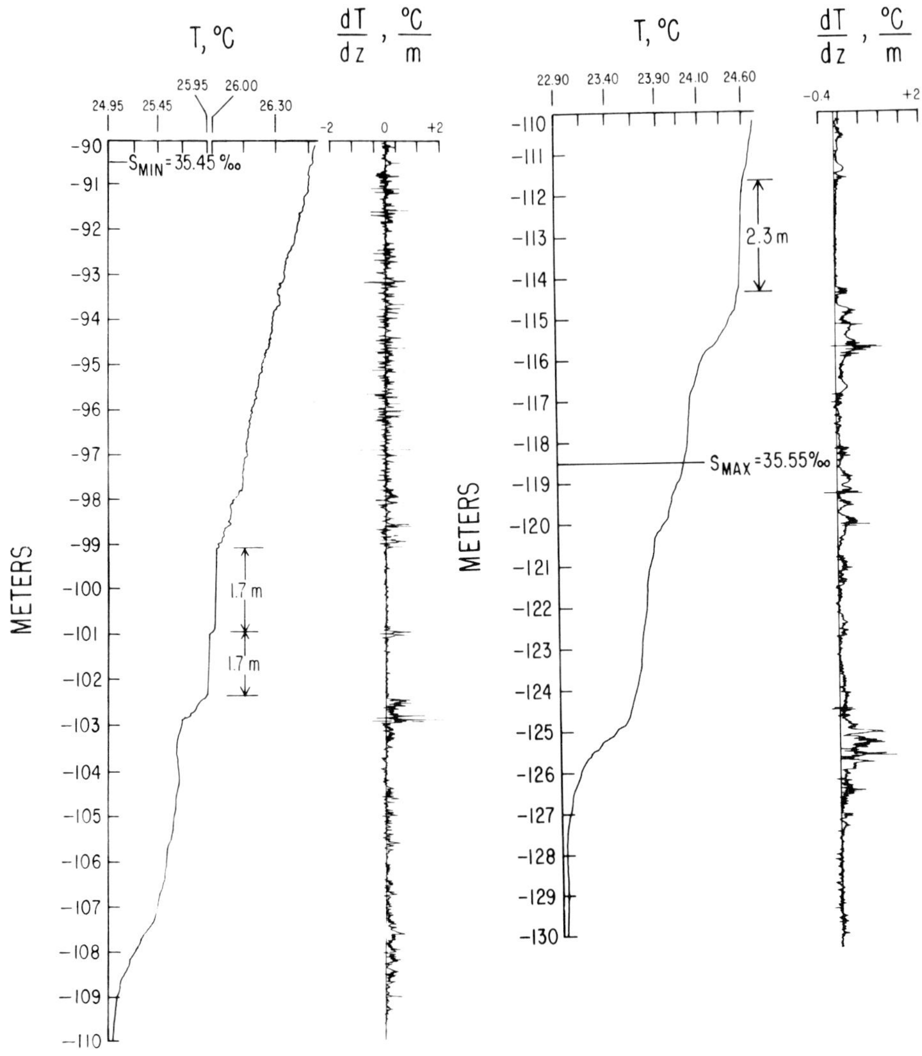

Fig. 3. Temperature and temperature gradient profiles through the high velocity core layer at 0^0 155^0W, from Gregg (1976). Microstructure patches at 100 m and 113 m are apparently fossil temperature turbulence remnants of much stronger previous turbulence and mixing, with ε_0 and χ_0 values shown in Fig. 2.

3. COMPARISON OF TOWED BODY AND DROPSONDE TEMPERATURE SPECTRA AT CORE DEPTHS.

Figure 4 shows a comparison between the Williams and Gibson (1974) towed body temperature spectra in the core at 1°N and 0° with the Gregg (1976) dropsonde spectra in the same depth range. The dropsonde spectra were computed

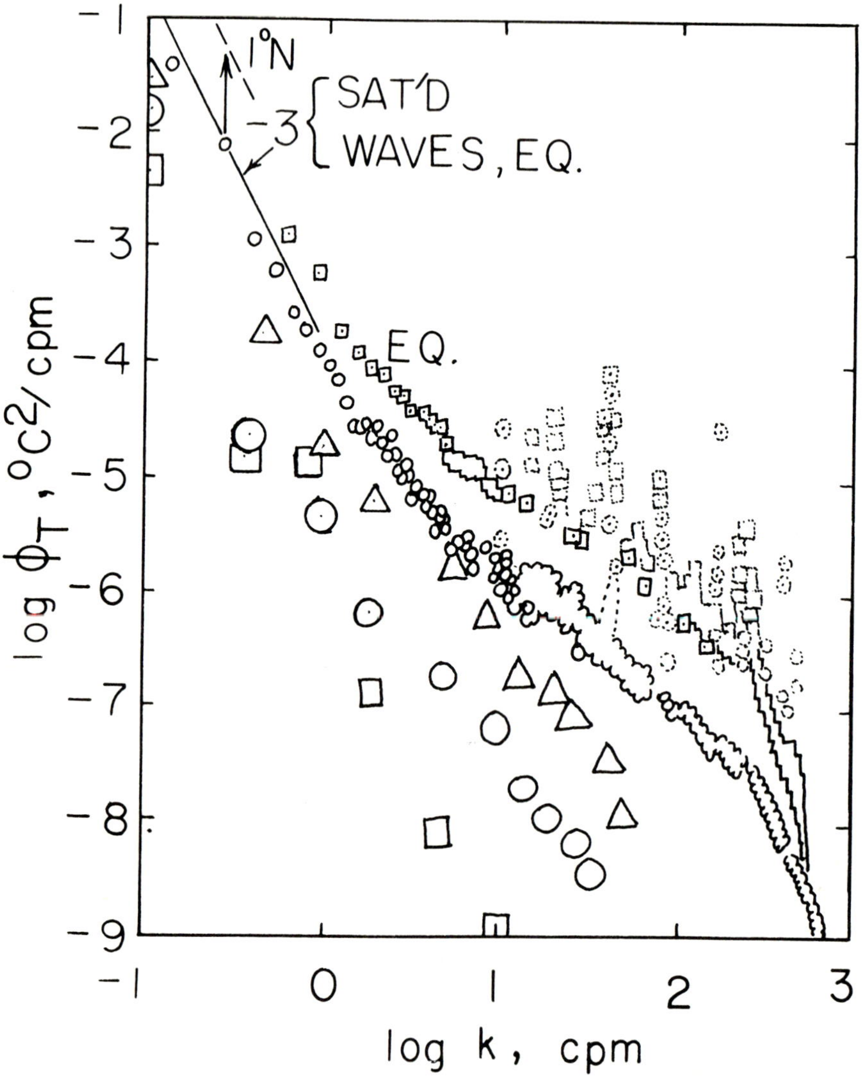

Fig. 4. Temperature spectra from tow body and dropsonde in the core layer of the Pacific equatorial undercurrent. Tow body: Williams and Gibson (1974) 150°W; small squares, at equator, 110 m; small circles, 1°N, 99 m. Dropsonde: Gregg (1976) 155°W; circles 90-99 m, triangles 99-108 m, squares 110-119 m.

from three 9 m records compared to two 350 m records for the towed body spectra, or about 4% as much data.

The dropsonde spectra show differences in amplitude between themselves of almost three decades at $k = 10$ cpm, much more than the difference between the average dropsonde spectral level and the towed body spectral levels at this wavenumber. Some of the towed body spectral bands are contaminated by line noise and are shown as dotted points. Both the 1°N and 0° towed body spectra show a prominent peak at 60 Hz with nearly the same amplitude. However, most spectral bands at $k < 1$ cpm and many for $k > 100$ cpm were dominated by signal and are shown by solid symbols.

The main differences between the towed body and dropsonde spectra occur at low and high wavenumber. The dropsonde spectra indicate diffusive cutoff wavenumbers of about $k = 2$ cpm or less, indicating a dissipation rate ε of less than 7×10^{-6} cm^2/s^3. This is less than the minimum necessary for turbulence to exist, which is about $25\,\nu N^2$ according to Gibson (1980) or 6×10^{-5} cm^2/s^3 if N is 1.4×10^{-2} rad/s. Therefore the microstructure observed by Gregg (1976) in the core is not turbulent but fossil turbulence at all scales.

This conclusion is also indicated by the low wavenumber dropsonde spectra, which fall well below the level for temperature fluctuations produced by saturated internal waves

$$(\phi_T)_{SAT'D\ WAVES} = (0.7 \pm 0.2)\ (\partial\overline{T}/\partial z)^2\ k^{-3} \tag{5}$$

discussed by Gibson (1980, 1983), where in (5) k is measured in radians/wavelength. The saturated internal wave spectrum of (5) is shown in Fig. 4 for equatorial values of $\partial\overline{T}/\partial z$ by the solid line with slope -3 at low k values. If the microstructure patches in a layer are closely spaced and completely turbulent the associated internal wave motion should be saturated, and the temperature spectrum should be given by (5). The equatorial towed body spectrum approaches the saturated wave spectral level very closely as is to be expected if small scale internal waves at k values of 1 cpm and less are in equilibrium with fully developed turbulence. From the high wavenumber cutoff at about $k = 200$ cpm the viscous dissipation ε was estimated to be about 0.1 cm^2/s^3 and the temperature dissipation rate χ was about 10^{-4} $^{o}C^2/s$. According to the Gibson (1980) fossil turbulence model, microstructure is fossil turbulence when

$$\varepsilon \le \varepsilon_o = (13/2)\chi_o\ N^2/(\partial\overline{T}/\partial z)^2 \ge \varepsilon_o' \equiv (13/2)\chi N^2/(\partial\overline{T}/\partial z)^2 \tag{6}$$

where equality shows the turbulence is just at the point of fossilization, indicated by o-subscripts. Substituting the measured values of χ, N and $\partial\overline{T}/\partial z$ in

(6) gives $\varepsilon_0' = 0.11\ cm^2/s^3$, which shows that the turbulence producing the measured microstructure was completely active but close to the fossilization point. Consequently, turbulence should exist for all wavelengths smaller than about a meter, and saturated waves should exist for scales somewhat larger than a meter, as observed.

The spectrum from the tow at 1^0N is more difficult to interpret. The viscous dissipation rate ε inferred from the diffusive cutoff was also about $0.1\ cm^2/s^3$, but χ was only about $10^{-5}\ {}^0C^2/s$ and $\partial\bar{T}/\partial z$ was $2.8 \times 10^{-3}\ {}^0C/cm$, so ε_0' from (6) is $4.6 \times 10^{-3}\ cm^2/s^3$. Because $\varepsilon > \varepsilon_0'$, if the turbulence were continuous one might infer that the turbulence is so active that buoyancy effects are unimportant, even at the largest scales of the turbulence. However, Williams (1974) found χ at 1^0N was extremely intermittent, with $\sigma^2_{\ln\chi} > 5$ compared to $\sigma^2_{\ln\chi} < 2$ for the equatorial data, so the 1^0N turbulence is probably quite patchy. The saturated wave spectrum for the 1^0N towed body data is shown by the dashed line which lies above the 0^0 line shown in Fig. 4 by a factor of 6, reflecting the larger $\partial\bar{T}/\partial z$ value at 1^0N from (5). Thus the 1^0N low wavenumber spectrum is subsaturated. However, because the χ record is very intermittent the subsaturated level of the 1^0N spectrum may only be a consequence of averaging a few very active patches ($\varepsilon \simeq 0.1\ cm^2/s^3$) with large regions of nonturbulent fossil microstructure accompanied by subsaturated internal wave motions. Note that the ε value inferred from the high wavenumber cutoff of the temperature spectrum reflects the maximum ε value in the record and may be larger than the average ε for the record by a factor of order (active fraction)$^{-1}$; possibly a very large factor of order 10 but probably less than 100.

4. INTERCOMPARISON OF TOWED BODY AND DROPSONDE χ VALUES IN THE MILE MIXED LAYER EXPERIMENT

Because the towed body and dropsonde ε and χ values shown in Figures 2 and 4 were measured at various longitudes and times it might be argued that some of the variability is not due to the intermittency of the turbulence process in the undercurrent but due to zonal variations or seasonality. Therefore it is useful to examine the results of an intercomparison test between a towed body and two different dropsonde systems during the MILE mixed layer experiment at Ocean Station P, 50^0N 145^0W in August-September 1977, a period during which the mixing processes and property distribution in the upper ocean were relatively stationary.

Figure 5 shows the results of the intercomparison of mean χ values as a function of depth in the upper 60 meters. Dillon (1982) used a rapidly deployable dropsonde to collect six profiles during periods of light winds (Cast A)

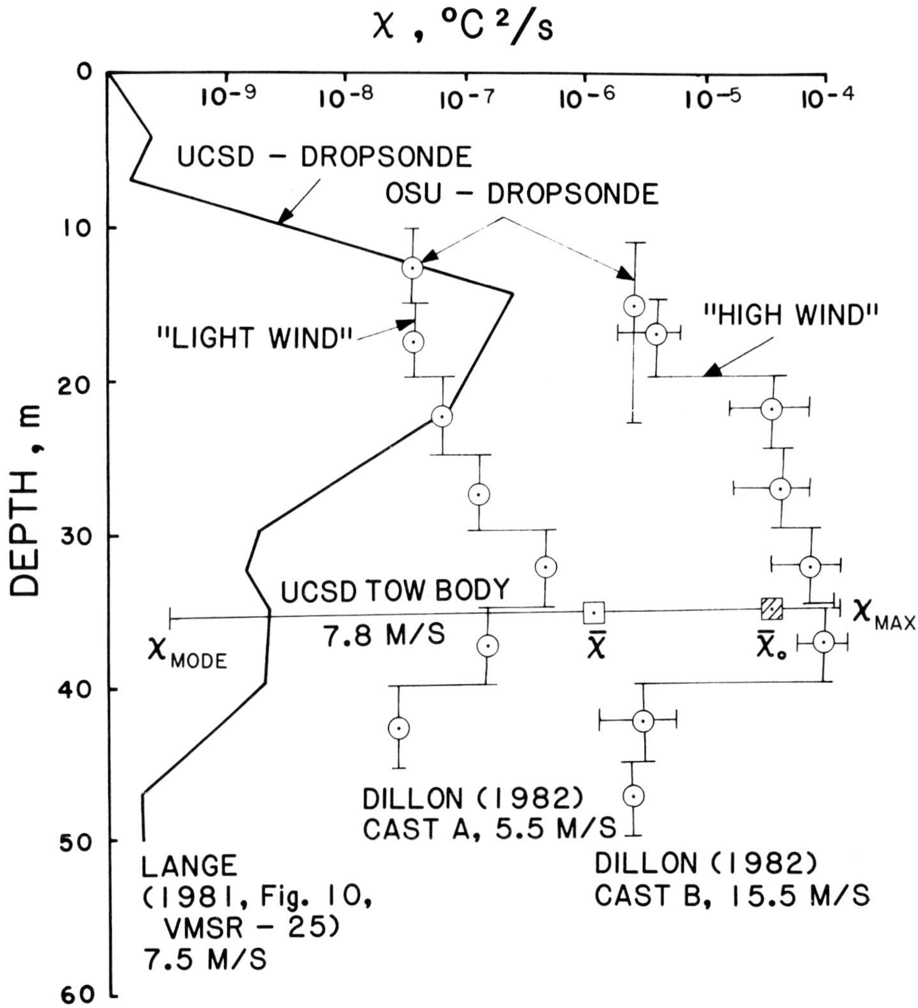

Fig. 5. Intercomparison of $\overline{\chi}(z)$ and $\chi(z)$ from dropsondes and tow body during MILE mixed layer experiment, 50°N 145°W. Dillon (1982): open circles; high (15.5 m/s) and low (5.5 m/s) winds; multiple profiles averaged over depth intervals shown, standard deviation interval shown for high wind case. Lange (1981): solid line, wind 7.5 m/s, single $\chi(z)$ profile VMSR-25. UCSD tow body: open square shows space average $\overline{\chi}$ in seasonal thermocline, bar shows range between χ_{mode} of measured lognormal probability density function and χ_{max}, the maximum χ in a patch, wind 7 m/s. $\overline{\chi}_0$ is the space average found if all patches are assumed to be at fossilization. Note the large discrepancy between Lange's single $\overline{\chi}$ profile and the average χ profile, particularly at the depth of maximum stratification at 30-40 m where intermittency is maximum.

and high winds (Cast B). The profiles were divided into 56 shorter records which were individually analyzed and the statistics tabulated. The shorter records were used to compute the $\overline{\chi}$ profiles shown in Fig. 5 by averaging over the vertical depth intervals shown. Standard deviations about the means are shown for the high wind case.

Both the high wind profile and the low wind profile show that $\overline{\chi}$ is a maximum at the depth of maximum vertical temperature and density gradient, or seasonal thermocline, which occurs at about 30-40 m. This is in sharp contrast with the pattern indicated by single dropsonde profiles reported by Lange (1981), using the same MSR dropsonde as Gregg (1976), an example of which is shown in Fig. 5. The isolated MSR profiles typically show minimum χ values in the seasonal thermocline even though the average $\overline{\chi}$ profiles indicated by Dillon (1982) are maximum at this depth with values 10^3-10^5 times larger.

The towed body average $\overline{\chi}$ in the seasonal thermocline was obtained using a microconductivity sensor to detect high frequency temperature fluctuations, as described by Gibson (1981), Washburn and Gibson (1982) and Washburn (1982), and is in good agreement with the Dillon (1982) results. The probability density function of χ measured by the towed body was close to lognormal with $\sigma^2_{\ln\chi} = 5.3$ and $\chi_{mode} = 4 \times 10^{-10}\ {}^oC^2/s$. Thus χ_{mode} is close to the Lange (1981) individual profile values as might be expected. This demonstrates the danger of treating an individual χ profile as representative of the mean profile, especially in a very intermittent layer such as the seasonal thermocline. The towed body average was based on about 6 km of data, of which only about 3% contained significant microstructure activity. The dominant microstructure patches were fossil turbulence at the largest scales. An upper bound on the space-time average χ was estimated by correcting each patch to its dissipation rate χ_o at fossilization, giving the $\overline{\chi}_o$ value shown in Fig. 5. Presumably the true space-time average lies somewhere between the space averages $\overline{\chi}$ and $\overline{\chi}_o$.

Figure 6 shows the individual microstructure patch values from the towed body and the Dillon (1982) high and low wind profiles compared to three individual χ profiles of Lange (1981), all in the 30-40 m depth range of the seasonal thermocline. The range of values is over 5 decades, demonstrating the extreme intermittency possible in such a strongly stratified layer.

Figure 7 shows a profile of $\sigma^2_{\ln\chi}$ as a function of depth estimated from the towed body as well as the Dillon (1982) values grouped by depth interval. The number of $\ln\chi$ values used to compute $\sigma^2_{\ln\chi}$ are shown. Although there is considerable scatter it seems clear that the intermittency is greater in the strongly stratified seasonal thermocline than in the mixed layer above or the less stratified layers below. Elliott and Oakey (1980) found χ was lognormal in the depth interval 30-130 m during GATE with $\sigma^2_{\ln\chi} = 3.8$, and this value is included in Fig. 7 for comparison.

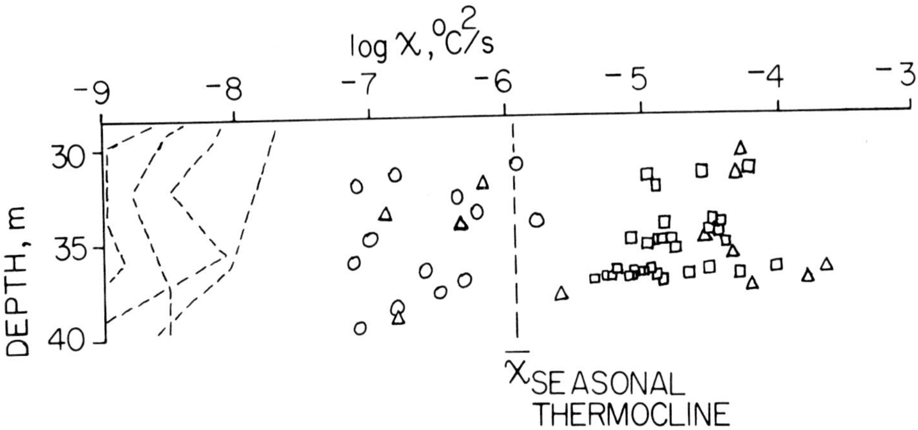

Fig. 6. Range of χ values in seasonal thermocline during MILE. Dillon (1982): triangles, high wind; circles, low wind. UCSD tow body: squares. Lange (1981) χ profiles: dashed lines. Note that individual profiles tend to underestimate $\bar{\chi}$ by large factors in this very intermittent layer (see Fig. 7).

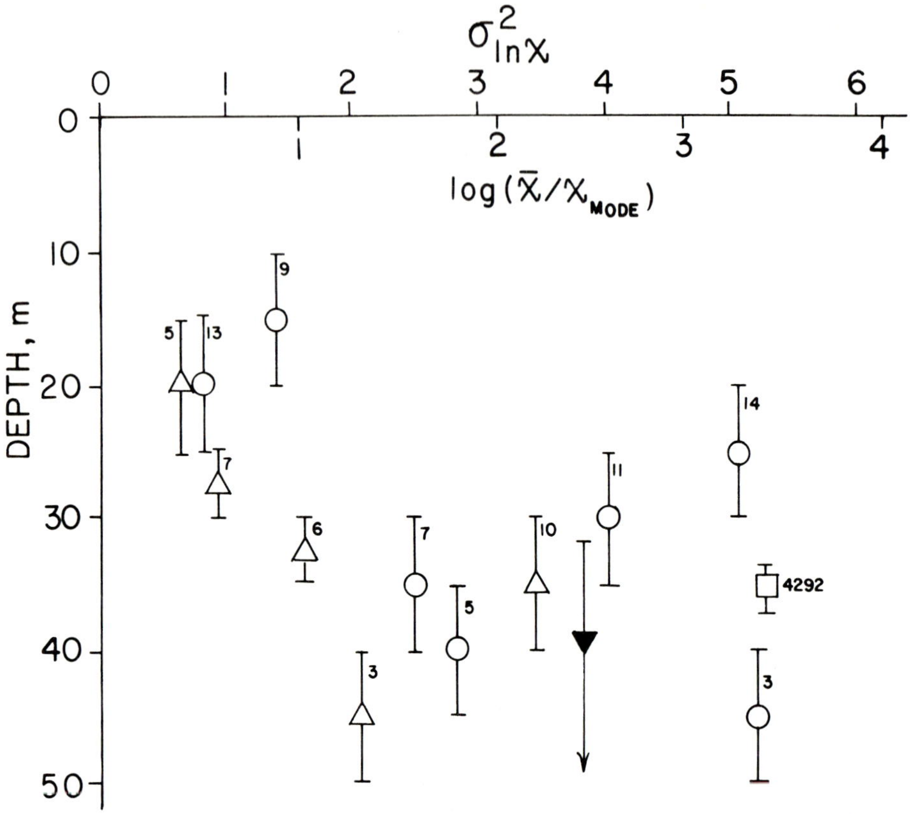

Fig. 7. Intermittency $\sigma^2_{\ln\chi}$ and probable undersampling error χ_{mean}/χ_{mode} = exp $(1.5\ \sigma^2_{\ln\chi})$ versus depth. Dillon (1982): circles and triangles as in Fig. 6, depth intervals and number of profiles averaged shown. UCSD tow body: square, 4292 1.1 m length samples averaged. Elliott and Oakey (1980): dark inverted triangle, 30-130 m depths, during GATE.

The undersampling error factor expected for a short data record, $\chi_{mean}/\chi_{mode} = \exp(1.5\ \sigma^2_{\ln\chi})$, is shown on the intermittency scale. For mixed layer depths down to 25 m the expected undersampling error factor is less than 10, at the seasonal thermocline it is over 1000, and there is perhaps some decrease in layers below.

5. ESTIMATES OF $\overline{\chi}$, DIFFUSIVITY AND HEAT FLUX IN THE EQUATORIAL UNDERCURRENT DEPTH ZONES FROM LOGNORMAL MODEL

Figure 8 shows all of the available measurements of χ in the core layers of

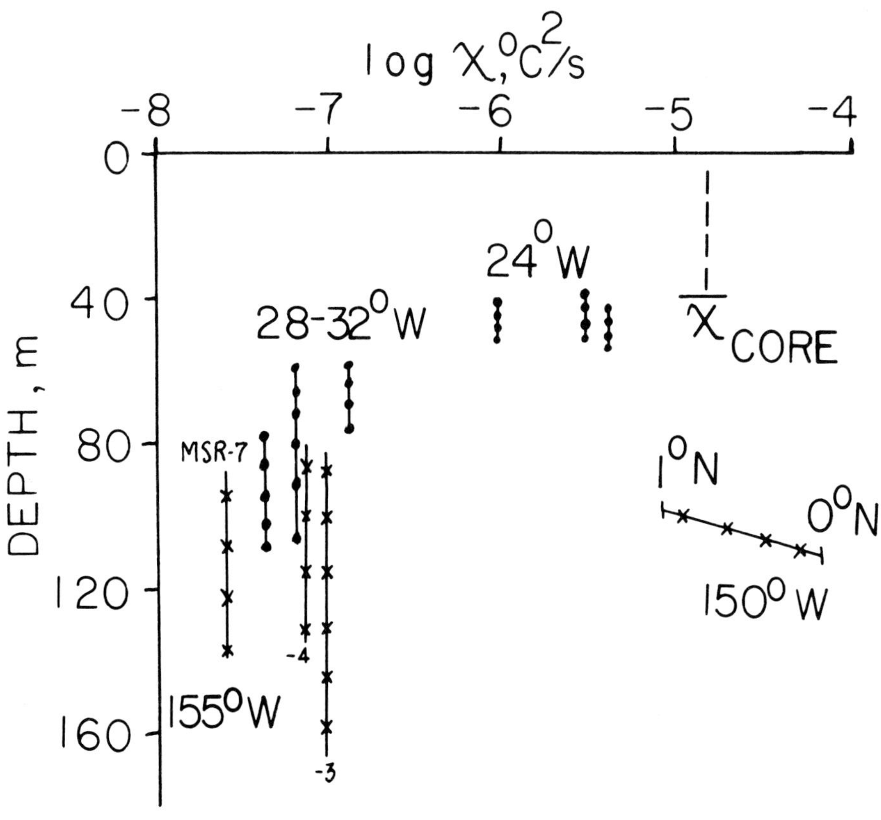

Fig. 8. χ measurements in the high velocity core of the equatorial undercurrent. Gregg (1976): crosses on depth intervals of MSR-3,4,7. Osborn and Bilodeau (1980): dots on depth intervals. Williams and Gibson (1974): crosses on depth interval. Longitudes are shown, latitudes are at the equator unless noted.

the equatorial undercurrent in the Atlantic and Pacific. The scatter is very wide, covering nearly 4 decades, but this range does not seem unreasonable in view of the scatter observed in the seasonal thermocline shown in Fig. 6. Treating each of the 11 samples shown as independent, and giving each equal weight gives a mean $\ln\chi$ value μ of -14.51 and $\sigma^2_{\ln\chi} = 6.89$. Therefore, assuming lognormality, $\overline{\chi}$ for the core should be about 1.6×10^{-5} $^{o}C^2/s$ based on this data, where $\overline{\chi} = \exp\mu \exp(\sigma^2_{\ln\chi}/2)$. This value is shown in Fig. 8.

Table 1 lists estimates of $\overline{\chi}$ for the three depth zones of the equatorial undercurrent indicated by the lognormal model. It should be emphasized that these mean values are based on a small number of samples which have variable degrees of statistical significance and do not take into account seasonal and zonal variability which may be substantial. The indicated mean χ values in the undercurrent depth zones are about an order of magnitude larger than the corresponding mean values in the seasonal thermocline shown in Fig. 5, which seems reasonable.

Table 1. Estimates of χ_{mean} in Equatorial Undercurrent Depth Zones - Lognormal Model

	Samples	$\mu = \overline{\ln\chi}$	$\sigma^2_{\ln\chi}$	$\overline{\chi} = \exp\mu\exp(\sigma^2/2)$
				$^{o}C^2/s$
Shear layer above core	6	-13.16	2.72	7.5×10^{-6}
Core layer	11	-14.51	6.89	1.6×10^{-5}
Shear layer below core	6	-14.81	2.05	1.0×10^{-6}

Figure 9 shows the vertical eddy diffusivity K indicated by the dropsonde values of Gregg (1976), Osborn (1980) and the present work, where

$$K = \overline{\chi}/2\,(\partial\overline{T}/\partial z)^2 \qquad (7)$$

using the Osborn and Cox (1972) model. The most striking difference is in the core layer where the present lognormal model gives $K = (2\text{-}25)\ cm^2/s$ compared to $(1.5-3) \times 10^{-2}\ cm^2/s$ inferred by Gregg (1976) and Osborn (1980).

The corresponding values of vertical heat flux Q are shown in Figure 10, where

$$Q = -\rho\, C_p\, K\,(\partial\overline{T}/\partial z) = -\rho\, C_p\, \overline{\chi}/2\,(\partial\overline{T}/\partial z) \qquad (8)$$

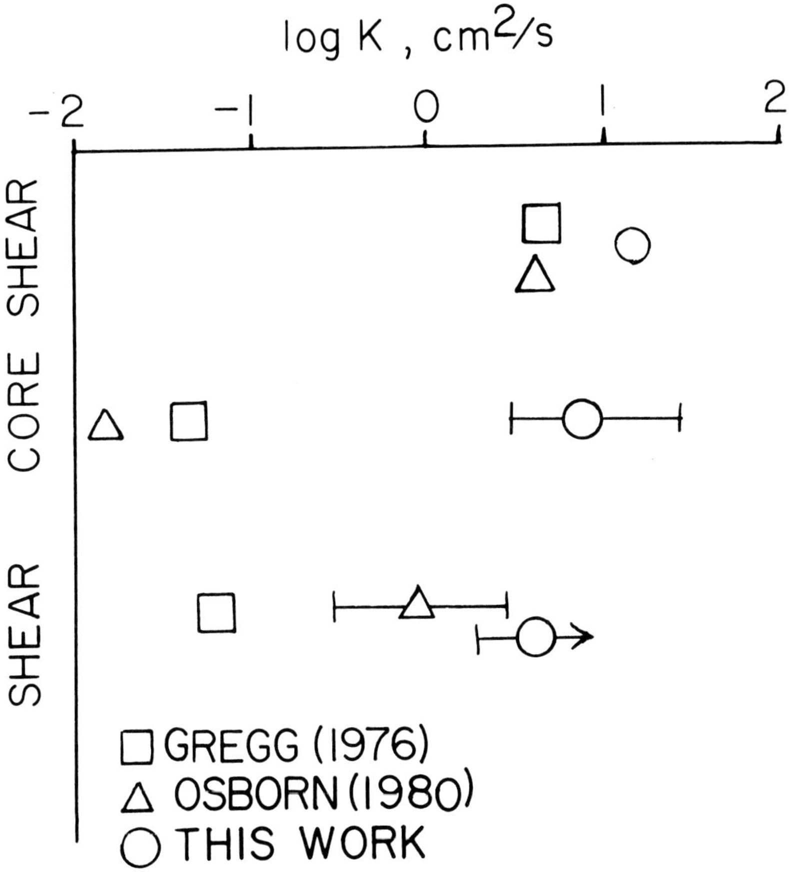

Fig. 9. Vertical diffusivity K in various depth zones of the equatorial undercurrent. Gregg (1976), squares. Osborn (1980), triangles. Average of all available values corrected for undersampling error using log-normal model, circles.

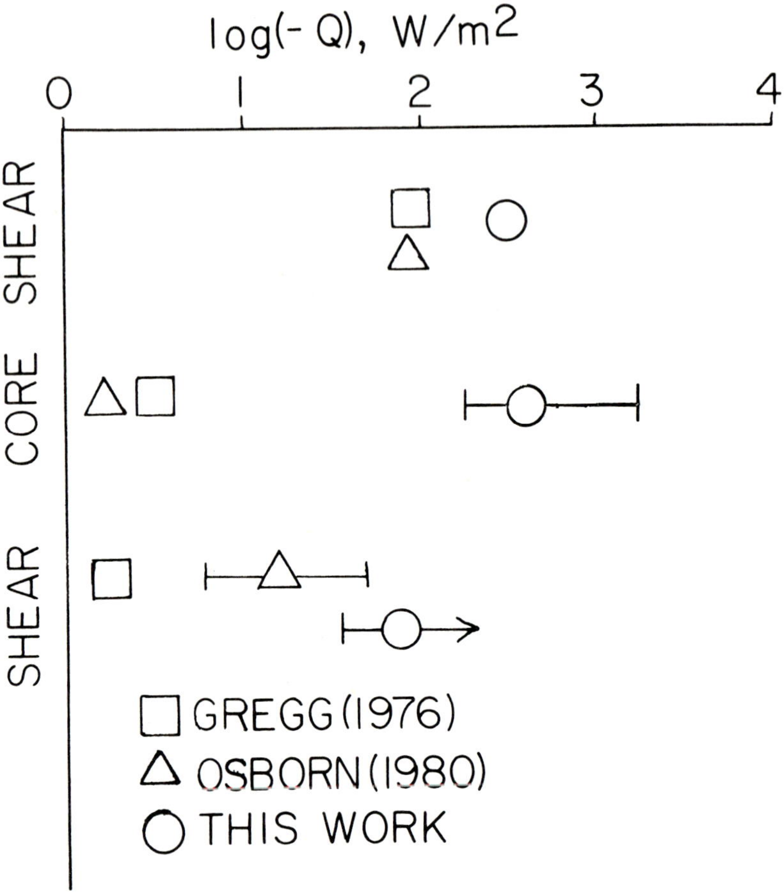

Fig. 10. Vertical heat flux Q in shear zones of equatorial undercurrent. Same symbols as Fig. 9.

ρ is density and C_p is heat capacity. Vertical temperature gradients were taken to be 1×10^{-3} °C/cm in the core and 5×10^{-4} °C/cm in the shear zones above and below in computing K and Q values in Figures 9 and 10.

The vertical heat flux in the core zone is larger than in the zones above and below, and is about 2 orders of magnitude larger than inferred by Gregg (1976) and Osborn (1980) and 1 order of magnitude larger than the upper bound of 10 W/m^2 proposed by Crawford (1982). From the present work, Q is about (0.2-2) kW/m^2 downward at the core which is larger than the net surface heat flux. Therefore it appears that the undercurrent may be collecting heat by turbulent diffusion from warmer surface layers north and south of the equator (see Fig. 1) and mixing the heat down through the core to deeper layers at rates 1-2 orders of magnitude higher than the vertical heat flux at latitudes outside the undercurrent influence.

6. SUMMARY AND CONCLUSIONS

Estimates of the true space-time average χ, K and Q values in the velocity core of the equatorial undercurrent have been made from all available microstructure data. Undersampling errors due to turbulent patchiness and intermittency are taken into account by assuming the temperature dissipation rate χ is lognormal. Such undersampling errors appear to be maximum in strongly stratified layers such as the undercurrent high velocity core and in the seasonal thermocline at higher latitudes. The range of χ values measured in the undercurrent core is very large, nearly 4 decades as shown in Fig. 8, but this range is less than the range observed in the north Pacific seasonal thermocline, which covers nearly 6 decades as shown in Fig. 6. Gregg (1976) and Crawford (1976, 1982) have suggested that towed body ε and χ values in the undercurrent must be contaminated by vibrational noise because their dropsonde values tend to be orders of magnitude smaller. This suggestion does not seem to be justified. Rather, it appears that dropsonde ε and χ values generally represent the mode of lognormal random variables with means several decades larger in stratified ocean layers such as the undercurrent core and the seasonal thermocline. For the tow body-dropsonde intercomparison in the seasonal thermocline the upper end of the range of χ values shown in Fig. 6 was measured from a dropsonde by Dillon (1982) rather than the towed body. No attempt was made to estimate the mean values of ε in the undercurrent depth zones because less data is available, although it seems likely that the towed body values in the core of order 10^{-1} cm^2/s^3 shown in Fig. 2 are closer to the true average than the dropsonde values of 10^{-4} - 10^{-5} cm^2/s^3 proposed as representative of the true average by Osborn (1980) without accounting for intermittency.

The much higher $\overline{\chi}$, K and Q values inferred in this paper call into question the conclusion reached by Gregg (1976, p. 1195) that "in view of the small volume of the undercurrent system, it is of negligible importance in the global balance of temperature fluctuations." Gregg's statement was based on his assumptions that his dropsonde χ values of 10^{-8} - 10^{-7} $^oC^2/s$ were representative of the undercurrent average in the upper 200 m and that these were not much different than average oceanic χ values in this depth range. However, the present analysis suggests $\overline{\chi}$ for the undercurrent is probably in the range 10^{-5} - 10^{-4} $^oC^2/s$, which may be 10^1 - 10^2 times larger than the upper 200 m ocean average, which may also be higher than Gregg (1976) assumes. Thus the undercurrent may constitute a very important factor in tropical heat and mass transport processes which affect planetary weather and climate.

It is clear that the present information about turbulence, mixing and diffusion in the equatorial undercurrent system is quite incomplete: fragmentary might be a better term. Much more data at all scales is needed to establish reliable values of the relevant parameters on which the present rather crude analysis depends. It is of particular importance to measure the intermittency of the turbulence and mixing at various depths, times and positions in order to be able to evaluate any microstructure sampling procedure or any particular data set. Profiles of $\sigma^2_{\ln\chi}$ and $\sigma^2_{\ln\varepsilon}$ are most desirable.

Acknowledgements. The author's attendance of the Liège colloquium was supported by a UNESCO travel grant. Preparation of this paper was supported by a grant from the UCSD Committee on Research.

REFERENCES

Belyaev, V.S., A.N. Gezentsvey, A.S. Monin, R.V. Ozmidov and V.T. Paka, 1975a. Spectral characteristics of small-scale fluctuations of the hydrophysical fields in the upper layer of the ocean. Journal of Physical Oceanography, 5: 492-498.

Belyaev, V.S., M.M. Lubimtzev and R.V. Ozmidov, 1975b. The rate of dissipation of turbulent energy in the upper layer of the ocean. Journal of Physical Oceanography 5: 499-505.

Crawford, W.B., 1976. Turbulent energy dissipation in the Atlantic equatorial undercurrent. Thesis, The University of British Columbia, Canada.

Crawford, W.R. and T.R. Osborn, 1980. Microstructure measurements in the Atlantic equatorial undercurrent during GATE. In: GATE-2, Equatorial and A-Scale Oceanography, Walter Düing, Ed., Pergammon, 285-308.

Dillon, T.R., 1982. Vertical overturns: a comparison of Thorpe and Ozmidov length scales. Journal of Geophysical Research, 87 (C12): 9601-9613.

Elliott, J.A., and N.S. Oakey, 1980. Average microstructure levels and vertical diffusion for Phase III, GATE. In: GATE-1, Equatorial and A-Scale Oceanography, Walter Düing, Ed., Pergammon, 273-293.

Gammon, R.H., J. Cline and D. Wisegarvern, 1982. Chlorofluoromethanes in the northeast Pacific ocean: measured vertical distributions and application as transient tracers of upper ocean mixing. Journal of Geophysical Research, 87 (C12): 9441-9454.

Gibson, C.H., 1980. Fossil temperature, salinity, and vorticity turbulence in the ocean. In: Marine Turbulence, J.C.J. Nihoul, Ed., Elsevier, 221-257.

Gibson, C.H., 1981. Buoyancy effects in turbulent mixing: sampling turbulence in the stratified ocean. AIAA Journal 19: 1394-1400.

Gibson, C.H., 1982. Alternate interpretations for microstructure patches in the thermocline. Journal of Physical Oceanography, 12:374-383.

Gibson, C.H., 1983. Internal waves, fossil turbulence, and composite ocean microstructure spectra. Submitted to Journal of Physical Oceanography.

Gregg, M.C., 1976. Temperature and salinity microstructure in the Pacific equatorial undercurrent. Journal of Geophysical Research, 81:1180-1196.

Gregg, M.C., 1977. Variations in the intensity of small-scale mixing in the main thermocline. Journal of Physical Oceanography, 7: 436-454.

Gregg, M.C., 1980. Microstructure patches in the thermocline. Journal of Physical Oceanography, 10: 915-943.

Helm, R., H.U. Lass and M. Sturm, 1980. Some peculiarities of the Atlantic equatorial undercurrent core structure and its variation in time and space. In: GATE-2, Equatorial and A-Scale Oceanography, Walter Düing, Ed., 249-259.

Lange, R.E., 1981. Observations of near-surface oceanic velocity strain-rate variability during and after storm events. Journal of Physical Oceanography, 11: 1272-1279.

Osborn, T.R., and C.S. Cox, 1972. Oceanic finestructure. Geophysical Fluid Dynamics, 3: 321-345.

Osborn, T.R., 1980. Estimates of the local rate of vertical diffusion from dissipation measurements. Journal of Physical Oceanography, 10: 83-89.

Osborn, T.R., and L.E. Bilodeau, 1980. Temperature microstructure in the equatorial Atlantic. Journal of Physical Oceanography, 10(1): 66-82.

Schedvin, J.C., 1979. Microscale temperature measurements in the upper ocean from a towed body. Thesis, University of California at San Diego.

Tully, J.P. and F.G. Barber, 1960. An estuarine analogy in the sub-Arctic Pacific ocean. J. Fish. Res. Bd. Canada 17 (1): 91-112.

Washburn, L., and C.H. Gibson, 1982. Measurements of oceanic temperature microstructure using a small conductivity sensor. Journal of Geophysical Research, 87(C6): 4230-4240.

Washburn, L., 1982. Horizontal observations of temperature microstructure within the seasonal thermocline. Thesis, University of California at San Diego.

Williams, R.B., and C.H. Gibson, 1974. Direct measurements of turbulence in the Pacific equatorial undercurrent. Journal of Physical Oceanography, 4: 104-108.

Williams, R.B., 1974. Direct measurements of turbulence in the Pacific equatorial undercurrent. Thesis, University of California at San Diego.

Wyrtki, K., 1982. An estimate of equatorial upwelling in the Pacific. Journal of Physical Oceanography, 11: 1205-1214.

ON THE INTERANNUAL WIND-DRIVEN RESPONSE OF THE TROPICAL PACIFIC OCEAN

Antonio J. Busalacchi,[1] Kensuke Takeuchi,[2] and James J. O'Brien

Mesoscale Air-Sea Interaction Group
Florida State University
Tallahassee, Florida 32306 (USA)

Present Affiliation: [1] Goddard Laboratory for Atmospheric Sciences
NASA/Goddard Space Flight Center
Greenbelt, MD 20771 (USA)

[2] Department of Geophysics
Hokkaido University, Sapporo, 060 (JAPAN)

ABSTRACT

A linear numerical model forced by ship-board wind estimates for each month from January, 1961 to December, 1978, has been used to study the interannual variability of the tropical Pacific Ocean. Model pycnocline variations at several stations are similar to the observed sea level fluctuations. El Nino events are depicted as periods when the pycnocline is persistently deep along the eastern boundary. Remotely forced equatorial Kelvin waves are responsible for this response. The character of each simulated El Nino is strongly dependent on the relation between zonal wind stress changes in the western and central equatorial Pacific. A rapid shoaling of the pycnocline in the western tropical Pacific during each El Nino is caused by westward-propagating Rossby waves. Interannual pycnocline displacements in the central equatorial Pacific are determined by the superposition of Kelvin waves excited to the west and first-mode Rossby waves generated to the east.

INTRODUCTION

The importance of the interannual variability of the tropical Pacific lies in its relationship with the El Nino phenomenon. In the strictest sense, El Nino is a term used presently to represent the anomalous occurrence and persistence of warm sea surface temperatures (SST) along the coast of Ecuador and Peru. It is used more loosely to represent anomalous high sea level and deep thermocline periods in the eastern Pacific consistent and coincident with the SST changes. An El Nino occurrence is not an isolated phenomenon, rather it is one facet of global-wide climate anomalies characterized by changes in the Southern Oscillation Index (Kidson, 1975). With respect to the tropical Pacific Ocean, one might expect that large perturbations to the mean state are not unique to the eastern boundary region. Fortuitously, there are several island stations in key areas of the tropical Pacific (Fig. 1) from which sea level time series can be used to monitor the interannual variability.

In the eastern Pacific, sea level changes at the Galapagos Islands are representative of the non-seasonal fluctuations at the equator and along the eastern boundary (Wyrtki, 1975; Hickey, 1975). In the western tropical

Pacific, there is a drastic decrease in sea level and shoaling of the thermocline during the El Nino year (Hickey, 1975; Wyrtki, 1975, 1977, 1979; Meyers, 1982). The largest drop in sea level occurs north of the equator. The sea level record at Truk Island is representative of the variability in this region and is one of the longest time series available. A lagged cross correlation analysis (Hickey, 1975) indicates a maximum negative correlation of sea level between the eastern and western Pacific occurs when the eastern Pacific leads by several months. In the central equatorial Pacific, sea level data from Canton Island and Christmas Island indicate an anomalous rise in sea level during the latter half of the El Nino year. The processes responsible for the phase relations between the eastern, central equatorial, and western Pacific are not understood.

Wyrtki (1975) has attributed El Nino to the excitation of Kelvin waves due to the interannual fluctuations of the southeast trade winds over the central equatorial Pacific. A statistical analysis by Barnett (1977) linked the variability of the zonal component of the southeast trades west of the dateline with sea level changes across the Pacific basin. Numerical and analytical models by Hurlburt et al., (1976) and McCreary (1976) simulated the excitation of an equatorially trapped internal Kelvin wave in response to changes in the zonal component of the wind stress. Upon impingement with the eastern boundary, the incoming Kelvin wave excited westward-propagating internal Rossby waves and poleward-propagating coastal trapped Kelvin waves. This is consistent with the theoretical work of Moore (1968) and Moore and Philander (1977). Kindle (1979) demonstrated that the magnitude, duration and timing of changes in the southeast trades over the central and western Pacific were important in determining the character of a simulated El Nino response. O'Brien et al., (1981) discussed these aspects of Kelvin waves and El Nino in a simpler physical context.

This study uses the results of a linear numerical model to analyze the interannual variability of the tropical Pacific Ocean. The model of Busalacchi and O'Brien (1980) is forced by monthly estimates of the surface wind field based on ship-board observations over the tropical Pacific for 1961-1978. The dense time and space coverage provided by such a model will be used to provide insight into the causality of the wind-driven response in different parts of the basin. Regions of important wind stress fluctuations will be identified and the relevance to sea level/pycnocline responses in the eastern, central equatorial and western Pacific will be addressed. Since the model is simple, the wind induced variability can be interpreted easily in physical terms. If the model results compare favorably with observations, physical interpretation of the model response can be applied to the observed oceanic response.

THE MODEL AND WIND DATA

The model utilized by Busalacchi and O'Brien (1980) is a $1^{1}/2$ layer reduced-gravity, linear transport model on an equatorial β-plane. The model geometry is an idealization of the tropical Pacific from 18°N to 12°S (Fig. 1). Open boundary conditions at the northern and southern boundaries allow coastal Kelvin waves to propagate freely out of the model domain. A complete discussion of the model and its equations are included in the authors' previous work.

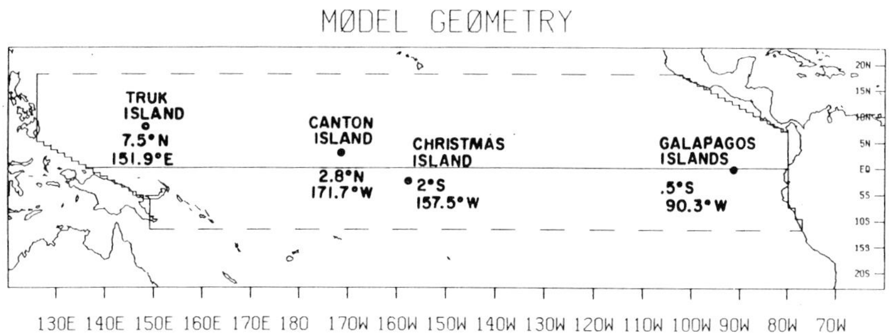

Fig. 1. Comparison of the model geometry with the tropical Pacific Ocean. The model basin extends from 18° N to 12° S and 126° E to 77° W. The dashed line represents an open boundary; all remaining boundaries are solid walls. Sea level data from the island stations indicated provide details of the interannual variability in different regions of the Pacific.

For the present study, the upper layer thickness, H, is taken to be 300 m. This depth is sufficient to prevent any surfacing of the density interface during an 18-year integration. With H = 300 m the first baroclinic phase speed is 2.45 m s^{-1}. This speed is the only dynamically important free parameter in the model. The Laplacian friction coefficient only cosmetically smooths the very high wave number numerical noise.

Details of the wind stress data base and a spectral analysis of that data for 1961-1970 were discussed by Goldenberg and O'Brien (1981). Monthly estimates of the surface wind stress based on ship observations over the tropical Pacific from January, 1961, to December, 1978, were subjectively analyzed onto a 2° by 2° mesh. The resulting data set, continuous in time and space, was input as a body force into the momentum equations of the model. Though the approach used in collecting and preparing the wind data is not the ultimate methodology for monitoring the wind field over the tropical Pacific Ocean, it has resulted in the only data set available with spatial and temporal resolution suitable for interannual modelling studies.

A new open boundary condition is implemented for this 18-year simulation. After 7 years of integration, the variant of the Hurlburt (1974) open boundary condition used in the previous study becomes unstable. The boundary condition at the northern and southern boundaries is replaced with an open boundary condition developed by Camerlengo and O'Brien (1980). This modification of the Orlanski (1976) open boundary treatment allows the model to be integrated for the full 18 years. Prior to the onset of any instability, calculations with the Hurlburt open boundary condition and those with the Camerlengo and O'Brien open boundary condition render the same solutions. The model is initiated using the results from year 4 of the seasonal calculation in Busalacchi and O'Brien (1980). Thus, spin-up transients are confined essentially to the first year of integration.

EASTERN PACIFIC RESPONSE

The 18-year period of the numerical simulation encompasses several notable eastern Pacific events of varying intensity. The strongest El Nino since 1957 occurred during 1972 (Quinn et al., 1978). The associated environmental changes resulted in the accelerated collapse of the Peruvian anchoveta fishery. Sea surface temperatures (SST) off the coast of Ecuador and Peru were up to 4°C warmer than the climatological monthly means (Fig. 2). Sea level was at its highest point in years. The sea level record at the Galapagos Islands (Fig. 3) is a useful indicator of this interannual variability along the eastern boundary (Wyrtki, 1975; Hickey, 1975). For the 1972 event, as opposed to previous Los Ninos, large anomalies were present during the middle of the year. Positive SST anomalies peaked in March and June. Sea level began to rise in late 1971 and reached a maximum in June. Consistent with several previous El Nino events another peak in SST and sea level occurred at the end of the year.

During 1965 and 1976, moderate El Nino conditions were present in the eastern tropical Pacific. Sea surface temperatures were 2°-3°C above normal. The 1965 event was characterized by warm SST and elevated sea level episodes at the beginning and end of the year. For the 1976 El Nino, SST was anomalously warm throughout the year. Sea level began to rise late in 1975 and peaked in June similar to the 1972 occurrence. A secondary sea level maximum was present late in the year but was much smaller relative to the first peak than in the 1972 event.

The weakest of the year-long warm events during this 18-year period were in 1963 and 1969. Positive SST anomalies were less than 2.5°C (Quinn et al. 1978). Once again sea level was elevated along the eastern boundary. However, for the very weak event in 1963 the sea level anomaly at the Galapagos Islands was not truly representative of the anomalies at coastal stations. Only at

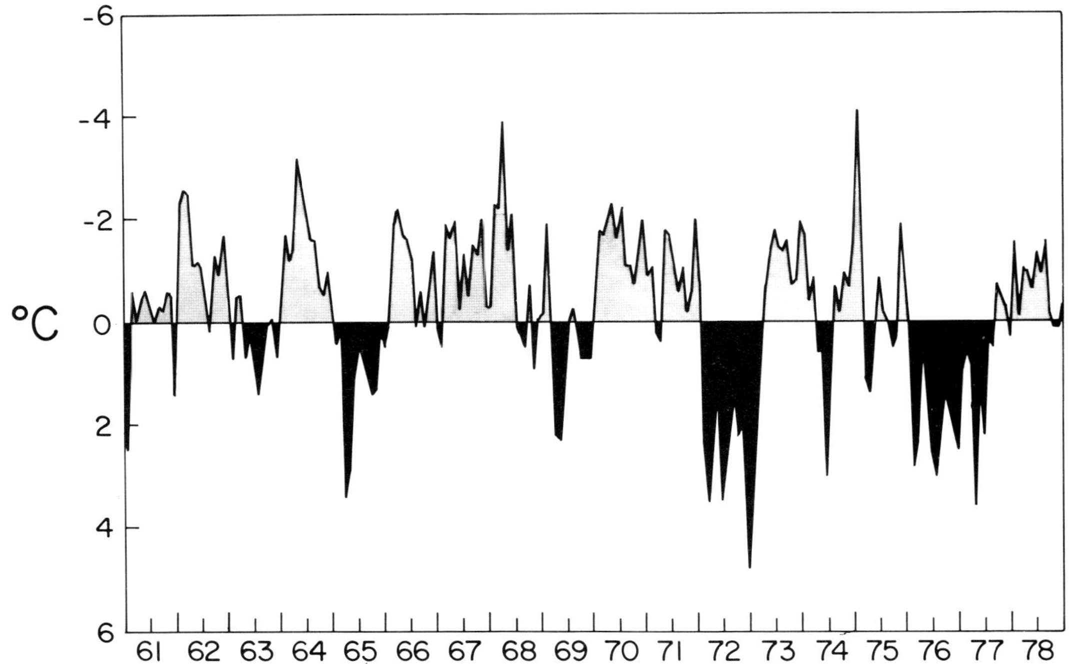

Fig. 2. Time series of departures from the long-term monthly mean shore temperatures at Talara, Peru, from 1961-1978, inclusive. Anomalously warm periods are shaded black.

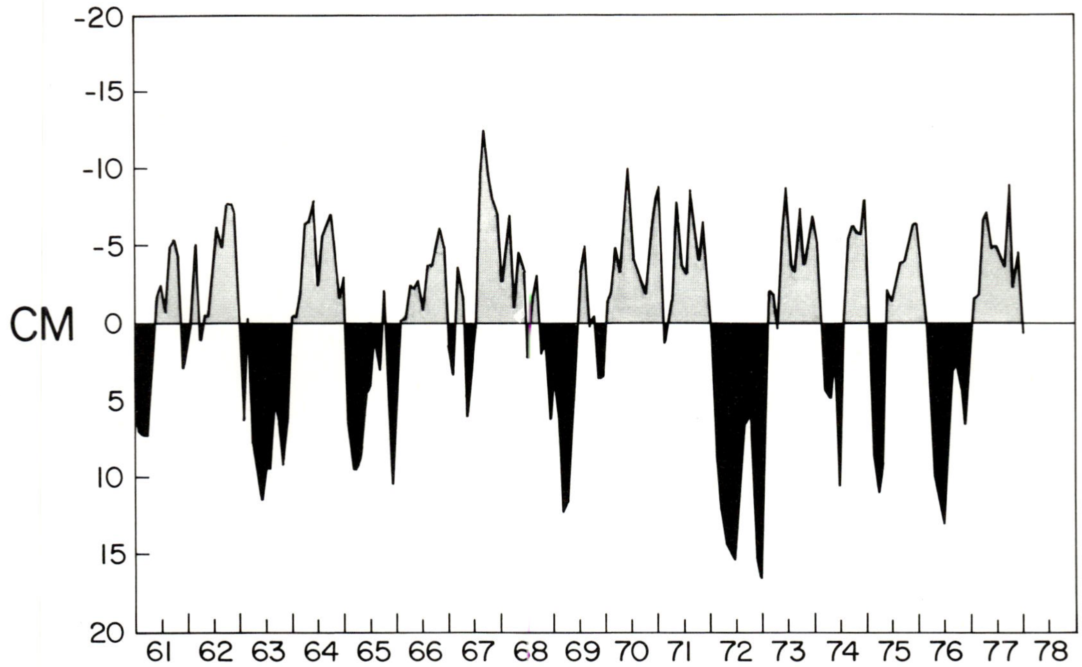

Fig. 3. Time series of observed sea level at the Galapagos Islands from 1961-1977, inclusive. Data for 1978 are not available. Periods when the mean monthly sea level is higher than the long-term annual mean are shaded black.

Galapagos was the sea level for 1963 higher than during the 1965 El Nino. At all the South American coastal stations monthly sea levels were higher in 1965.

Another period worthy of attention is 1975 even though it was not an El Nino year. In 1974 it was noted that there was an ongoing decrease in the Southern Oscillation Index (Wyrtki et al., 1976). Based on previous studies linking a low Southern Oscillation Index with El Nino (Quinn, 1974), a weak El Nino was predicted for 1975. An expedition mounted to observe this event found anomalous conditions present off the coast of Ecuador and Peru in February and March of 1975. The distribution of warm SST anomalies, changes in thermocline depth, and elevated sea levels were similar to the onset phase of previous El Nino episodes. These conditions did not persist long enough and were not large enough in magnitude for 1975 to be recognized as an El Nino year. By April and May the development of the warm episode had been aborted and conditions were returning to normal.

Interspersed among the El Nino episodes were periods of cool SST and depressed sea level. In this study of the linear wind-driven Pacific response the dynamics responsible for pycnocline and sea level changes during El Nino will be the same as in these non-El Nino events. Most of the following discussion of the model results will focus on the El Nino incidents since they often dominate the interannual variability, have important climatic implications, and have been the subject of previous observational and theoretical studies.

The sea surface and pycnocline fluctuate 180° out of phase in the model formulation used. Meyers (1979) has shown, using sea level and interfacial depth profiles, that the central and eastern Pacific act in a manner compatible with the dynamics implicit in this reduced-gravity approach. Since the model does not contain thermodynamics the results cannot give a direct indication of SST change. In general, there is no linear relation between upper layer thickness, sea level, or thermocline depth and SST. Yet, in the eastern tropical Pacific, warm (cold) SST anomalies are significantly correlated with elevated (depressed) sea level (Barnett, 1977; Enfield and Allen, 1980). Modelling efforts including mixed-layer physics will provide much needed information on such relations between thermocline depth, vertical velocity, and SST.

A time series of the pycnocline height anomaly (PHA) at a location representing the Galapagos Islands (Fig. 4) depicts the vertical displacement of the model pycnocline throughout the 18-year integration. The linear trend, resulting from a mean background wind stress, has been removed. Upwelling and downwelling bursts of relatively short duration constitute the variability in the entire record. The model pycnocline is the deepest during 1963, 1965,

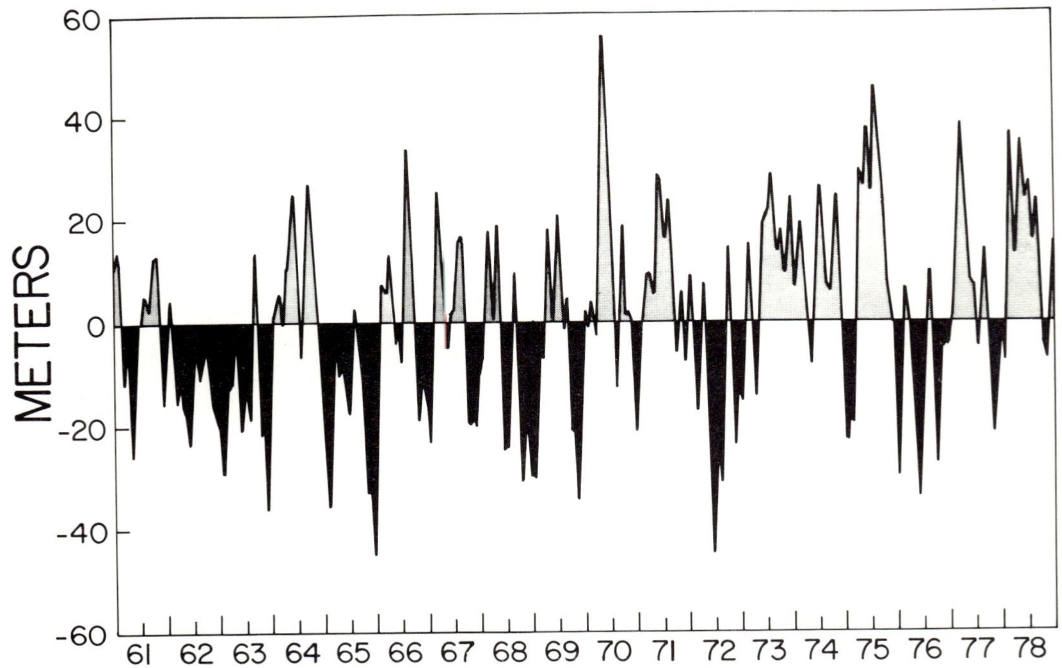

Fig. 4. Model pycnocline height anomaly (PHA) at the Galapagos Islands. Periods when the pycnocline is deeper than the 18-year mean are shaded black.

1969, 1972, and 1976; the five El Nino years. The double-peak El Nino signature found in the sea level record (Fig. 3) is present in the model results for 1965 and 1969. During these years the upper layer becomes deepest at the beginning and end of the El Nino year. The onset of the 1972 El Nino is preceded by a deepening of the pycnocline during the second half of 1971. The pycnocline reaches its maximum depth in June followed by a secondary maximum at the end of the El Nino year. The 1976 El Nino is characterized by downwelling from mid-1975 to early 1976, maximum pycnocline depths in June, and a deep pycnocline again in the latter half of the year. The weak event of 1975 is represented by a downwelled pycnocline at the beginning of the year followed by a rapid upwelling leaving a shallow pycnocline for the remainder of the year. The timing of major upwelling and downwelling events is similar even though there are periods when the relative amplitude of observed sea level change and pycnocline displacement do not agree. Changing the phase speed of the model by ±30% changes the phase of the eastern boundary response by less than one month. When comparing with monthly mean observations this shift in phase is not significant. A major difference between the model pycnocline fluctuations and observed sea level is the presence of more high frequency variability in the pycnocline time series. The cross correlation of pycnocline variability and observed sea level less the seasonal signals is maximum at zero lag with a correlation of 0.52 which is significant at the 99% level. The degrees of freedom used in the significance test were determined by dividing the total record length by the integral time period needed to obtain two independent realizations (Davis, 1976). The number of equivalent degrees of freedom was 51.

The observed sea level and model pycnocline time series for the Galapagos Islands have been subjected to a 12-month, running-mean filter (Fig. 5). This provides a more graphic representation of the interannual variability during the 18-year period. The similarities and differences between the model and observations are clearly displayed. The filtered data indicate a tendency for the sea level and the model pycnocline to be elevated or depressed for spans of 10-30 months. Discrepancies between the model and observations include the phase of the response from mid-1961 through 1962, the relative amplitudes of the response from late-1967 to early-1968, and the amplitude in 1972. The cross correlation between these two filtered time series was .61.

The model pycnocline variability at the location of Talara, Peru (Fig. 6) is very similar to that at the Galapagos Islands, except the maximum cross correlation (r = .98, significant at 99% level) occurs when the signal at Talara lags Galapagos by 1-3 days. This lag increases when time series further poleward along the coast are cross correlated with the Galapagos time series. For example, the pycnocline time series at the location of Callao,

Peru (12°S) lags Galapagos by approximately one week and is highly correlated. The great degree of similarity between these time series attests to the large areal extent of the variability near the eastern boundary. The poleward increase in lag indicates a poleward propagation of information. Smith (1978), Enfield and Allen (1980), and Romea and Smith (1982) have reported on an observed poleward propagation along the coast. Knox and Halpern (1982) have observed eastward propagation along the equator from 152°W to 91°W.

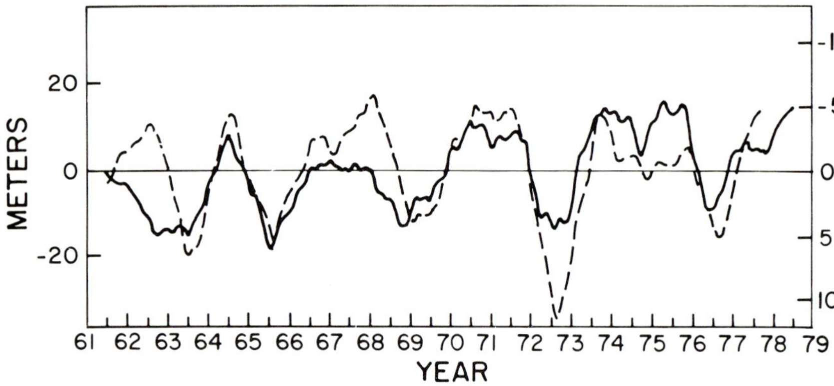

Fig. 5. Comparison of the model pycnocline height anomaly (solid) and observed sea level (dashed) at the Galapagos Islands filtered by a 12-month running-mean. The pycnocline displacement scale is in meters and sea level changes are expressed in cm.

The pycnocline variability at the Galapagos Islands and Talara consists of discrete upwelling and downwelling events with a duration of several months. What is the mechanism responsible for these changes in depth of the pycnocline? A cross correlation of the pycnocline variability at the equatorial eastern boundary with all points west along the equator implies that the effects of internal Kelvin and Rossby waves are important. The lag structure of the pycnocline variability along the equator is illustrated in Fig. 7. The most striking aspect is the linear increase in lag away from the eastern boundary for high correlations. The solid line represents an eastward phase speed of 2.45 m s^{-1}, corresponding to the phase speed for an internal, equatorially-trapped Kelvin wave in this model. As a result, 50% of the variability at the eastern boundary is accounted for by the variability along the Kelvin wave characteristic in the central Pacific.

The linear increase in lead away from the eastern boundary (dashed line) represents the westward propagation of the lowest order Rossby waves excited

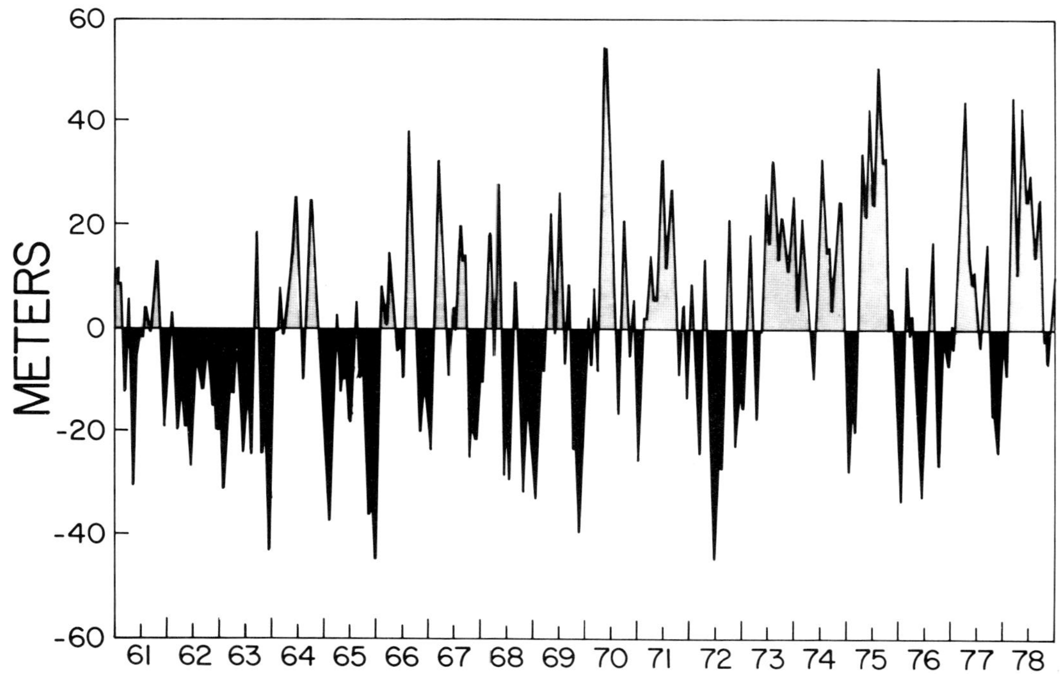

Fig. 6. Model pycnocline height anomaly at a location corresponding to Talara, Peru.

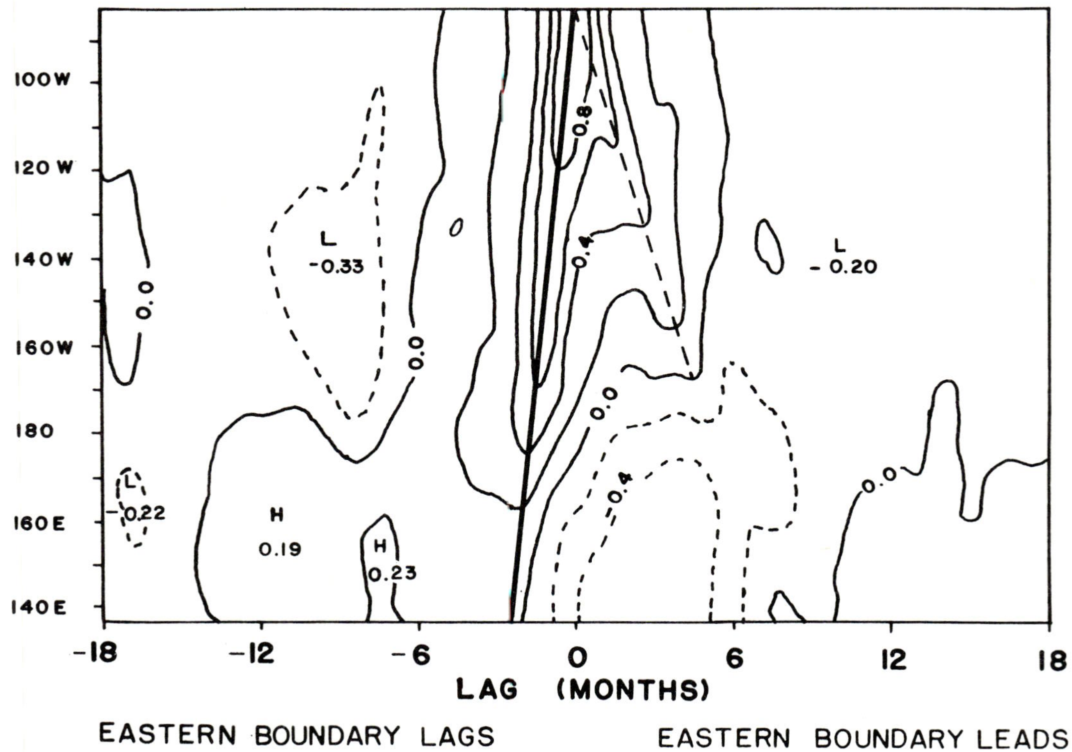

Fig. 7. Contours of lagged cross correlation between the pycnocline variability at the equatorial eastern boundary and all points west. The slope of the solid line indicates an eastward propagation representative of equatorially trapped Kelvin waves. The solid line represents the westward propagation of Rossby waves excited at the boundary.

at the coast. These waves were excited by the impingement of Kelvin waves or local changes in the wind stress. These waves have a noticeable effect on the pycnocline variability westward to 160°W. The slope of the dashed line represents a phase speed of .82 m s^{-1}, one-third the speed of the Kelvin wave. There is also a notable negative correlation when the eastern boundary leads the western boundary by several months.

Away from the boundaries an equatorial Kelvin wave can only be generated by temporal fluctuations of the zonal wind stress symmetric about the equator. A test calculation was performed in which the model was only driven by the zonal wind stress for the period mid-1973 through 1977. This was intended to test the premise that the interannual events in the eastern part of the basin are remotely forced by Kelvin waves. The time interval was chosen because it contained several large changes in depth of the pycnocline. The resulting pycnocline response along the eastern boundary was essentially the same as the total response to zonal and meridional winds. The cross correlation between the zonal wind stress solution and the total forced case was .99 at zero lag.

The pycnocline variability at the eastern boundary of this linear model is the integrated response to the zonal equatorial wind stress fluctuations to the west. The evolution of significant events, e.g., El Nino occurrences, can be traced back in time and space along the Kelvin wave characteristics. The location of the forcing responsible for each event has been identified in this manner.

We now describe the character of El Nino events from 1961-1978 depicted in this model as periods when the pycnocline was persistently deep. The period 1964-1965 is a good time frame within which to study the transformation from a non-El Nino regime to an El Nino regime. The semiannual pycnocline variability at the eastern boundary previously discussed by others (Meyers, 1979; Kindle, 1979; Busalacchi and O'Brien, 1980) has been associated with Kelvin wave signals excited between 180° and 120°W. Evidence for this seasonal cycle is seen throughout the model integration (Figs. 4,6). The semiannual variability is typical of a non-El Nino year in the 1960's. From January, 1964, to April, 1964, the pycnocline in the eastern two-thirds of the equatorial Pacific is being upwelled (Fig. 8). This is a result of an upwelling Kelvin wave front excited during an October, 1963 to March, 1964 intensification of the easterlies between 180°-160°W. The downwelling at the western boundary through June, 1964, is caused by a downwelling Rossby wave front excited by the same intensification of the easterlies between 180°-160°W. Another large amplitude upwelling Kelvin wave front is excited in the interior from July through September. Towards the end of a non-El Nino year a downwelling Kelvin wave front is excited in the central equatorial Pacific. A simulated seasonal El Nino along the eastern boundary is

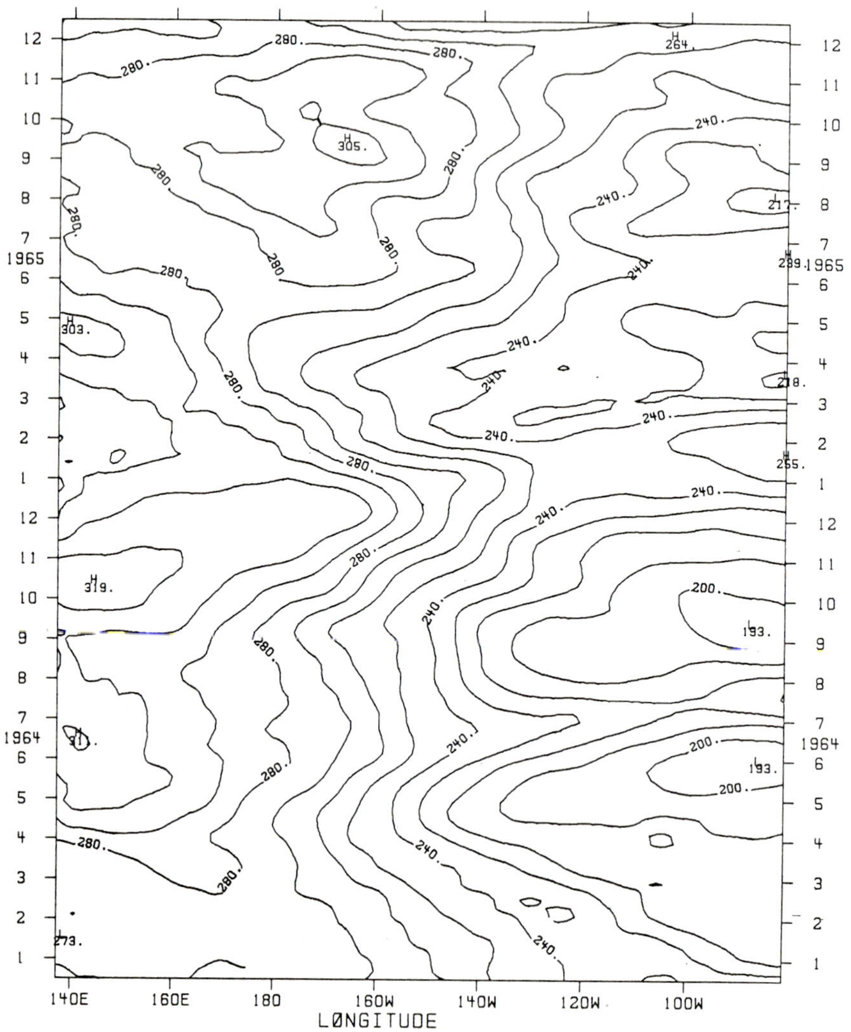

Fig. 8. An X-T section along the equator for the upper layer thickness (ULT) during 1964-65. The phase speed of the downwelling Kelvin wave responsible for triggering the 1965 El Nino in the model is indicated by the slope of an imaginary line connecting a point at the western boundary for November, 1964, with a point at the eastern boundary for February, 1965. Such a line also indicates the end of the equatorial downwelling associated with the passage of the Kelvin wave.

associated with this downwelling Kelvin wave front. Though the amplitude and precise timing of these upwelling signals may vary amongst non-El Nino years, there is often an upwelling of the pycnocline early in the year. The upwelling is important because it causes the pycnocline to be upwelled following the seasonal downwelling of the pycnocline during the austral summer.

The changes in pycnocline depth that typify the 1965 El Nino event have their roots in the western Pacific during the latter half of 1964. A 0.4 dynes cm^{-2} relaxation of the easterly wind stress between the western boundary and the dateline from August-November, 1964, excites a downwelling Kelvin wave front. The slope of the downwelling signal in Fig. 8 represents the inverse phase speed of an internal Kelvin wave. The equatorial pycnocline is depressed as the wave front propagates eastward; terminating at the eastern boundary in January, 1965. The timing of this wave is such that it greatly enhances the normal seasonal downwelling during the Southern Hemispheric summer. The same sequence of events preceeds the 1969 El Nino.

In addition to the downwelling Kelvin wave excited near the western boundary, the semiannual upwelling signals of 1965 are much weaker than those of non-El Nino years. As a result, there is no mechanism to raise the deeper than normal downwelled pycnocline present in early 1965. The upwelling signals are weaker than non-El Nino years because the intensification of the interior easterlies is not as strong. This cessation of the normal semiannual variability of the zonal wind stress also occurs during 1969. Therefore, over a one-year period of time the pycnocline is deeper than the long term mean.

Towards the end of 1965 (Fig. 8), the seasonal relaxation of the interior easterlies excites a downwelling Kelvin wave front. This downwelling Kelvin wave is the cause of the downwelling at the eastern boundary at the beginning of 1966. Though this Kelvin wave is associated with the seasonal signal, it is atypical in the sense that it was excited at a time when the pycnocline was deeper than normal. Hence, there is a double-peak downwelling signature during 1965.

The excitation of a large amplitude downwelling Kelvin wave west of the dateline, which then enhances the normal seasonal downwelling, and the concomitant cessation of the semiannual interior zonal wind stress is the scenario that describes the 1965 and 1969 El Nino events. The minor El Nino of 1963 is solely due to weakened seasonally varying zonal wind stress over the central equatorial Pacific. There were not any large amplitude downwelling Kelvin waves excited during late 1962.

The events of the 1970's were somewhat different than in the previous decade. Beginning in October 1971, there was a relaxation or weakening of the trade winds west of the dateline (Fig. 9). This band of wind relaxation progressed eastward (Wyrtki, 1977) to 160°W during the entire El Nino year.

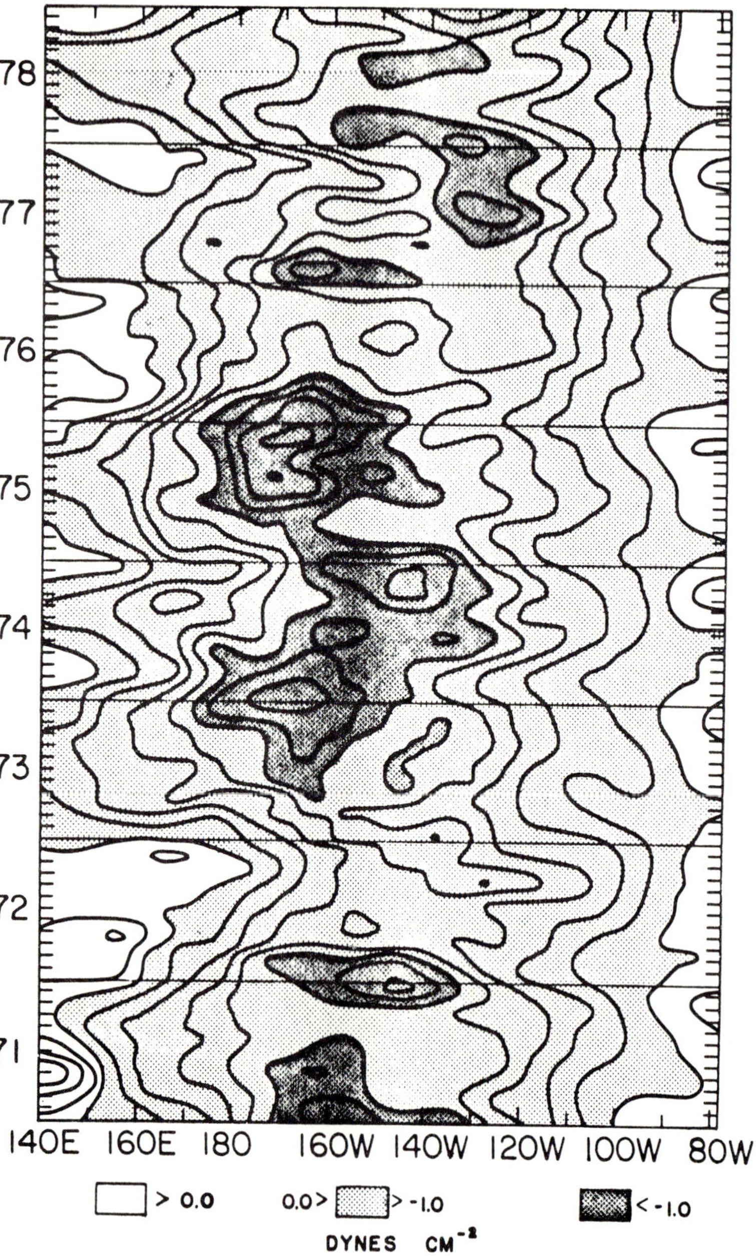

Fig. 9. Zonal wind stress along the equator from 1971-1978. White areas indicate westerly flow. Time periods in which the stress is greater than 1.0 dynes cm^{-2} are shaded the darkest. Contour interval is .2 dynes cm^{-2}.

Concurrent with the initial weakening in the west there was an intensification of the central equatorial easterlies, 130°W-170°W, from mid-October until the end of 1971. The combined effect of weakening winds in the west and intensifying winds in the central Pacific yielded average pycnocline depths in the east at the beginning of 1972 (Fig. 4). From January through April, at 120°W-180°, there was a decrease in the zonal wind stress of more than .6 dynes cm^{-2}. Thereafter, the winds in the central equatorial Pacific remained weak. This weakening of the easterlies excited a downwelling Kelvin wave front responsible for the deep model pycnocline (Fig. 4) corresponding with observed elevated sea level (Fig. 3) along the eastern boundary in June. During the latter half of the year an intensification-relaxation sequence at 100W-140W was responsible for the second downwelling peak at the end of 1972. Recovery from the 1972 El Nino began with strengthening trades in January of 1973.

The trade winds west of the dateline began decreasing in September of 1974 and continued until January, 1975. The winds in the central Pacific did not change appreciably. The resulting downwelling Kelvin wave front induced the deep pycnocline in the east at the beginning of 1975. As opposed to 1972, the trades in the western and central Pacific did not continue to weaken. An intensification of the easterlies beginning west of the dateline in January and east of the dateline in March produced a rapid recovery from the downwelling at the beginning of the year. A persistently deep pycnocline was not present at the eastern boundary in 1975. Consistent with observations, the event in early 1975 never matured into an El Nino.

The onset of the 1976 El Nino began when the winds west of 160°E started decreasing in October of 1975. An eastward progression of this weakening transpired similar to the wind changes in 1972. A more rapid decrease in the trades between 140°W and the dateline during the first four months of 1976 was again responsible for the large downwelling peak in June of the El Nino year. Continued weakening between 160°W and 140°W produced the last downwelling pulse of the El Nino year. The zonal wind stress remained weak up until December. Recovery from the 1976 El Nino began with a short period of intensified easterlies between 140°W and 180° (Wyrtki, 1979). The subsequent upwelling Kelvin wave front produced shallow pycnocline depths along the eastern boundary during the first half of 1977.

Comparisons of the simulated El Nino events of 1972 and 1976 with those of 1965 and 1969 yield an important difference. The model pycnocline variability during Los Ninos of the 1960's were comprised of pycnocline depth maxima at the beginning and end of the El Nino year. In the middle of the El Nino year the pycnocline was not significantly deep. On the contrary, during the 1972 and 1976 El Nino events the maximum pycnocline depth was attained in June. The reasons for this difference bear out the uniqueness of each El Nino.

The central equatorial easterlies during the 1970's had a significantly greater impact on the eastern boundary variability than in 1965 or 1969. The downwelling pulses at the beginning of 1965 and 1969 were due to weakening trades in the western Pacific during the four preceding months. During this same period the winds in the central Pacific did not intensify. Similar conditions were present at the beginning of 1975 and 1976; years in which the pycnocline was deep in January. A deep pycnocline did not exist at the beginning of 1972 because the easterlies in the central Pacific intensified in late 1971 while the winds west of the dateline were changing from easterly to westerly. From January to April of 1972 and 1976 the central equatorial easterlies decreased markedly. This resulted in the deep pycnocline along the eastern boundary in June. During the first four months of 1965 and 1969 the central equatorial trade winds did not weaken noticeably. Hence, the pycnocline was not deep in mid-1965 or mid-1969. In 1975 the trade winds intensified and were anomalously strong for the entire year resulting in a rapid upwelling from an initially deep pycnocline.

WESTERN PACIFIC RESPONSE

Since Wyrtki (1975) proposed a scenario of El Nino it has been thought that, due to strong easterlies, sea level in the western tropical Pacific was deserving of attention in the year preceding El Nino. However, as Meyers (1982) recently reported, and, as shown in Fig. 10, the sea level change at Truk Island is much more dramatic in El Nino years themselves. The sea level record at Truk Island is presented because it is representative of the interannual changes in the western Pacific and is one of the longest time series available.

In every year classed as an "El Nino year" for the period 1961-1978, a significant drop of the sea level is observed in the latter half of the year. Sustained significantly high sea levels are not necessarily observed in the year previous to every El Nino. The decreases in sea level at Truk during El Nino years have a common pattern. They begin in late winter or spring (February-May). In most cases (1963, 1969, 1972 and 1976) an initial sea level minimum is attained in the middle of the year (June-September). The lowest sea level, as low as 20 cm below the annual average, is reached at the end of the year. This is followed by a rapid recovery within the next few months, usually from January to April of the following year.

Wyrtki (1979) examined the sea levels at widely scattered islands in the western Pacific Ocean for the 1976 El Nino event. Widespread low sea levels in the western tropical Pacific for the second half of the year were evident. The maximum sea level decrease was within a zonal band at 5°N-10°N. This indicates that Truk Island is located in an advantageous position to monitor

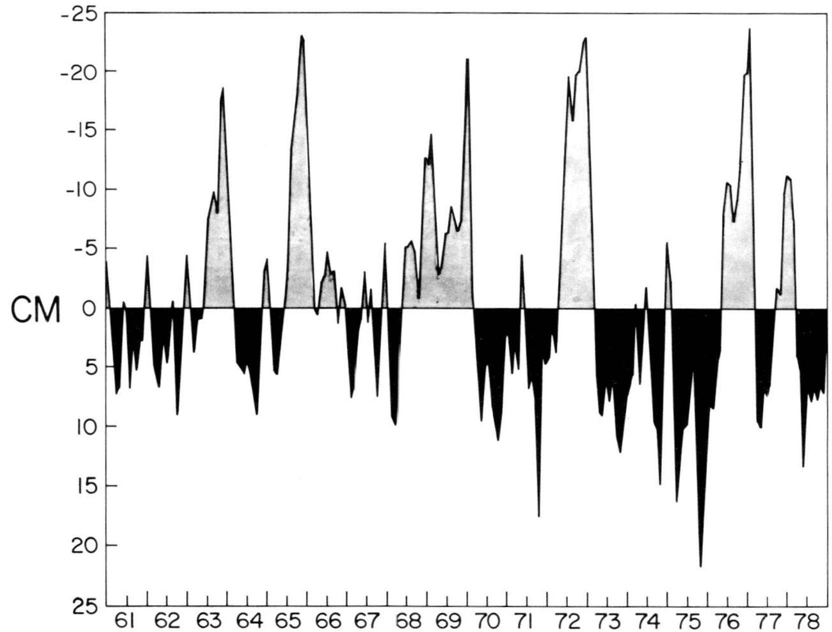

Fig. 10. Time series of observed sea level at Truk Island from 1961-1978, inclusive.

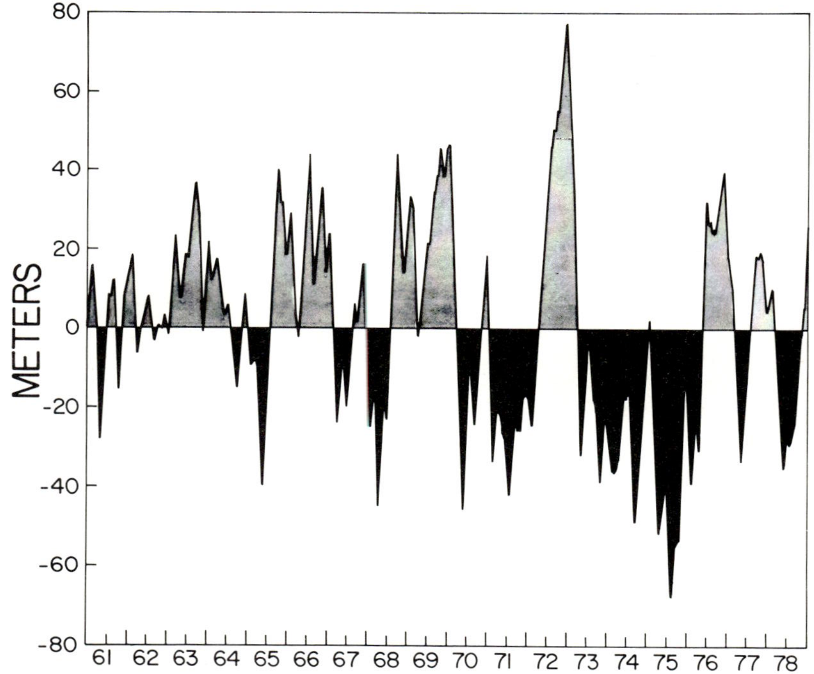

Fig. 11. Model pycnocline height anomaly for 1961-1978 at a location corresponding to Truk Island.

the sea level change during El Nino years. The sea level excursions in the Southern Hemisphere are much smaller.

The time series of the model pycnocline displacement at the location of Truk Island is depicted in Fig. 11. The time series is characterized by several shallow pycnocline events. Except for 1966, these events occur during the latter half of an El Nino year and correspond well to the sea level time series at Truk (Fig. 10). Note again a shallow pycnocline (positive PHA) corresponds to low sea level. As expected from this comparison, the correlation between the observed sea level and the model pycnocline at Truk has a maximum as high as 0.76 at zero lag.

The spatial distribution of the anomalous model pycnocline depths in an El Nino year is also in agreement with the observations. Fig. 12 shows the difference in depth of the model pycnocline for the western Pacific between October, 1976 and October, 1975. The spatial structure of the change in pycnocline depth agrees with the observations shown by Wyrtki (1979, Fig. 7) except for the region close to the northern open boundary of the model. The maximum pycnocline rise is also found around 5°N-10°N. A smaller maximum is found in the southern hemisphere near the western boundary.

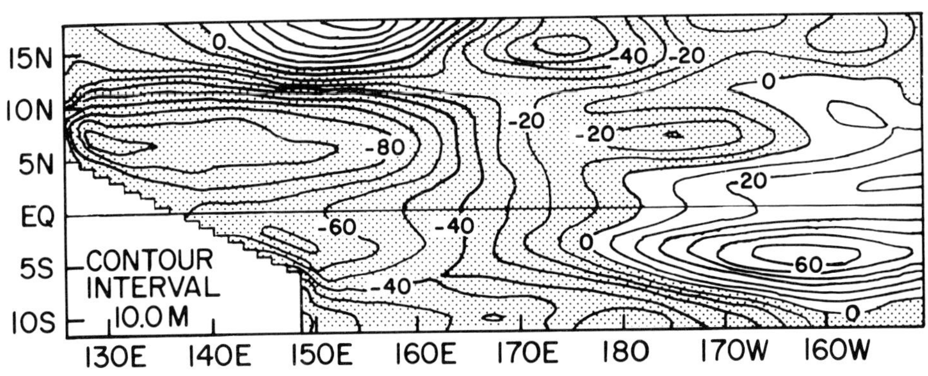

Fig. 12. The change in upper layer thickness from October, 1975 to October, 1976 in the western portion of the model domain. The stipled regions indicate where the pycnocline was shallower in October, 1976. Contour interval is 10 m.

A Y-T diagram of the model pycnocline at 160°E for the period 1971 to 1977 (Fig. 13) is convenient for examining the meridional distribution of the shallow pycnocline during El Nino years. The longitude 160°E is selected to avoid the effects of the western boundary in the Southern Hemisphere. It is not considered inadequate to choose 160°E as being representative of the

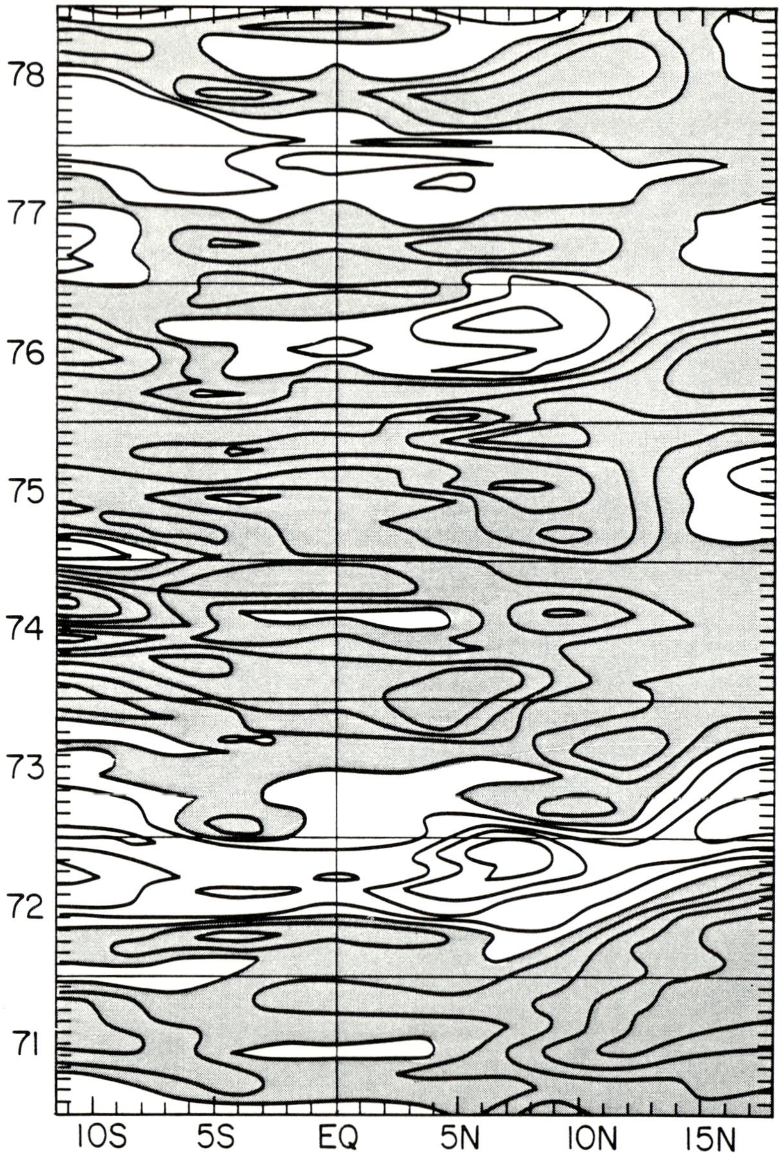

Fig. 13. A Y-T section of the pycnocline height anomaly along 160°E for 1971 to 1978. White regions indicate the model pycnocline is shallower than the 18-year mean. Contour interval is 15 m.

western tropical Pacific since the model pycnocline variability at 7.5°N, 160°E, has a similar time series to that of Truk Island.

In both the 1972 and 1976 cases the maximum positive pycnocline anomalies take place around 7°N. Anomalies in the Southern Hemisphere are smaller for both El Nino events. Also common to both events are smaller positive anomalies found in both hemispheres 3-4 months prior to the main peaks. The latitudes of the smaller anomalies are 4° to 6° from the equator depending on the event and the hemisphere, but always closer to the equator than the maximum anomalies. These earlier periods of shallow pycnocline depth are more symmetric to the equator than the latter ones and represent the initial shallowing of the pycnocline at Truk Island. The maximum positive anomaly at 7°N corresponds to the period when the pycnocline was shallow at Truk near the end of the El Nino year. This pattern of pycnocline depth change is not as evident in the El Nino events of the 1960's. For example, during the 1965 El Nino the maximum positive pycnocline anomalies are more symmetric to the equator. However, it can be stated that the Y-T pattern of the shallow pycnocline anomaly near 7°N late in the year is typical of an El Nino year for the 18-year record studied here.

A matrix of lagged cross correlations between Truk Island and all points along 7.5°N (Fig. 14) shows the pycnocline variability at Truk lagging that to the east. The inclination of maximum positive correlations represents a westward propagation of the pycnocline signal at approximately 40 cm s^{-1}. This compares favorably with the 35 cm s^{-1} phase speed for third latitudinal mode Rossby waves in this calculation. Projection of the meridional pycnocline structure onto the free modes of the system indicates the dominant response at Truk is due to third-mode Rossby waves. Depending on the event, the variability at Truk may also be influenced by first, second, and fourth mode Rossby waves.

As discussed in the previous section, relaxation of the easterlies takes place in the central and western Pacific during El Nino. It is suspected that the wind changes that are responsible for high sea level in the eastern equatorial Pacific through the generation of equatorial Kelvin waves may also be responsible for low sea level in the western tropical Pacific through the generation of Rossby waves at the same time.

McCreary (1977) showed that the onset of westerlies, which is equivalent to the relaxation of easterlies in linear theory, generates an eastward-propagating, downwelling, equatorial Kelvin wave front and a westward-propagating, upwelling, Rossby wave front which has minimum pressure on both sides of the equator. Therefore, it seems quite reasonable to attribute the decrease of sea level in the western tropical Pacific to the relaxation of the easterlies. This would also explain the negative correlation between the

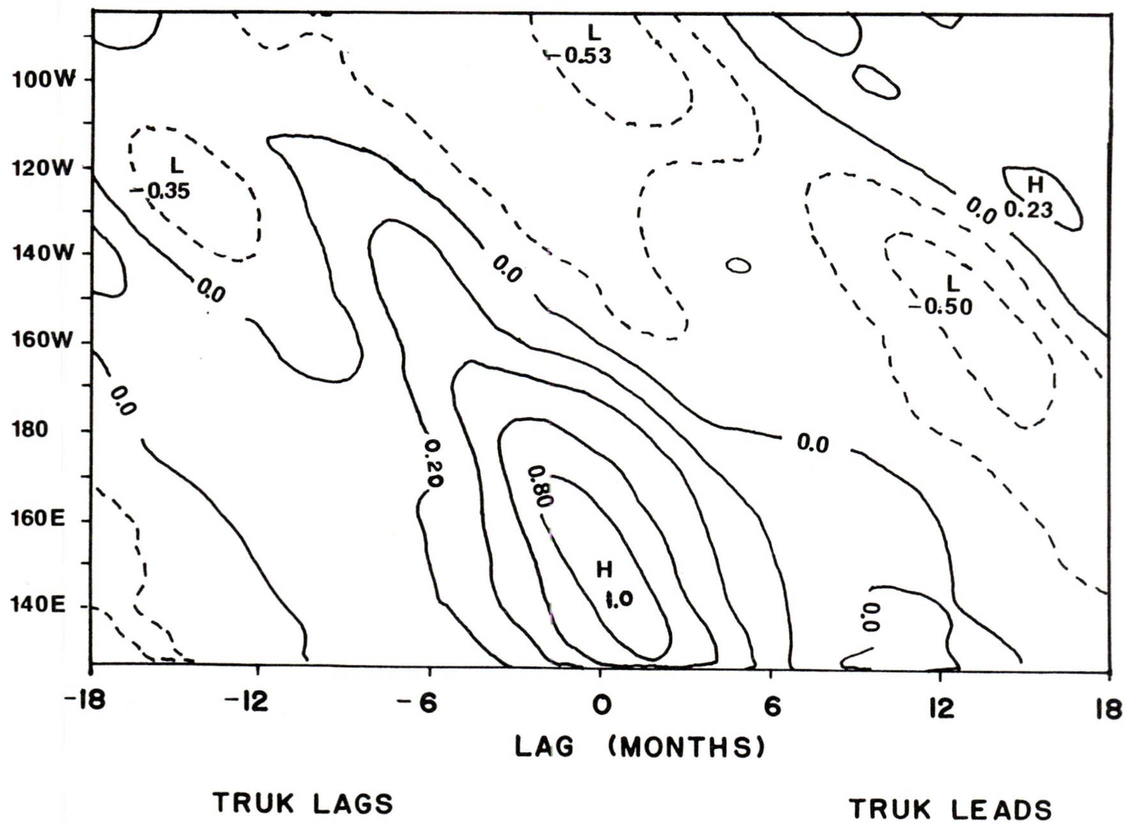

Fig. 14. Contours of lagged cross correlation between the model pycnocline variability at Truk Island and all points east and west along 7.5°N.

equatorial eastern boundary and western boundary variability (Fig. 7). The maximum cross correlation between the non-seasonal model pycnocline signal at Truk and Galapagos is also negative and occurs when Galapagos leads by a few months.

In McCreary's study, the wind stress and the resultant sea level distribution was symmetric to the equator. The observed sea level and the model pycnocline fluctuations, however, are not symmetric to the equator. This suggests that the relaxation of the easterlies may not be symmetric to the equator. To judge if this is true the meridional distribution of the wind stress variability is examined next.

Only the zonal component of the wind stress is considered because it is more important in forcing low frequency motion in the equatorial ocean. This is supported by the test calculation when the model was forced only by the zonal wind stress. At Truk the pycnocline response to zonal forcing was very similar to that for the full model calculation forced by both zonal and meridional components. The meridional structure of the zonal wind stress variability is obtained by averaging the annual anomalies of the zonal wind stress between 160°E and 140°W for 1971 to 1978 (Fig. 15). The longitudinal range is chosen to include the major wind stress changes in the El Nino years. The width of each wind event is 30°-40° of longitude. The area west of 160°E is omitted because long Rossby waves generated there have no influence eastward.

The winds are predominantly easterly for the range 160E-140W, so positive values in Fig. 15 indicate weakened easterlies. The seasonal variations are obvious and are basically out of phase between the hemispheres; the weakest easterlies occur in the second half of each year in the Northern Hemisphere and in the earlier half of each year in the Southern Hemisphere. In 1972 and 1976, the weakening of the easterlies in the Northern Hemisphere was larger and the area of weakening (5°N-10°N) was closer to the equator. The easterlies at the equator were also weaker than usual beyond March.

Does such a distribution of decreasing easterlies cause the pressure field distribution found in Fig. 13? The response of the ocean to an idealized wind forcing event is calculated to determine this. An eastward wind stress which has a Y-T pattern shown in Fig. 16a, constant in the zonal direction between two meridions, and zero outside of it is assumed. The width of the region is taken to be 3300 km, corresponding to about 30° of longitude. This idealized forcing is a crude approximation of the relaxation and recovery phase of the northeast trades during the 1972 and 1976 Los Ninos. Initially the ocean is assumed to be at rest. The Y-T distribution of pressure along the meridion at the west end of the forcing region is calculated and shown in Fig. 16b. The geostrophic balance is assumed for the zonal momentum equation. Forcing

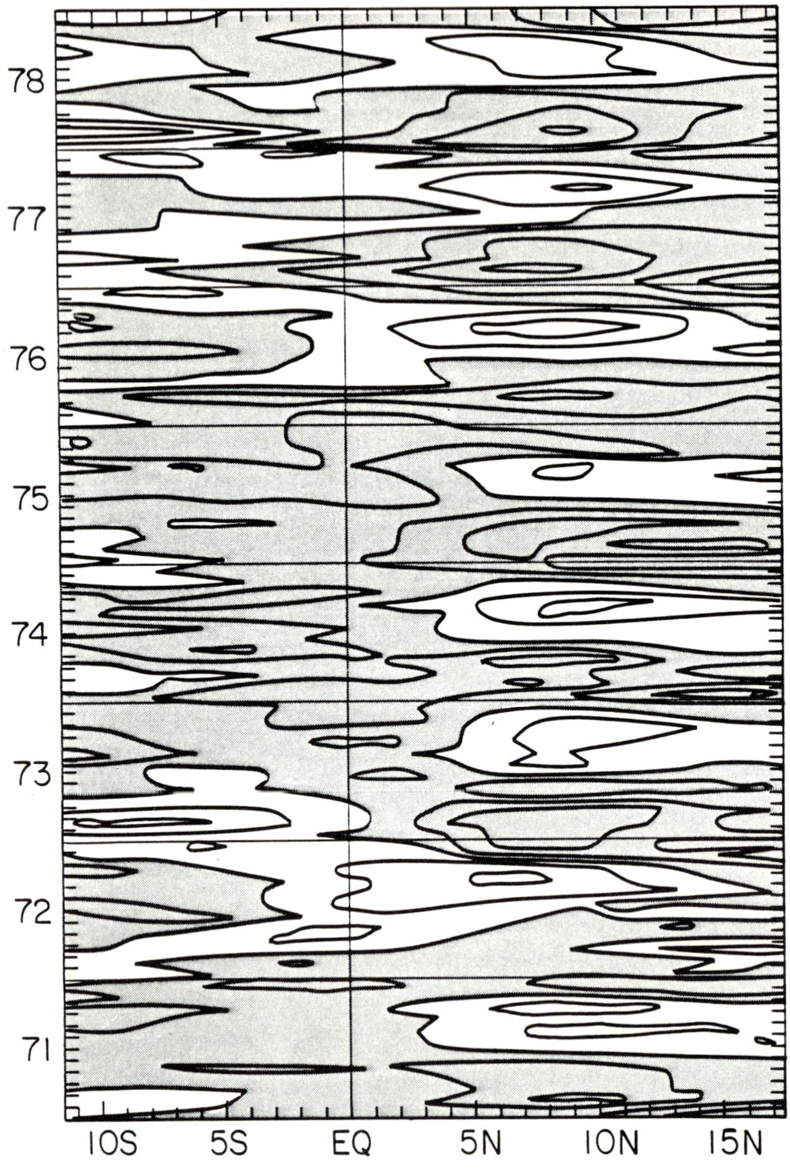

Fig. 15. A Y-T diagram of the zonal wind stress averaged over 160°E-140°W for 1971-1978. White regions indicate the easterlies are weaker (westerly anomaly) than the long-term annual mean. Contour interval is .25 dynes cm^{-2}.

functions for equatorial Rossby modes are calculated analytically and integrated numerically along the characteristics of each mode.

The response to the idealized forcing (Fig. 16b) has some similarities with the modelled response of the western tropical Pacific in El Nino years (Fig. 13). The response has a peak in each hemisphere although the wind forcing is mostly in the Northern Hemisphere. The largest response is north of the equator. The minimum pycnocline depth appears around 7°N, about 1.5 months after the wind forcing minimum. This supports the idea which attributes sea level anomalies in the western tropical Pacific in the latter half of the El Nino year to the weakening of easterlies centered north of the equator.

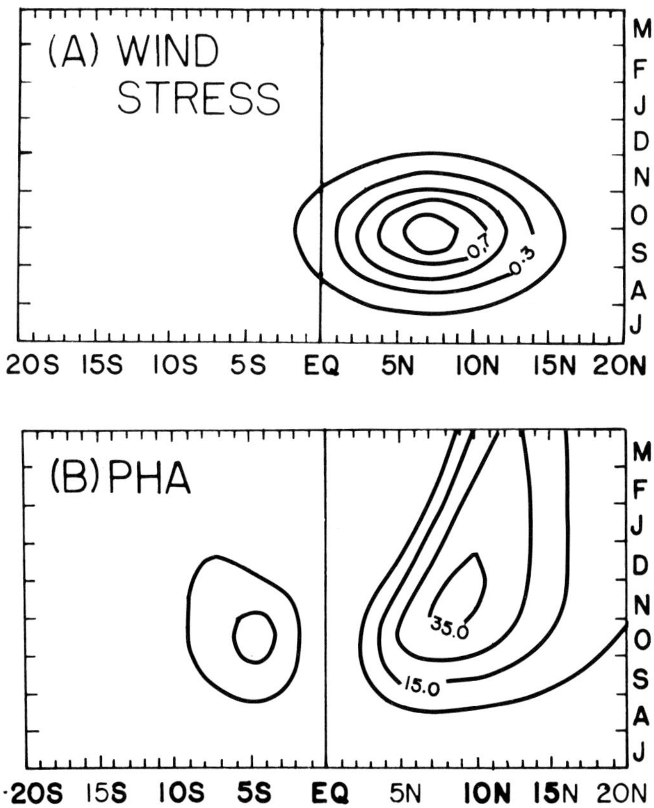

Fig. 16. Idealized model of the response of the tropical ocean to the westerly wind stress anomaly confined within two meridians 30° apart. a) Y-T diagram of the wind stress forcing function. Contour interval is .2 dynes cm^{-2}. b) Y-T diagram of the pycnocline response along the western edge of the forcing region. Contour interval is 10 m.

The response of the model ocean is fairly sensitive to the location of the wind forcing. If the center of the forcing shifts northward to around

10°N, similar to the location for the center of normal seasonal weakening, the low-latitude response becomes considerably smaller and the maximum response shifts northward more than the shift of the wind forcing (Fig. 17). In this case the forced motion is more like mid-latitude Rossby waves than equatorial Rossby waves. Also, it may explain why the relaxation of the easterlies found in the latter half of most years produces only a small response in the low latitudes of the western tropical Pacific except in El Nino years.

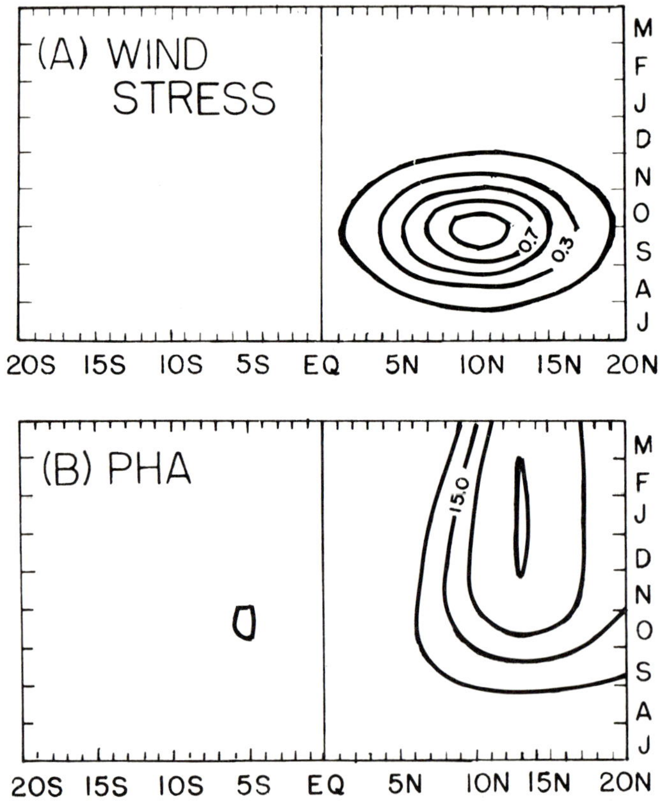

Fig. 17. Same as Fig. 16 except the forcing is shifted northward by 3°.

Four or five months prior to the main weakening of the northeast trades, weakened easterlies exist about the equator in both 1972 and 1976 (Fig. 15). Since the wind anomalies are maximum near the equator, the response of the ocean should be symmetric to the equator. Late in the El Nino year the relaxation of the northeast trades is followed by strong westward anomalies. The switch from eastward to westward anomalies takes place very rapidly. This rapid recovery of the easterlies, along with the earlier eastward anomalies near the equator, might provide a more complete description of the typical sea level change at Truk Island during an El Nino event, i.e., a two-step drop in sea level during the El Nino year with lowest sea level occurring at the end of the year followed by a rapid rising.

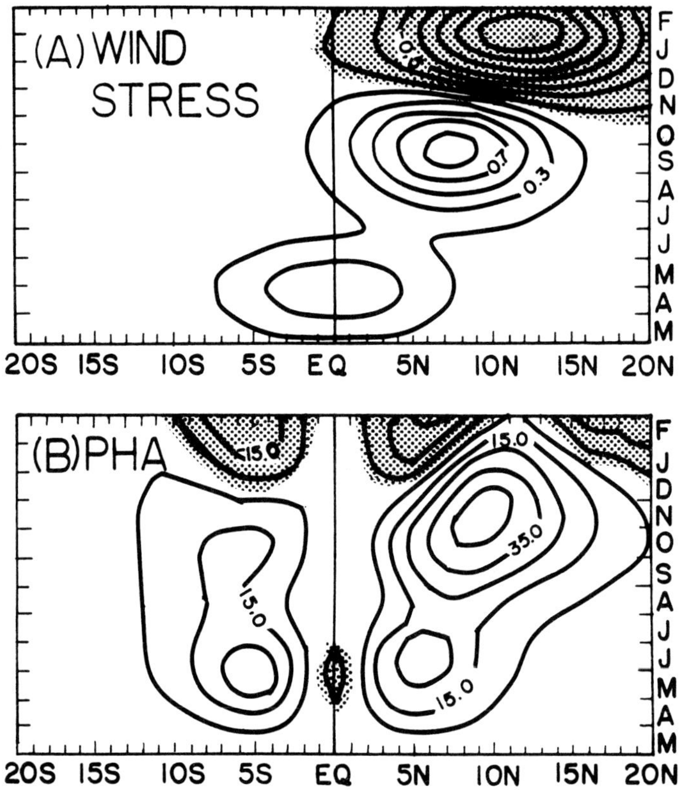

Fig. 18. Similar to Figure 16 but the forcing function is now preceeded by an equatorial relaxation sequence and followed by an intensification of the trades (stipled) late in the year.

Another idealized forcing calculation is performed in which there are minimum easterlies at the equator in April, minimum northeast trades in September, followed by a rapid strengthening yielding anomalously strong northeast trades in the following January and February (Fig. 18a). The resulting pycnocline response (Fig. 18b) is initially symmetric about the equator with minimum pycnocline depths at $\pm 5^\circ$ in May-June. The remainder of the response is not symmetric to the equator whereby the pycnocline depth minimum is located at 9°N in November. This compares favorably with the shallow pycnocline depths of the 1972 and 1976 events depicted in Fig. 13. The pycnocline response (Fig. 18b) at 7.5°N is also similar to that at Truk Island during an El Nino year (Fig. 11). There is an initial shoaling of the pycnocline from March to May, followed by a few months during which there is a slight increase in pycnocline depth, further rising of the pycnocline until October-November, and a final rapid increase in pycnocline depth.

CENTRAL EQUATORIAL RESPONSE

The two previous sections have demonstrated that for 1961-1978 the variability in the eastern tropical Pacific may be influenced by equatorially trapped Kelvin waves and the western tropical Pacific may be influenced by westward-propagating Rossby waves. In the central equatorial Pacific there are not many long-term observations. The sea level data at Canton and Christmas Islands are among the few available. Unfortunately, the time series are not continuous and do not cover the entire 18 years of interest. Furthermore, data is missing during several El Nino events. Nonetheless, the most significant features are high sea levels at both islands during the second half of an El Nino year. Smaller high sea level peaks are often found at the begining of an El Nino year. A lagged cross correlation with the model pycnocline time series is impossible because of the data gaps in the sea level record. A zero lag cross correlation can be performed if the sea level observations are concatenated and then correlated with the corresponding points in time for the model pycnocline record. The zero lag cross correlations are 0.40 at Canton and 0.67 at Christmas Island.

Instead of comparing the observed sea level and model pycnocline variations, it is of interest to know the relationship of these variations at Christmas and Canton to those in the eastern and western Pacific discussed in the previous sections. This may aid future analyses of sea level records when complete time series become available. The location of these two islands are opportune for two reasons. First, longitudinally the islands are in a region where there are important wind stress changes to the east and west. The sea levels at these islands may therefore be affected by both Kelvin waves generated west of the islands and Rossby waves generated east of the islands. Second, latitudinally these islands are situated where both Kelvin waves and first-mode Rossby waves have relatively high amplitudes in the pressure field.

For the baroclinic phase speed chosen in this study the equatorial radius of deformation is 327 km. At the latitudes of Christmas, 1°59'N, and Canton, 2°48'S, sea level variations due to an equatorial Kelvin wave are 80% and 64% of that at the equator, respectively. At low frequencies, the first-mode equatorial Rossby wave has its maximum pressure field amplitude at ±3.6° from the equator. The amplitude of Rossby wave pressure variations at Christmas and Canton are 80% and 95% of the maximum, respectively. Therefore, it is suggested that the sea level variations at these islands are strongly influenced by both Kelvin and first-mode Rossby waves. It is impossible to separate Kelvin and Rossby signals from the observed sea levels at these islands. Fortunately, the model pycnocline variability is similar at times with the observed sea level. It is possible to get some hint of the contributions of Kelvin and Rossby waves to the sea level variations by projecting the model pycnocline structure onto the Kelvin and Rossby wave components.

In Fig. 19 and 20, Kelvin and first-mode Rossby wave components of the model pycnocline are plotted for the 1965 El Nino event at Canton and Christmas. Most of the variations of the model pycnocline at both islands are attributed to Kelvin waves and first-mode Rossby waves. The pycnocline variability at Christmas Island contains more Kelvin wave signal than at Canton. This can be explained by considering the locations of the two islands. Since Christmas Island is located closer to the equator than Canton, if the amplitudes of the Kelvin waves at the longitudes of these islands are the same, Christmas Island would experience a stronger Kelvin wave signal than would Canton. In addition, the longitudinal location of these islands can also lead to stronger Kelvin signals at Christmas since the Kelvin waves generated between the longitudes of Canton and Christmas only reach Christmas Island. By similar reasoning, Canton Island is likely to receive larger Rossby signals than Christmas if the Rossby waves generated between the islands add constructively.

For the 1965 El Nino, the downwelling peak at the beginning of the year at both islands is primarily due to the Kelvin wave component. The shallow pycnocline in March-April is greatly influenced by the upwelling Rossby wave front discussed in the previous section. Another deep pycnocline episode occurs in the second half of the year. However, in this latter downwelling event, the first-mode Rossby wave component shares the role of depressing the pycnocline. At Christmas Island, the downwelling peaks of the Kelvin wave component and first-mode Rossby wave component have a slight time lag and result in a double peak in the pycnocline variability. This feature is also present in the observed sea level record.

The temporal variation of the Kelvin wave component at these two islands is similar at a slight time lag. The variability at Galapagos Island is also similar at a lag of about 1.5 months. The downwelling peak of the Kelvin wave component at the beginning of the year does not change much in amplitude as it passes Canton, Christmas, and Galapagos. The peak later in the year grows in amplitude as it passes Canton, Christmas, and Galapagos. This suggests the forcing region of the Kelvin wave front responsible for the deep pycnocline in the beginning of the year is west of the forcing region for the Kelvin wave front excited later in the year. This is common among many El Nino events in

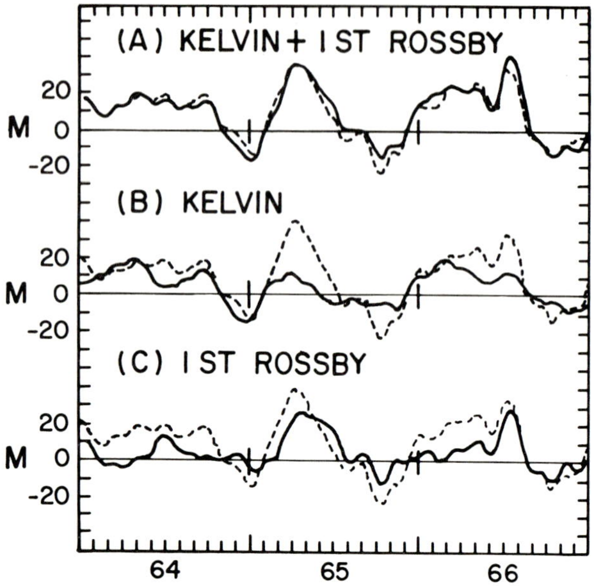

Fig. 19. Kelvin and first-mode equatorial Rossby wave components (solid) of the model pycnocline signal (dashed) at Canton Island for 1964 to 1966. a) Kelvin plus first-mode Rossby wave compared with the pycnocline signal. b) Kelvin wave compared with the pycnocline signal. c) First-mode Rossby wave compared with the pycnocline signal.

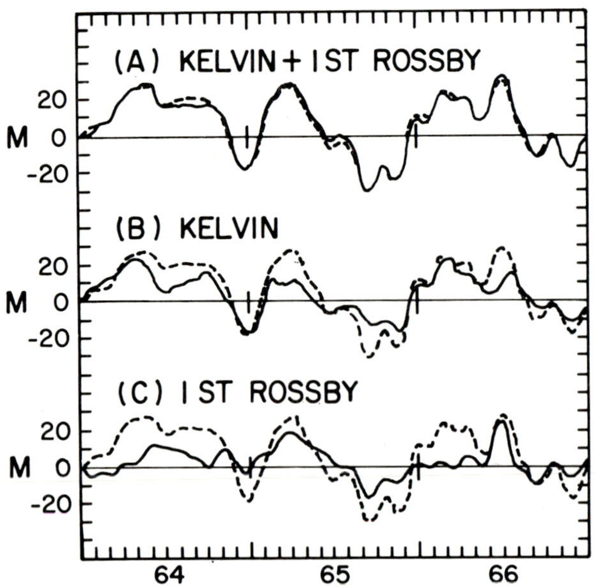

Fig. 20. Same as Fig. 19 except at Christmas Island.

this 18-year period and consistent with the previous identification of the forcing regions.

The 1972 El Nino event (Fig. 21) provides another interesting example of the relationship between the effects of the Kelvin waves and the Rossby waves. In this event the Kelvin wave component exhibits two downwelling peaks: at the beginning and middle of the year. The initial downwelling Kelvin wave signal of 1972, which emanates in the west, has little influence on the pycnocline depth at Galapagos because of intensified easterlies in the central Pacific. This intensification, from mid-October until the end of 1971, is responsible for a downwelling Rossby wave response at Canton at the beginning of 1972. The Kelvin wave peak later in the year is not found in the total pycnocline signal at Canton because the Kelvin wave component is canceled out by an upwelling first-mode Rossby wave contribution. At Galapagos, this second Kelvin wave signal is no longer affected by a Rossby wave and thereby contributes to a deep pycnocline in June. The large downwelling at the end of 1972 is mainly a Rossby wave response.

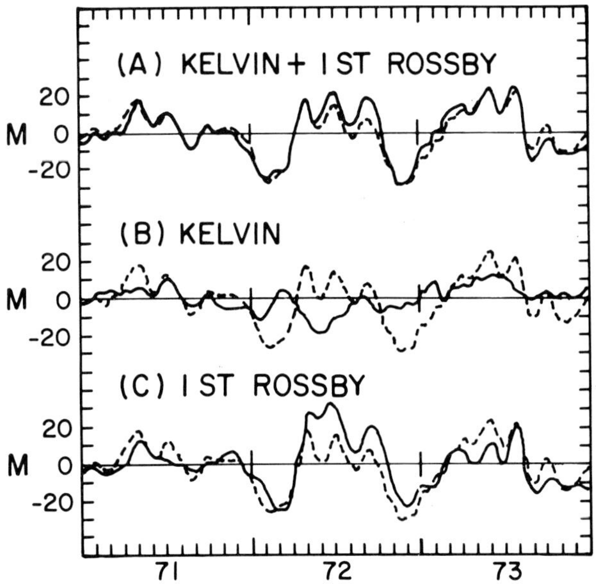

Fig. 21. Same as Fig. 19 except from 1971-1973.

It appears there may be great difficulty in interpreting the sea level data at Canton and Christmas Islands due to the interplay between Kelvin waves and Rossby waves from one event to the next. This is also evident if the 18-year pycnocline records for all points along a longitude representative of Canton or Christmas Island are correlated with the model pycnocline variability at Galapagos. At 2°-3° from the equator the influence of the Kelvin waves have been reduced enough that the resulting maximum positive correlation when the central Pacific leads Galapagos is comparable to the maximum negative correlation due to Rossby waves when the central Pacific variability lags Galapagos. At times the sea level data of these islands can be used to discuss the passage of Kelvin waves. However, the present study suggests a strong influence of the first-mode equatorial Rossby waves at these islands. In order to fully understand the processes responsible for the observed sea level variability at these islands, it is necessary to know the sea level and wind stress variability to the east and west or the meridional structure of the sea level variability.

SUMMARY AND CONCLUSIONS

The results of a linear, wind-driven, numerical model have been used to study the interannual variability of the eastern, central equatorial, and western Pacific for a span of 18 years. The model was forced by monthly estimates of the observed surface wind field over the tropical Pacific for 1961-1978. Such a model provides a straightforward way of assessing the basin-wide response to the overlying wind field in a manner that would not be possible by studying the wind field variability. Of interest during the period of study were several Los Ninos. The character of these events in the model results and the nature of the forcing were described.

The model pycnocline variability in the eastern Pacific was induced by internal equatorially trapped Kelvin waves excited by fluctuations in the zonal wind stress at the equator in the western and central Pacific. The model pycnocline variability at the Galapagos Islands was remarkably similar to the observed sea level variability there. Cross correlation between the model pycnocline variability and observed sea level indicated a maximum significant correlation at zero lag. One important discrepancy between the model pycnocline time series and observed sea level was the presence of more high frequency fluctuations in the pycnocline record. This is to be expected since the temporal resolution of the model pycnocline time series is much greater than the monthly sea level record. Furthermore, in this single mode calculation the time scales of the upwelling and downwelling episodes in the east are indicative of the time and spatial scales of the wind changes to the west. The addition of slower higher order vertical modes may lead to longer duration events. Another possible reason for this discrepancy may be that the wind estimates used have more power at high frequencies than the actual wind field.

The five Los Ninos years of 1963, 1965, 1969, 1972 and 1976, were depicted as years when the pycnocline was anomalously deep. Los Ninos of 1965 and 1969 have been classified as being the strongest for the 1960's. The variability of the southeast trades on both sides of the dateline was the cause of these events in the model. The onset of El Nino was due to the excitation of a large amplitude downwelling Kelvin wave front west of 180°. This Kelvin wave front was excited by a 0.4 dynes cm^{-2} relaxation of the easterly wind stress during the latter half of the year preceding El Nino. The relaxation in 1964 and 1968 was so large that the zonal wind stress changed from easterly to westerly. The timing of this Kelvin wave is such that it enhanced the remotely forced seasonal downwelling at the eastern boundary.

Besides the initial downwelling impulse, the El Nino year was characterized by the persistence of a downwelled pycnocline. The pycnocline remained deep because the seasonal intensification of the central Pacific easterlies was not as strong as during non-El Nino years. The upwelling Kelvin waves excited were small in amplitude. Hence, the pycnocline was not significantly upwelled. This lack of a normal seasonal upwelling kept the pycnocline deeper than normal throughout the El Nino year. Towards the end of the El Nino year the semiannual variability of the interior easterlies was reestablished. The resulting downwelling Kelvin wave front generated towards the end of the year created the second major downwelling pulse of the El Nino year. This completed the double-peak downwelling signature characteristic of the 1965 and 1969 El Nino events.

In contrast, the minor El Nino of 1963 was only influenced by the wind stress variability east of 180°. During 1962 there was not a large relaxation of the wind field west of the dateline. After the seasonal downwelling at the beginning of 1963 there was a cessation of the semiannual intensification of the southeast trades. The pycnocline remained depressed throughout 1963 and did not recover until 1964.

Anomalous changes in the wind field leading up to the model expression of the 1972 El Nino began with an eastward progression of weakening trade winds west of the dateline in late 1971. A nearly simultaneous intensification of the central equatorial easterlies negated any influence from the west. This resulted in mean annual pycnocline depths at the eastern boundary in early 1972. From January through April a significant decrease of the zonal wind stress between 120°W and 180° excited a downwelling Kelvin wave front responsible for deep pycnocline depths at the eastern boundary in June. Observations indicated unusually high sea level for the same period. Late in the year, a relaxation of the easterlies at 100°W-140°W created a second downwelling episode, smaller relative to that in June. A second peak in sea level was observed at the end of the year. Intensification of the trade winds in January of 1973 led to a recovery from the 1972 El Nino.

Wind changes and their effects associated with the 1976 El Nino were similar to 1972. An eastward progression of weakening trade winds began west of 160°E in late 1975 and generated a downwelling Kelvin wave front. As opposed to late 1971, the winds in the central Pacific did not intensify. Thus, the model pycnocline was deep along the eastern boundary at the beginning of 1976. A decrease in the equatorial easterlies between 140°W and 180° during the first several months of 1976 was once again responsible for the deep model pycnocline at the eastern boundary in June of the El Nino year. Additional weakening of the wind stress in the central Pacific produced a final downwelling burst later in the year. The recovery phase began with intensified easterlies in December.

There was one major difference between the simulated Los Ninos of the 1970's and those of 1965 and 1969. For the 1960's events, observed sea level was highest and the model pycnocline was deepest towards the beginning and end of the El Nino year in the eastern Pacific. During the 1972 and 1976 El Nino occurrences, elevated sea level and maximum pycnocline depth were in June. Common to all these events was a weakening of the trade winds west of the dateline late in the preceding year. The model pycnocline was deep in June of 1972 and 1976 because the central equatorial easterlies had decreased considerably from January to April. The pycnocline was not deep in mid-1965 or mid-1969 because the trade winds in the central Pacific did not weaken during the first four months of the year. Thus, during the 1970's the easterlies in the central equatorial Pacific had more of an active role in the evolution of El Nino than in the previous decade. This in turn implies there are important differences in the location and timing of the wind changes responsible for each El Nino.

The season in which an anomalously deep pycnocline occurs might have important ramifications for determining the size of anomalous SST. For the El Nino events of the 1960's the model pycnocline was anomalously deep during the periods of seasonal warming in the Southern Hemisphere. For the El Nino events of the 1970's the pycnocline was deepest during the cool SST season and SST anomalies were larger than in the previous decade. Does a depression of the thermocline during the auroral fall and winter result in a larger positive SST anomaly than a thermocline depression in summer? Modelling efforts including thermodynamics will provide the information needed to understand relations between thermocline depth, vertical velocity, and SST.

The El Nino sequence of initial wind relaxation west of the dateline followed by weakening or quiescent winds in the central Pacific did not apply to 1975. This event began when winds west of the dateline decreased in late

1974 similar to the type of relaxation prior to an El Nino. A downwelling Kelvin wave was excited and resulted in a deep pycnocline at the eastern boundary in early 1975. The similarity with an El Nino event stopped there. After the initial weakening, the winds intensified and remained anomalously strong for the remainder of the year. At the eastern boundary, there was a rapid upwelling and the model pycnocline remained shallow. Hydrographic observations in the eastern Pacific indicated El Nino type conditions were present at the beginning of 1975 but soon deteriorated (Wyrtki et al., 1976).

The interannual variability in the western tropical Pacific was characterized by a shoaling of the model pycnocline and an observed drop in sea level in all El Nino years. Maximum changes in model pycnocline depth and observed sea level occurred at 5°N-10°N. Truk Island at 7.5°N, 151.9°E is in a strategic position to monitor these fluctuations. The model pycnocline time series at Truk was highly correlated at zero lag with the observed sea level record. A majority of the pycnocline variability at Truk was due to westward-propagating, third latitudinal mode Rossby waves.

The shallow pycnocline and low sea level at the end of the El Nino year was attributed to an upwelling Rossby wave front excited by a decrease in the trade winds to the east. Related wind changes at the equator were responsible for downwelling at the eastern boundary in the middle and end of the El Nino year. This accounts for the out of phase relation between the eastern and western tropical Pacific during the El Nino simulations.

The meridional structure of the zonal wind stress variability between 160E and 140W indicated that the largest perturbation to the wind field during El Nino was not symmetric about the equator and involved the northeast trades. The seasonal weakening of the northeast trades was larger in area and closer to the equator (5°N-10°N) during the 1972 and 1976 Los Ninos. A simple analytical model forced by an idealization of this weakening yielded an upwelling Rossby wave response in both hemispheres west of the forcing with maximum upwelling at 7°N, 1.5 months after the minimum wind forcing. When the idealized forcing consisted of decreasing southeast trades symmetric about the equator at the beginning of the El Nino year, followed by the El Nino weakening of the northeast trades, and a subsequent recovery of the northeast trades, the response at 7°N-8°N was in qualitative agreement with the pycnocline variability at Truk during the 1972 and 1976 events. The oceanic response is strongly dependent on the location of the weakened trade winds. This may account for differences between the seasonal and nonseasonal response in the western tropical Pacific. When the center of the northeast trade wind forcing was shifted to 10°N, similar to the normal location of the maximum seasonal weakening, the low-latitude response was considerably smaller and the maximum response was displaced northward more than the shift of the forcing.

In the central equatorial Pacific, limited sea level records were available for Canton Island and Christmas Island. This is an area for which there are important wind changes to the east and west during El Nino. These islands are therefore subject to the influence of equatorially trapped Kelvin waves excited to the west and long Rossby waves generated to the east. Since these islands are located 2°-3° from the equator, the Kelvin wave amplitude is reduced and the Rossby wave amplitude is increased relative to that at the equator. Decomposition of the model pycnocline variability at these islands indicated that almost all of the signal was due to Kelvin waves and first latitudinal mode Rossby waves. The superposition of Kelvin and Rossby waves during El Nino changes from one event to another depending on the location and timing of the forcings. Analysis of sea level variability in the central equatorial Pacific is therefore difficult because Kelvin wave signals generated in the west and ultimately causing pycnocline displacements in the east may be masked temporarily by a superposition with Rossby waves of opposite sign.

Even though all Los Ninos events are unique, a few generalizations are in order. The results of the numerical model demonstrate that interannual changes of the zonal wind stress in the central and western Pacific are very important in determining the sea level/pycnocline responses of El Nino. For most Los Ninos during the period of study the onset phase was linked to perturbations to the zonal wind field in the western Pacific. How the El Nino event progressed and matured was subsequently determined by the nature of the wind field in the central Pacific. Recently, it was noted (Keen, 1982) that there is often a greater incidence of tropical cyclone pairs in the western Pacific prior to El Nino. A succession of these disturbances, at times penetrating east of 180°, results in strong eastward perturbations to the wind field several times per month. In the future it will be necessary to determine how these high frequency disturbances affect the wind field monthly means. It would be of interest to know how the model results would change if there was a greater temporal resolution of the wind field.

The results of this numerical simulation have also provided some information regarding monitoring the tropical Pacific wind field. In view of the correlations between the model results and observations at several locations, important fluctuations in the low-latitude surface wind field must have been large enough in space and time to have been resolved by ship-board estimates. Admittedly, there are several problems with this method of observing the wind field. Yet it is the only data set available suitable for interannual modelling studies. Nonetheless, this study has demonstrated a definite need for continuous monitoring of the detailed structure of the entire tropical Pacific wind field.

With respect to real-time wind monitoring, the simulations of the El Nino events for 1961-1978 indicate that a predictive capability for El Nino will not be straightforward. At the present time, persistence of anomalous conditions in the tropical atmosphere cannot be predicted. An example of why this is important is the aborted event of 1975 when the winds in the western Pacific initially weakened in a manner similar to the onset of previous El Nino events. Shortly thereafter the winds abruptly intensified and remained strong. Monitoring a significant wind relaxation without knowledge of a subsequent intensification may lead to an erroneous El Nino prediction. Furthermore, equal weight must be given to all regions when monitoring the tropical Pacific wind field. As illustrated by the conditions during the first several months of 1972, an anomalous decrease in the trade winds in the western Pacific may be followed by an intensification of the winds in the central Pacific. The net result of such circumstances being no change in the eastern Pacific.

The wind-driven modelling results presented here have shown that a significant amount of the observed interannual sea level variability in the eastern, central equatorial, and western Pacific can be accounted for by linear theory when reasonable estimates of the wind field are provided. Internal Kelvin and Rossby waves play important roles in determining the variability throughout the tropical Pacific. Drastic shoaling of the pycnocline in the western Pacific during El Nino may be induced by Rossby waves excited by changes in strength and position of the northeast trades. In the eastern tropical Pacific, remotely forced Kelvin waves generated in the central and western equatorial Pacific are responsible for the El Nino pycnocline displacements. By definition El Nino is associated with warm SST but at present the cause and effect relation of a Kelvin wave induced deep thermocline and SST is not understood.

ACKNOWLEDGMENTS

This work was supported by the National Science Foundation under grants ATM7920485 and OCE8119052. Computations were performed at the National Center for Atmospheric Research and the Florida State University Computation Center. We gratefully acknowledge the many persons involved in the wind data analysis. Special thanks to Pat Teaf for manuscript preparation. Drafting services were kindly provided by Dewey Rudd and Roberta Scott. Contribution number 187 of the Geophysical Fluid Dynamics Institute, The Florida State University, and contribution number 15 to PEQUOD.

REFERENCES

Barnett, T. P., 1977. An attempt to verify some theories of El Nino. J. Phys. Oceanogr., 7: 633-647.
Busalacchi, A. J., and J. J. O'Brien, 1980. The seasonal variability in a model of the Tropical Pacific. J. Phys. Oceanogr., 10: 1929-1951.
Camerlengo, A. L., and J. J. O'Brien, 1980. Open boundary conditions in rotating fluids. J. Comp. Phys., 35: 12-35.
Davis, R. E., 1976. Predictability of sea surface temperature and sea level pressure anomalies over the North Pacific Ocean, J. Phys. Oceanogr., 6: 249-266.
Enfield, D. B., and J. S. Allen, 1980. On the structure and dynamics of monthly mean sea level anomalies along the Pacific coast of North and South America. J. Phys. Oceanogr., 10: 557-578.
Goldenberg, S. B., and J. J. O'Brien, 1981. Time and space variability of tropical Pacific wind stress. Mon. Wea. Rev., 109: 1190-1207.
Hickey, B., 1975. The relationship between fluctuations in sea level, wind stress and sea surface temperature in the equatorial Pacific, J. Phys. Oceanogr., 5: 460-475.
Hurlburt, H. E., 1974. The influence of coastline geometry and bottom topography on the eastern ocean circulation. Ph.D. dissertation, Florida State University, 104 pp.
Hurlburt, H. E., J. C. Kindle and J. J. O'Brien, 1976. A numerical simulation of the onset of El Nino. J. Phys. Oceanogr., 6: 621-631.
Keen, R. A., 1982. The role of cross-equatorial tropical cyclone pairs in the Southern Oscillation, Mon. Wea. Rev., in press.
Kidson, J. W., 1975: Tropical eigenvector analysis and the southern oscillation, Mon. Wea. Rev., 103: 187-196.
Kindle, J. C., 1979. Equatorial Pacific Ocean variability - Seasonal and El Nino time scales. Ph.D. dissertation, Florida State University, 134 pp.
Knox, R. A., and D. Halpern, 1982. Long range Kelvin wave propagation of transport variations in Pacific Ocean equatorial currents, J. Mar. Res., 40: 329-339.
McCreary, J., 1976. Eastern tropical response to changing wind systems: with application to El Nino. J. Phys. Oceanogr., 6: 632-645.
McCreary, J., 1977: Eastern ocean response to changing wind systems, Ph.D. dissertation, University of California, San Diego.
Meyers, G., 1979. Annual variation in the slope of the 14 C isotherm along the equator in the Pacific Ocean. J. Phys. Oceanogr., 9: 885-891.
Meyers, G., 1982. Annual and interannual variation in sea level near Truk Island, J. Phys. Oceanogr., 12: 1161-1168.
Moore, D. W., 1968. Planetary-gravity waves in an equatorial ocean. Ph.D. thesis, Harvard University, 201 pp.
Moore, D. W., and S. G. H. Philander, 1977. Modeling of the tropical oceanic circulation. The Sea, Vol. 6, E. Goldberg, et al., Eds., Wiley-Interscience, 319-361.
O'Brien, J. J., A. Busalacchi and J. Kindle, 1981. Ocean models of El Nino. In: Resource Management and Environmental Uncertainty: Lessons from Coastal Upwelling Fisheries. M. H. Glantz and J. D. Thompson, Eds., Wiley-Interscience, 159-212.
Orlanski, I., 1976. A simple boundary condition for unbounded hyperbolic flows. J. Comp. Phys., 21: 251-269.
Quinn, W. H., 1974: Monitoring and predicting El Nino invasions, J. Appl. Meteor., 13: 825-830.

Quinn, W. H., D. O. Zopf, K. S. Short, and R. T. W. Kuu Yang, 1978. Historical trends and statistics of the Southern Oscillation, El Nino and Indonesian droughts. Fish. Bull., 76: 663-678.
Romea, R. D., and R. L. Smith, 1982. Further evidence for coastal trapped waves along the Peru coast. Submitted to J. Phys. Oceanogr.
Smith, R. L., 1978. Poleward propagating perturbations in currents and sea levels along the Peru coast. J. Geophys. Res., 83: 6083-6092.
Wyrtki, K., 1975. El Nino - the dynamic response of the equatorial Pacific Ocean to atmospheric forcing. J. Phys. Oceanogr., 5: 572-584.
Wyrtki, K., Sea level during the 1972 El Nino, J. Phys. Oceanogr., 6: 779-787, 1977.
Wyrtki, K., The response of sea surface topography to the 1976 El Nino. J. Phys. Oceanogr., 9: 1223-1231, 1979.
Wyrtki, K., Stroup, E., Patzert, W., Williams, R., and Quinn, W., Predicting and observing El Nino, Science, 191: 343-346, 1976.

ON WESTWARD PROPAGATION OF SEA-SURFACE TEMPERATURE ANOMALIES IN THE EQUATORIAL PACIFIC

ERIC B. KRAUS and HOWARD P. HANSON

Cooperative Institute for Research in Environmental Sciences
University of Colorado/NOAA, Box 449, Boulder,
Colorado, USA 80309

ABSTRACT

Sea-surface temperature anomalies in the tropical Pacific, both in the Eastern Pacific coastal areas and in the Central Pacific, have been the focus of much attention in recent years, the former because of impacts on South American fisheries and the latter as a possible modulator of the North American climate. The westward spreading of warm anomalies from the South American coast occurs at speeds of around three times the mean advection rate; the mechanics of this spreading must therefore be associated with dynamic phenomena which propagate westward substantially faster than the surface current. Moreover, the strength of the anomalies suggests that ocean-atmosphere interactions cannot be ignored as a feature of the overall process.

The westward spreading is discussed here in terms of a simple mixed-layer model in the context, first, of a non-rotational equatorial channel model and, second, of the dynamics of the equatorial beta plane. Model results indicate substantial enhancement of westward propagation due primarily to the effect on the mixed layer of anomalous air-sea heat fluxes. Limitations and implications of the model are discussed, and the need for finite-amplitude (numerical) investigations is indicated.

1. INTRODUCTION

The so-called Southern Oscillation, specified by a see-saw between the surface pressure anomalies in the region of Indonesia and Northwest Australia on the one hand, and those in the Southeast Pacific (Tahiti, Rapa, Easter Island) on the other, is best interpreted as a symptom or index of a global energy relaxation cycle. It seems that it involves a quasi-cyclical voiding and filling of a large oceanic and atmospheric reservoir of thermal and available potential energy. Though relatively large, the signal associated with the cycle is obscured to some extent by non-linear interactions associated with other energy sources and sinks and by the general climatic noise. As a result, no two sequences or episodes are exactly identical. The similarities are, however, sufficiently pronounced to permit the construction of a "typical" sequence from the composite of a number of cycles. This was done by Rasmusson and Carpenter (1982) for the tropical Pacific (30N-30S) sea surface temperature and wind fields. The characteristic sequence of events in the Western Pacific has been discussed by Donguy and Herrin (1980). Among many other relevant diagnostic studies, one might quote Julian and Chervin (1978), Horel and Wallace (1981) or Pazan and Meyers (1982).

One particular link in the whole sequence of events is the appearance of pronounced warm sea surface temperature anomalies throughout the Central Equatorial Pacific. The climatological importance of these warm equatorial anomalies has been discussed by Horel and Wallace (loc. cit.). On the basis of analytical computations by Hoskins and Karoly (1981), they demonstrated the existence of long wave trains in the upper atmosphere that appear to be forced in the first instance by the anomalous (latent) heat flux from the equatorial ocean. These waves not only modify the tropospheric pressure patterns in a rather persistent manner, but they also can contribute to the transport of atmospheric potential energy in the meridional direction.

The appearance of sea surface temperature (SST) anomalies in the Central Equatorial Pacific is commonly (but not always) preceded by the appearance of a pronounced warm anomaly in the Eastern Pacific along the South American coast -- the so-called El Niño. After the high temperatures are observed in the surface waters off Peru, the anomaly seems to spread westward at a rate of about 0.5-1.0 m s^{-1} (Rasmusson and Carpenter, loc. cit.). The positive anomalies continue to increase in the Central Equatorial Pacific for 3-6 months following their maximum off the Peru coast. During the final stages of such a warm episode, a negative SST anomaly propagates westward from the South American coast at about the same rate as the earlier warm anomaly. It is with this westward propagation of sea surface temperature anomalies that this paper is essentially concerned.

It seems unlikely that the westward propagation of the anomaly is simply due to advection. The westward drift of the surface water above the eastern setting Equatorial Undercurrent has a mean velocity of only about 0.25 m s^{-1} or less (Bruce Taft, personal communication), which is substantially less than the spreading rate of the anomaly. It is also very shallow. At a depth of about 20 m the water tends to move eastward with the Undercurrent. The amount of heat which can be advected westward is correspondingly small. If one assumes that the heat loss to the atmosphere which would be associated with a warm SST anomaly of, say, 3C amounts to no more than 40 W m^{-2}, one finds that the heat reservoir in the top 20 m will have been practically exhausted over a distance of 1500 km. The distance from the coastal areas off Peru to the dateline is more than 10,000 km. It follows that warm anomalies in the Central and Western Pacific cannot have been kinematically advected there from the South American region. Their existence and westward propagation must be due to dynamic or thermodynamic effects.

Temperature anomalies can be propagated by waves. In particular, they can travel with the phase speed of an "odd" Rossby wave, provided there is a sufficient meridional temperature gradient across the Equator. In general, the meridional temperature gradient is too weak to account for the observed phenomenon, which can involve anomalies of more than 3C. Anomalies can be propagated westward also, however, by gravity waves and by "even" Rossby waves, which produce westward moving undulations of the isotherms. The interpretation of the westward propagating sea surface temperature changes simply in terms of the phase speed of these waves leaves, however, also a number of problems unanswered.

Firstly, the longitudinal extent of the anomalies along the Equator changes with time. The phenomenon seems to correspond to the spreading of a warm surface layer from the coastal regions of South America until it covers all the Eastern and Central Equatorial Pacific. Subsequently something like a cold front appears to move through the same area until the more common equatorial cold tongue has been restored. While it may be possible to account for these phenomena in terms of groups of Rossby waves, we are not aware that this has been done so far; it may be difficult to do so because part of the group may be found to carry energy eastward.

A second difficulty arises from the fact that the vertical advection velocity associated with the waves must become zero at the sea surface. Surface temperature changes can therefore be produced only either by horizontal advection or by non-adiabatic processes. In particular, if the propagating temperature anomalies are to be associated with westward propagating even Rossby waves, concomitant changes in the vertical heat flux convergence and mixing have to be invoked. It is not possible to do this in the framework of linear wave theory (e.g., Moore and Philander, 1977) and the stipulation of a structure with a variable surface mixed layer becomes convenient in this case.

When the mixed layer is deep, entrainment tends to be reduced and the heating by solar radiation of equatorial surface layers is then enhanced. When the mixed layer is shallow, entrainment tends to be more vigorous and the entrained cold water is mixed into a thinner layer with enhanced cooling. These processes imply negative feedback and increased dissipation of the energy of waves associated with vertical displacements at the bottom of the mixed layer. Other feedbacks associated with air-sea interaction processes can independently produce a westward propagation of surface temperature anomalies. One such feedback could be caused by changes in the stress and the surface heat flux associated with the passage of the wind over the anomalies. It is with this process that the present paper is concerned.

We model the ocean's response to stress and surface heat flux variations which are associated with a steady easterly wind in the presence of variations in the sea-air temperature difference. The mean zonal wind along the Equator, typically of the order 5 m s^{-1}, exerts a stress of some 0.04 kg m^{-1} s^{-2} (Katz et al., 1977; O'Brien and Goldenberg, 1982). This corresponds to an oceanic friction velocity of about 0.6 cm s^{-1}. At these relatively low wind speeds, the friction velocity is very sensitive to changes in the air-sea temperature difference, as shown in Fig. 1. The easterly wind, in passing over the eastern boundary (Fig. 2, right) of a patch of warm equatorial water, increases the stress, causing divergence, upwelling and also increased wind stirring and entrainment. The resulting cooling will cause the boundary to shift westward. On the westward side of the patch the stress must decrease as the wind moves from a warmer to a colder surface, causing divergence and downwelling. Entrainment is suppressed, and a new shallower mixed layer is formed and heated by solar radiation, producing a gradual westward extension of the anomaly.

Changes in the flux of latent and sensible heat greatly reinforce the mechanical process. As cold and relatively dry air moves in over a warmer surface along the eastern boundary, the water below will be cooled by enhanced conduction and evaporation. The resulting convection also can reinforce the entrainment. The opposite happens at the western boundary where the air moves from a warmer to a colder surface.

The response of the ocean to these processes is discussed here based on, first, results of a simple non-rotational channel model including mean currents and wind stress variations and, second, the behavior of Rossby waves on the equatorial beta plane as modified by heat flux feedbacks. Based on the heuristic descriptions of the feedback processes outlined above, one might expect the speed of westward propagating sea surface temperature anomalies to be increased due to the air-sea heat and momentum exchange. The strength of the ocean's response would seem to be directly related to the strength of the air-sea coupling; a less obvious, but highly important, parameter measures the strength of the vertical coupling within the ocean (which controls the vertical mixing rate). In linear equatorial wave theory this is a friction coefficient or eddy viscosity; in the mixed layer model it is the entrainment rate. These considerations anticipate results below. A more subtle aspect of the ocean's

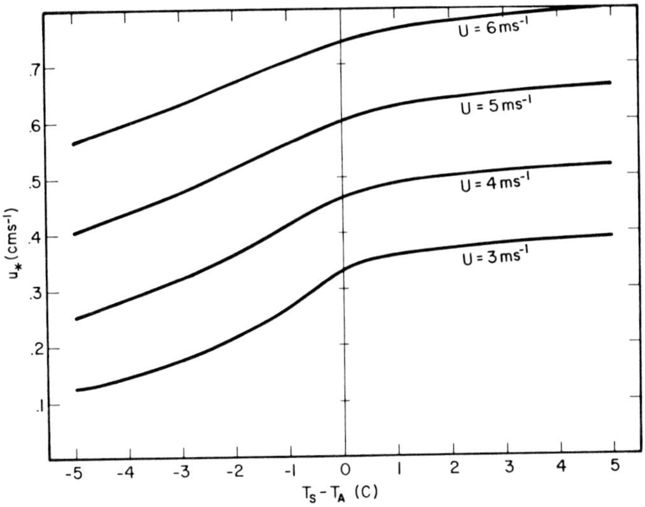

Figure 1. Variation of oceanic friction velocity vs. sea-air temperature difference, for various wind speeds, after Kondo's (1975) parameterization. From Kraus and Hanson (1983).

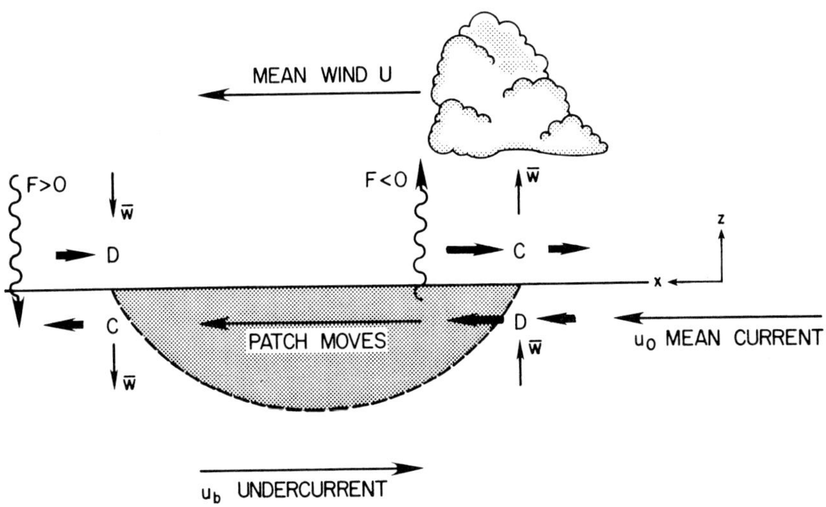

Figure 2. Schematic diagram of air-sea interaction associated with an anomolous patch of warm water. Arrows and symbols represent perturbation convergence (C), divergence (D), and associated vertical motion $\overline{w}$, and heat fluxes (F). From Kraus and Hanson (1983).

response involves the interplay of the two quantities which determine the upper-ocean heat content; i.e., the upper ocean temperature (the SST in a mixed layer model) and the mixed layer depth. Combined, they determine the relevant pressure variable appearing in the momentum budgets. However, assumptions involved in their analytic combination are shown here to mask a response that is significant to westward SST anomaly propagation.

2. A NON ROTATIONAL MODEL

2.1 The Basic State Flow

We first consider the efficacy of the processes discussed above in modifying SST anomaly propagation by neglecting cross-equatorial displacements, and focus attention on forces and velocities in the equatorial plane without rotation or meridional motion.

The layered structure we use consists of a two-layer ocean with a well-defined surface mixed layer. This convenient approximation is never exactly realized along the Equator, where shear turbulence tends to produce a gradual transition from the shallow surface layers to the Undercurrent (Hayes, 1981; Crawford and Osborn, 1979). The last quoted authors have established, however, that most of the shear-generated turbulence is dissipated again at the base of the mixed layer. Because of this they found that a vertically quasi-uniform temperature distribution tends to exist in the surface layer above. In our model the lower layer is infinitely deep with a constant temperature. Balance between the zonal pressure gradient and frictional forces allows for an Undercurrent to be driven eastward with a zonally-uniform velocity u_b. The mixed layer zonal velocity u, depth h and temperature T are assumed to vary. The constant lower layer temperature is used as a reference; thus T represents the temperature difference between the two layers. The flux of buoyancy downward through the sea surface (F*) represents the sum of the latent, sensible and radiative fluxes; in general at the Equator, the sea is heated and $F^* > 0$. At the interface between the two layers, the entrainment velocity W* is found following Niiler and Kraus (1977):

$$W^* = \frac{2Mu_*^3}{C^2 - S(u-u_b)^2}\left[1 - h/L^*\right] \qquad (h/L^* < 1)$$
$$\equiv 0. \qquad (h/L^* \geq 1) \qquad (1)$$

The symbol C represents the velocity of the long internal gravity waves in the layer, and is proportional to its heat content (hT), and

$$L^* = -2Mu_*^3/F^* \tag{2}$$

is a generalized Monin-Obukhov length. The empirical factors M and S measure the fraction of mechanical energy which, after local dissipation, remains available for layer stirring by the work of the shear stresses at the surface and the layer interface, respectively.

A surface layer can be maintained only if $L^* < 1$. It follows that the factor M has to be fairly large to get a reasonably deep mixed layer when the wind is weak. Following Turner (1969) we set $M = 8$, rather on the high end of values proposed by various authors. We also set $S = 0.1$, which implies that 90% of the turbulence which is produced by shear at the bottom of the mixed layer is dissipated and that 10% is converted into potential energy by entrainment of denser water. Davis et al. (1981) derived values of S from the Mixed Layer Experiment results in the North Pacific which varied between 0. and 0.67. The high value was associated with the onset of a severe storm, and the low values corresponded to periods of relatively low wind. It is the latter case which is relevant for the equatorial problem considered here. We note that, inasmuch as there is really no sharp velocity jump between the top of the Equatorial Undercurrent and the surface layer, the entire formulation is somewhat unrealistic. We have simply taken M and S to be adjustable parameters, which are fitted to provide an adequate representation of the basic state.

Assuming hydrostatic balance, the equations for volume, heat and momentum conservation can be written

$$\frac{\partial h}{\partial t} = -\frac{\partial (uh)}{\partial x} + W^* \tag{3}$$

$$\frac{\partial T}{\partial t} = -u\frac{\partial T}{\partial x} + \frac{F^*/\alpha g - W^*T}{h} \tag{4}$$

$$\frac{\partial u}{\partial t} = -u\frac{\partial u}{\partial x} - \alpha g\,\frac{\partial (Th)}{\partial x} + \frac{u_*^2 - W^*(u-u_b) - R(u-u_b)^2}{h} \tag{5}$$

where α and g are the thermal expansion coefficient ($3. \times 10^{-4} C^{-1}$ and gravity ($9.81\ m\ s^{-1}$), respectively. The entrainment velocity W* can be eliminated between the last two equations, and, since $C^2 = \alpha g h T$, the equations reduce to

$$\frac{\partial(C^2)}{\partial t} = -\frac{\partial(uC^2)}{\partial x} + F_* \tag{4'}$$

$$\frac{\partial}{\partial t}\left[(u-u_b)h\right] = -\frac{\partial}{\partial x}\left[u(u-u_b)h\right] - h\frac{\partial(C^2)}{\partial x} + u_*^2 - R(u-u_b)^2. \tag{5'}$$

The factor R in the momentum equations (5,5′) measures the efficacy of loss of easterly momentum from the upper layer by the downward radiation of internal waves.

Rather than integrating Eqs. (1-5) with appropriate eastern and western boundary conditions, we find the basic state in the interior, which can then be perturbed by traveling waves, by setting

$$h = h_o\{1 + h_1(x/h_o) + h_2(x/h_o)^2 \cdots\} \tag{6}$$

with analogous formulae for u,T and W*. Taking the x-axis directed westward from an mid-oceanic origin, we assume for $h_o = 70$ m, $u_o = 0.25\ m\ s^{-1}$, and $T_o = 4.$ C at $x = 0.$ We further stipulate $u_b = -1.2\ m\ s^{-1}$, $u_* = 0.6\ cm\ s^{-1}$ (corresponding to a mean surface wind of about $5\ m\ s^{-1}$ under neutral conditions - Fig. 1) and $F^* = 3.\ 10^{-8} m^2\ s^{-3}$ (corresponding to a mean surface heat input of about 50. $W\ m^{-2}$). These values give $W_o^* = 2.10^{-6}\ m\ s^{-1}$ at $x = 0.$

As long as W* remains finite the layer continues to deepen. In the present case, if the wind and surface flux are assumed to remain constant everywhere, the layer deepens asymptotically from $h_o = 70.$ m at $x = 0.$ to $h_o = 120.$ m at $x = \infty$. On the other hand, a wind speed only 20% lower could not maintain 70.m deep mixed layer at all, and a new layer about 50.m deep would form. This new interior mixed layer would warm up rapidly because the entrainment would be small and the heat input would occur over a shallower layer.

A scale analysis of Eq. 5 in steady state shows that the first and fourth terms are more than an order of magnitude smaller that the wind stress term for realistic values of the mean (smoothed) thermocline slope. With $h_x = 10$ m/1000 km $= 10^{-5}$ and a mean temperature gradient $T_x = 0.3\ 10^{-6}$ K m^{-1} outside the boundary region (see, e.g. Fuglister, 1960), the ratio of $g\alpha h_o(Th)_x/u_*^2 = 0.3$. Though this is not insignificant, it does indicate that the second term, which represents the pressure force, remains considerably smaller than the surface stress. It follows that the input of momentum by the wind must be balanced in this model largely by internal wave radiation losses at the base of the mixed layer. On the basis of this consideration, we stipulate $R = 1.2\ 10^{-5}$. Use of the numerical values of the constants as derived above allows the equations to be solved for the first-order longitudinal variations in u, T, h and W* (i.e., u_1, etc. in (6)). Such a calculation yields a mixed layer which deepens, warms and decelerates toward the west, in agreement with observations. The model places rather too large an emphasis on the internal wave radiation as a frictional mechanism (the factor R being large); this is due to the model's confined geometry and neglect of upwelling (Kraus and Hanson, 1983). In any event, the linearization below transforms the processes at the base of the mixed layer into the usual Newtonian cooling and Rayleigh friction.

2.2. Feedback

The steady basic state of the previous section, found by introducing observational prejudice into the equations (1-5) in order to evaluate the necessary empirical factors, provides an environment in which waves can propagate and interact with the atmosphere as described in the Introduction. Our choice of an atmospheric model is made on the basis of simplicity and in order to isolate the ocean as much as reasonably possible. Large-scale circulation changes in the atmosphere are not included; description of such phenomena requires a dynamical model of the atmosphere. An example of a simple atmosphere-ocean model with atmospheric dynamics has been discussed by Lau (1981). We limit our attention here to the thermodynamic changes associated with a variable air-sea temperature difference. Due to its lower thermal inertia, the atmosphere responds to variable heating on much shorter time scales than the ocean; we therefore ignore effects of changing heat storage in the atmosphere. The atmospheric temperature is presumed to adjust over distances of hundreds of kilometers to heating at its lower boundary due to ocean temperature changes (the distance being an adjustable parameter). This adjustment can be interpreted symbolically as

$$U_A \frac{\partial\theta}{\partial x} = \frac{C_T}{H_A} |U_A| \; (1+1/B_o)(T'-\theta) \tag{7}$$

where C_T is the surface exchange coefficient and U_A and H_A are the zonal wind velocity and the atmospheric boundary layer depth. B_o is a Bowen ratio to be interpreted here as the net amount of latent heat flux into the atmosphere below H_A which is condensed locally. The combination of dimensional parameters yields a length scale γ , which is the adjustment scale for θ . Note that because of the wind speed term on the right side of (7) and the velocity term on the left, γ is not necessarily positive. Atmospheric temperature changes proportional to exp(ikx) take the form

$$(T'-\Theta) = fT' \tag{8a}$$

where

$$f = k/(k-i\gamma) \tag{8b}$$

is the sea-air transfer function.

In order to estimate the effects on the ocean of the feedback, the forcing terms $(F^{*\prime}, u_*')$ in the perturbation equations are set proportional to the air-sea temperature difference perturbations using (8b). The heat flux term in the thermodynamic equation (4) becomes

$$F^{*\prime}/(\alpha g h_o T_o) = \zeta_s f(T'/T_o) \tag{9a}$$

where T′ is the sea-surface temperature perturbation and the time scale for the mixed layer to adjust to surface heat flux changes can be estimated from

$$\zeta_s = \frac{\rho_A c_p}{\rho c} C_T |U_A| (1+1/B_o)/h_o . \tag{9b}$$

The friction velocity perturbations appear as both the wind stress (Eq. 4) and mechanical energy input (Eq. 1). They are found via linearization of the parameterization in Fig. 1 to be

$$u_*' = \mu_s f(T'/T_o) \tag{10a}$$

where the changes in surface stability scale changes in friction velocity by the factor

$$\mu_s = T_o \frac{\partial u_*}{\partial (T'-\Theta)} . \tag{10b}$$

The derivative is averaged about neutral conditions in Fig. 1, and T_o is taken from the previous section.

The feedback processes discussed in the Introduction are therefore formulated in terms of the simple transfer function (8b). A more complex atmospheric model would produce a transfer function analogous to f and another relating large-scale changes in U_A to T. The latter would appear in (5) with (10a) and likely dominate.

2.3. Irrotational Waves

A perturbation analysis of Eqs. (1-5), using the basic state and feedback derivations above, can be formulated as a 3x3 matrix eigenvalue problem; the determinant represents the dispersion relation for the system. The non-linearities in the system (1-5) lead to a large number of terms, many of which can be eliminated as small in a post-hoc analysis. The important terms can then be reduced to

$$\frac{\partial h}{\partial t^*} + \zeta_E h + \frac{\partial u}{\partial x} - \zeta_s fT = 0 \tag{11}$$

$$\frac{\partial T}{\partial t^*} + 2\zeta_s fT - \zeta_E h = 0 \tag{12}$$

$$\frac{\partial u}{\partial t^*} - \mu_E u + c_o^2 \left(\frac{\partial T}{\partial x} + \frac{\partial h}{\partial x} \right) + \mu_s fT = 0. \tag{13}$$

Here the time derivative is Doppler shifted with respect to the mean current $\partial/\partial t^* \equiv \partial/\partial t + u_o \partial/\partial x$ and T and h are non-dimensionalized $(T;h) = (T'/T_o;h'/h_o)$. The drag term μ_E is associated with the (linearized) internal wave radiation, and ζ_E is the effect of entrainment on h and T. The perturbation entrainment rate has been included in (11) and (12); this accounts for the feedback in (11) and the factor of 2 in (12). As before, (11) and (12) can be combined to give

$$\frac{\partial (c^2)}{\partial t^*} + \frac{\partial u}{\partial x} + \zeta_s fT = 0. \tag{14}$$

In the absence of any feedback or drag, the system reduces to

$$\frac{\partial^2 (c'^2)}{\partial t^{*2}} - c_o^2 \frac{\partial^2 (c'^2)}{\partial x^2} = 0 \tag{15a}$$

and the two gravity waves are Doppler shifted

$$\omega/k = u_o \pm C_o. \tag{15b}$$

With feedback and drag terms, it is necessary to solve the three equations (11-13). Without feedback, initial disturbances in the 3 equation set, in addition to generating the gravity waves above, are advected downstream at a rate u_o (Kraus and Hanson, 1983).

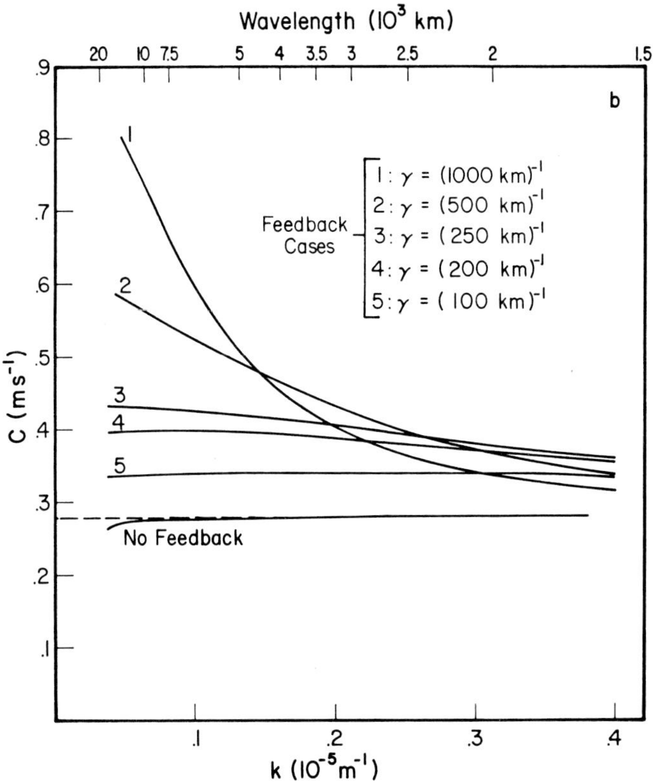

Figure 3. Speed of "wave" due to oscillation between mixed-layer depth and temperature, for various atmospheric adjustment scales. From Kraus and Hanson (1983).

The presence of feedback decouples the continuity and thermodynamic equations and a third "wave" appears due to the oscillation between T and h. Figure 3 shows the resulting phase speed (real part of ω/k) for a variety of adjustment scales γ. The air-sea interaction enhances the propagation speed of the low wavenumber disturbances by as much as a factor of 2 over the mean current for certain adjustment scales. Examination of the details of this enhanced propagation reveals that it is almost entirely due to the heat-flux feedback terms; the wind-stress terms have relatively little effect in this linear model. The drag terms at the mixed layer interface, however, cause decay of initial disturbances on time scales of weeks for the enhanced propagation.

Air-sea interaction therefore causes propagation rates for sea surface temperature anomalies which are increased over the mean advection speed. The heuristic description of the mechanism in the Introduction is borne out in the context of a simple channel model without rotation, although the dissipation rates are rather too large. This may be due to the importance placed on the internal wave drag (the factor R) in the present formulation; numerical calculations are required to resolve this issue. The role of these processes on the equatorial beta-plane can be investigated analytically; first results are described in the following section.

3. ROTATIONAL EFFECTS

The key to the enhanced propagation in the previous section was the decoupling of the mixed layer depth and temperature; it follows that studies which consider only the layer heat content as a variable (i.e. use only Eq. 14 in some form) cannot reproduce this result of air-sea interaction. Although a number of models of ocean circulation on the equatorial beta plane has appeared, each has used variant of Eq. 14, with feedback included, in a few cases, by replacing the temperature in the last term with the heat content. This despite the fact that surface fluxes depend on the surface temperature. Therefore it seems useful to extend the results of the previous section to include rotational effects.

Introducing the meridional direction y (and concomitant motion v) increases the complexity of the model to fourth order. We report here results from a simple formulation which retains the essential aspects of the new physical processes introduced above and neglects potential complexities. In particular, the interfacial drag terms in the momentum equations and the mean current are ignored, and only the heat-flux feedback is considered.

The model therefore represents a single baroclinic mode of the classical theory (e.g., Moore and Philander, 1977) with an equivalent depth $h_E = \alpha T_o h_o = c_o^2/g = 8.4$ cm; this is roughly the third baroclinic mode. The shear terms are not included in the entrainment rate (Eq. 1). The basic equations are (1-4) and the v-momentum equation; with some manipulation they become

$$\frac{\partial h}{\partial t} + \zeta_E h + \frac{\partial u}{\partial x} + \frac{\partial v}{\partial y} - \zeta_s fT = 0 \tag{16}$$

$$\frac{\partial T}{\partial t} + 2\zeta_s fT - \zeta_E h = 0 \tag{17}$$

$$\frac{\partial u}{\partial t} - \beta yv + C_o^2 \left(\frac{\partial T}{\partial x} + \frac{\partial h}{\partial x} \right) = 0 \tag{18}$$

$$\frac{\partial v}{\partial t} + \beta yu + C_o^2 \left(\frac{\partial T}{\partial y} + \frac{\partial h}{\partial y} \right) = 0 \tag{19}$$

and, as before, (16) and (17) combine to give

$$\frac{\partial (C^2)}{\partial t} + \frac{\partial u}{\partial x} + \frac{\partial v}{\partial y} + \zeta_s fT = 0. \tag{20}$$

(The non-dimensionalization and linearization of T and h imply $C'^2 = C_o^2(T + h)$.) Using (18), (19) and (20) without feedback and defining

$$y^* = (\beta/C_o)^{\frac{1}{2}} y \tag{21}$$

leads to the meridional structure represented by the Hermite functions in the variable y*; for order m the dispersion is

$$\omega^2 - C_o^2 k(k-\beta/\omega) - (2m+1)\beta C_o = 0 \tag{22}$$

where $m \geq 1$ for the (westward propagating) Rossby wave modes. In the previous section it was seen that the feedback is effective only at low wavenumbers; this is the case also with rotation. Figure 4 shows the no-feedback dispersion diagram for frequencies of interest. Its features have been described by Moore and Philander (1977). The Kelvin wave (m = -1) has phase speed C_o.

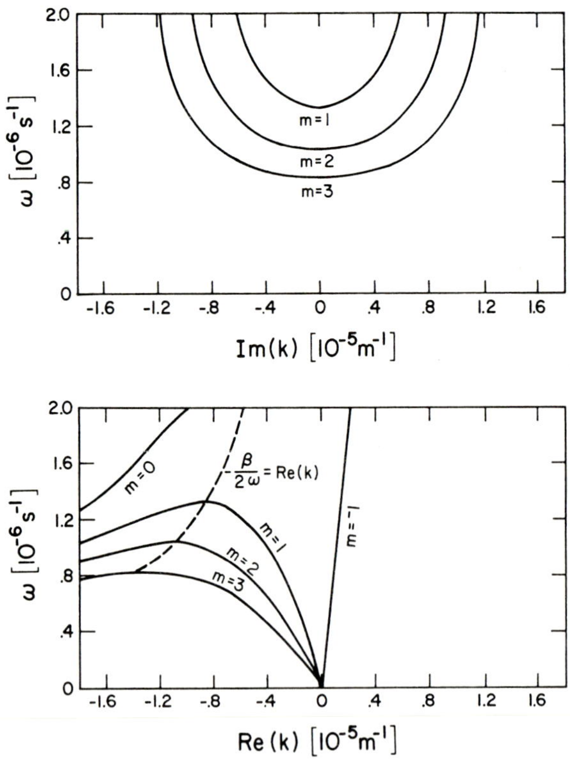

Figure 4. Low-frequency regime of waves on the equatorial beta plane for an equivalent depth of 8.4 cm.

In order to assess the effects of rotation of these waves, the four equations (16-19) must be solved. This presents no great difficulty due to the simple form of (17); the results can be cast in the classical form by defining (Hanson, 1983)

$$C_F = C_o(1+Z)^{-\frac{1}{2}} \tag{23a}$$

where the net effect of the feedback is contained in the complex factor

$$Z = \frac{-\zeta_E\zeta_s f/\omega}{\omega + i(\zeta_E+2\zeta_s f)}. \tag{23b}$$

The dispersion of the Rossby waves is found by solving (22) with C_o replaced by C_F. It should be noted that C_F is still the Kelvin wave speed; the form of Z implies that it has become dispersive. Note also that the "equivalent depth" -- the vertical separation constant in the three-dimensional problem -- is now frequency dependent. This is a post-hoc justification of our use of a layered model; expansion into vertical modes will not work for the feedback problem.

It is at this point in the analysis that the sign of the atmospheric adjustment scale becomes important. In the previous section all quantities were conveniently defined as positive toward the west without ambiguity. With rotation, waves propagate in both directions and (16-20) are formulated in the usual right-handed coordinate system. The easterly trade winds therefore imply that $\gamma < 0$ (see 7). In fact, from the heuristic discussion of the feedback mechanism given in the Introduction, one might expect waves traveling upwind to be slowed. This can be easily tested for the Rossby waves by presuming a westerly wind, so that $\gamma > 0$.

Figure 5 shows phase speeds for the low wavenumber m = 1 Rossby waves without feedback (dashed), for the feedback scales in Fig. 3, and for one case with $\gamma > 0$. The low wavenumber phase speed limit for the m = 1 wave in this case is $(\omega/k) = 0.30$ m s^{-1}; feedback enhances this by about 50% for an atmospheric adjustment scale of -500 km (i.e., easterly wind). The implied westerly case decelerates the wave speed, as expected. Comparison of Figs. 3 and 5 indicates that rotation constrains the model's response somewhat. In the non-rotational case, the low-wave number disturbances were free to propagate as fast as the match of (thermal exchange) scales allowed; with rotation the Rossby dynamics competes with this exchange and the low-wavenumber propagation enhancement is decreased.

If the phase speeds in Fig. 5 are superimposed on the mean current of Section 2.1, one arrives at a propagation rate between about 0.5 - 1.0 m s^{-1} for SST anomalies in the Equatorial Pacific, the exact rate depending on the wavenumber and the atmospheric adjustment scale. This is the range of propagation speeds deduced by Rasmusson and Carpenter (1982) from the composite of warming events.

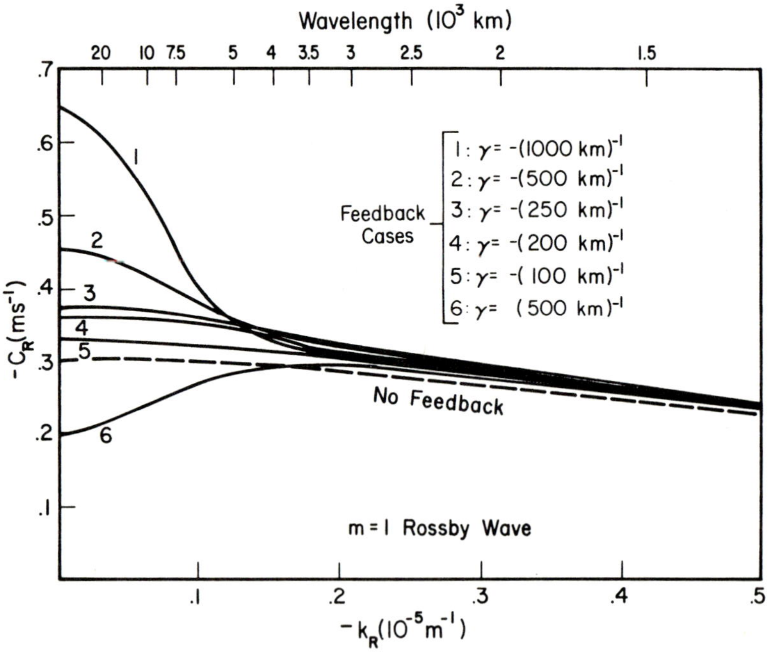

Figure 5. Phase speed of low wavenumber m = 1 Rossby wave vs. wavenumber for various atmospheric adjustment scales.

4. DISCUSSION

The results above indicate the potential importance of the role played by air-sea interaction in the westward propagation of equatorial SST anomalies. In the linear model used here, the dominant process is the sea-air heat exchange; in a non-linear model it is to be expected that the stress coupling through variable surface stability (Fig. 1) would become important also. For several reasons, non-linear modeling should be considered relevant to the study of this phenomenon.

For example, in reality the mixed layer depth may not be a continuous variable. A new shallower mixed layer will be formed whenever $h > L^*$ that is, when there is not enough mechanical energy available for entrainment of colder dense water. The formation of a new shallower mixed layer could considerably accelerate the warming to the west of a warm anomaly and therefore the rate of anomaly propagation. In the linear analysis here, we have only considered anomalies which had everywhere and at all times a mixed layer depth h which remained smaller than the generalized Monin-Obukbov length L^* . In other words we did not consider the intermittent cessation of the entrainment process. In a realistic, finite amplitude model this intermittent cessation would have to be considered.

Another reason for numerical modeling concerns the realistic description of the air-sea interaction processes themselves. The stress variability with sea-air temperature difference is clearly non-linear (Fig. 1). Moreover, our highly simplified atmospheric model neglects non-linearities associated with variable irradiance in addition to atmospheric circulation changes.

One important aspect of the enhanced propagation speeds shown in Fig. 5 (discussed more fully in Hanson, 1983) that is revealed in this linear study concerns the response of the mixed layer heat content (C^2) to the heat-flux feedback. Analysis of the components of the heat content shows that the enhanced propagation at the very low wavenumbers is almost entirely due to a temperature response, with the mixed-layer depth showing very little change. Toward increasing wavenumbers, where the propagation rate is less, a mixed layer depth response begins to dominate with the temperature response becoming smaller. The heat content thus changes in two distinct regimes: a slower thermal response and a faster pseudo-mechanical response. It is the thermal response that is associated with the increased sea-surface temperature anomaly propagation rates in the linear model. The decoupling of h and T (in Eqs. 16 and 17) allows these separate responses; a model using only the heat content equation (20) would not exhibit this characteristic. That the thermal response occurs on large space and long time scales makes it all the more relevant for climate studies.

In the present study we have simply tried to show that air-sea interaction can cause or contribute to the displacement of thermal boundaries. To do so, we have considered a constant wind in the free atmosphere. This is not entirely consistent with our allowing for adjustment of the atmospheric temperature field. In fact, it was the observed or stipulated changes in the atmosphere which largely generated the present interest in the study of the Equatorial Pacific region. There is reason to believe that the atmospheric changes are at least as energetic as those observed in the ocean. The two media interact. A complete understanding of the phenomenon will eventually require the development of a realistic, truly interactive ocean-atmosphere model. We consider this an ultimate goal, toward which the present work leads.

ACKNOWLEDGEMENT

The studies published in this paper were supported by the National Science Foundation with Grant ATM 80-23334 and by the National Oceanic and Atmospheric Administration as part of the Equatorial Pacific Ocean Climate Studies program.

REFERENCES

Crawford, W.R. and T.R. Osborn, 1979: Microstructure measurements in the Atlantic Equatorial Undercurrent during GATE. Deep-Sea Res., 26A (GATE Suppl. II), 285-308.

Davis, R.E., R. DeSzoeke and P. Niiler, 1981: Variability in the upper ocean during MILE. Part II: Modeling the mixed layer response. Deep-Sea Res., 28A, 1453-1475.

Donguy, J.R. and C. Henin, 1980: Climatic teleconnections in the Western South Pacific with El Niño phenomenon. J. Phys. Oceanogr., 10, 1952-1958.

Fuglister, F.C. 1960: Atlantic Ocean Atlas of Temperature and Salinity Profiles and Data from the International Geophysical Years of 1957-1958. WHOI Atlas Series, V. 1. Woods Hole, Mass., 209 pp.

Hanson, H.P., 1983: Equatorial waves in the presence of sea-air heat exchange. To appear in J. Phys. Oceanogr.

Hayes, S.P., 1981: Vertical fine-structure in the Eastern Equatorial Pacific J. Geophys. Res., 86, 10983-10999.

Horel, J.D. and J.M. Wallace, 1981: Planetary-scale atmospheric phenomena associated with the Southern Oscillation. Mon. Weath. Rev., 109, 813-829.

Hoskins, B.J. and D.J. Karoly, 1981: The steady linear response of a spherical atmosphere to thermal and orographic forcing. J. Atmos. Sci., 38, 1179-1196.

Julian, P.R. and R.M. Chervin, 1978: A Study of the Southern Oscillation and Walker Circulation. Mon. Weath. Rev., 106, 1433-1451.

Katz, E.J. and collaborators, 1977: Zonal pressure gradient along the

Equatorial Atlantic. J. Mar. Res., 35, 293-307.

Kondo, J., 1975: Air-Sea bulk transfer coefficients in diabatic conditions. Boundary-Layer Meteorol., 9, 91-112.

Kraus, E.B. and H.P. Hanson, 1983: Air-Sea interaction as a propagator of equatorial ocean surface temperature anomalies. J. Phys. Oceanogr., 13 , (1) in press.

Lau, K-M., 1981: Oscillations in a simple equatorial climate system. J. Atmos. Sci., 38, 248-261.

Moore, D.W. and S.G.H. Philander, 1977: Modeling of the tropical ocean circulation. In The Sea, V. 6, ed. Goldberg et al. Wiley Interscience, pp. 319-361.

Niiler, P.P. and E.B. Kraus, 1977: One-dimensional models of the upper ocean. In: Modeling and Prediction of the Upper Layers of the Ocean, ed. E.B. Kraus. Pergamon. pp. 143-173.

O'Brien, J.J. and S.B. Goldenberg, 1982: Tropical Pacific Wind-Stress Atlas, Florida State University Press, Tallahassee.

Pazan, S. and G. Meyers, 1982: Interannual fluctuations of the Tropical Pacific wind field and the Southern Oscillation. Mon. Weath. Rev., 110, 587-600.

Rasmusson, E.M. and J.H. Carpenter, 1982: Variations in tropical sea-surface temperature and surface wind fields associated with the Southern Oscillation/El Niño. Mon. Weath. Rev., 110, 354-384.

Turner, J.S., 1969: A note on wind-mixing at the seasonal thermocline. Deep-Sea Res., 16 (Suppl.), 297-300.

VARIABILITY OF COASTAL ZONES IN LOW LATITUDES
(with application to the Somali Current, the Gulf of Guinea and the El Nino Current)

P. DELECLUSE[1] and S.G.H. PHILANDER[2]

Abstract

Southerly winds in low latitudes drive very intense currents along the western coast and much slower currents along the eastern coast. Just under the surface layer, the currents are opposite to the wind in the eastern part (they flow upwind, following the pressure gradient force); they are in the same direction as the surface flow along the western coast. These different properties are due to the Ekman drift in the surface layer, which is maximal within 3° off the equator and drives strong zonal currents that give outer conditions to coastal currents. Non linearities intensify the western currents but have small effects on the eastern side. When the wind relaxes, southward current appears along the coast. The relevance of these results for the Somali Current, the Gulf of Guinea and the eastern Pacific currents is examined.

1. INTRODUCTION

Coastal currents in low latitudes present very different aspects along the western side and along the eastern side. Eastern boundary currents are broad and slow. Their variability is high because they hardly persist a few weeks as they are rapidly dispersed by long westward propagating Rossby waves. During most part of the year, southerly winds blow over the eastern part of equatorial oceans ; in February - March, the winds weaken and warm southward currents have been noticed at this period, along the western coast of Africa (Hisard, personal communication) and along the coasts of Peru and Ecuador. These currents flow upwind during the wind relaxation.

Along the western side of the ocean, very intense currents exist. In the Indian ocean, for instance, the southwest monsoon drives a coastal current that presents some peculiar characteristics as shown in Figure 1, during the FGGE experiment in 1979 (Duing, Molinari and Swallow, 1980 ; Brown, Bruce and Evans, 1980).

1. Laboratoire d'Océanographie Physique / MNHN, Paris, France.
2. Geophysical Fluid Dynamics Laboratory / NOAA, Princeton, U.S.A.

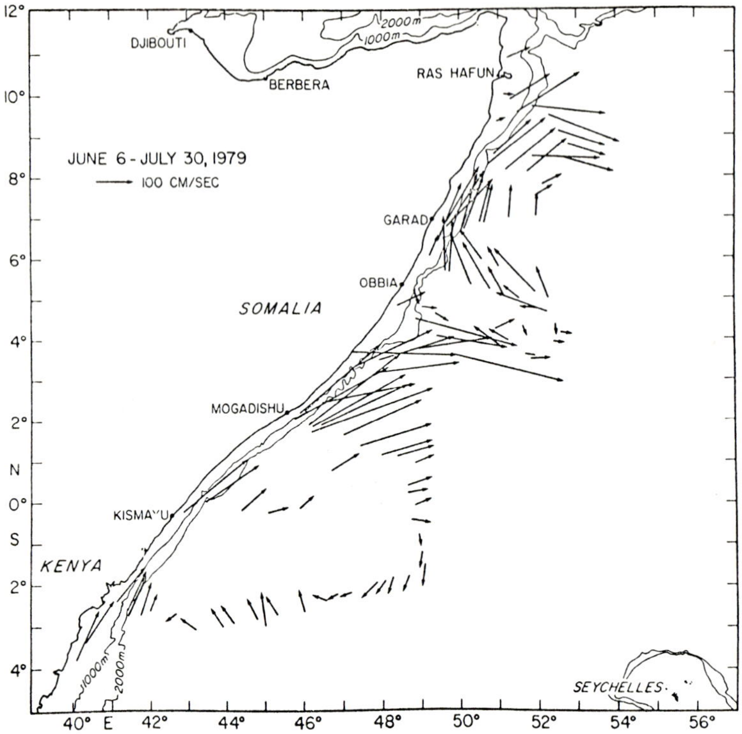

Fig. 1 : The distribution of surface currents between 6 June and 30 July 1979 off the eastern coast of Africa. Current arrows are centered on the point of observation. After Duing, Molinari and Swallow (1980).

The southern branch of the current veers offshore around 4° N (and this separation is marked by an upwelling) ; then, the coastal current seems to be part of an anticyclonic gyre situated between 4° N and 10° N and a second upwelling appears at 10° N. The monsoon winds increase from the equator to 12°N but there is no special features of the wind around 4° N.

In the following part, we try to explain why the Somali current leaves the coast at 4° N and why the eastern and western coasts exhibit such different behaviours. Part II describes the linear response of an equatorial ocean to a southerly wind stress; part III extends the results to a non linear ocean and to the relaxation of the wind. Finally, in part IV, the Somali current is examined in view of the previous results and the eastern boundary currents are discussed. A description of the numerical model used in this paper is presented in the appendix. For a more detailed discussion of some of the results described here, the reader is referred to Philander and Delecluse (1983)

2. THE LINEAR RESPONSE TO A SOUTHERLY WIND STRESS

Away from the coast, a southerly uniform wind stress over the ocean drives a northward current at the equator, an eastward current north of the equator and a westward current south of it. The linear problem, in a shallow water ocean, has been investigated by Moore and Philander (1977). The Ekman transport T^y/f (where T^y is the meridional wind stress component and f is the Coriolis parameter) grows quickly as the equator is approached. The Ekman transport becomes singular at the equator where the balance is ensured by the pressure gradient force :

$$P_y = \frac{T^y}{H}$$

The maximum of the Ekman transport (which corresponds to the maximum of zonal currents) is obtained within a distance equal to the radius of deformation off the equator. In a stratified model, the response is modified as the wind driven response presents a vertical profile different from the oceanic stratification structure. In the following study, only spin-up motions above the thermocline are considered. Long term equilibrium of an ocean in response to a southerly wind stress have been described in Philander and Pacanowski (1981) and Cane (1979) In the numerical approach, surface winds drive currents confined in the first layer (25m) and subsurface currents are situated between the first layer and the thermocline. The wind stress acts as a body force in the first layer.

A uniform southerly wind stress drives cross equatorial flow downwind, depresses the thermocline north of the equator and raises it south. A strong pressure gradient force exists just under the surface, which drives southward return current at the equator, westward current north of the equator and eastward current south of it, at right angle to the pressure force. This meridional circulation is depicted in Figure 2 where a multilevel numerical model, described in the appendix, is driven by a uniform southerly wind stress, acting on the first 25 meters. The time required to reach an equilibrium is estimated by:

$$T = (\beta N H)^{-1/2} = 2 \text{ days}$$

where H is the thermocline depth, N is the mean Brünt Vaïsälä frequency for the surface layers and β is the variation of the Coriolis parameter with latitude. The zonal currents are situated within a distance L off the equator :

$$L = (N H \beta^{-1})^{1/2} = 200 \text{ km}$$

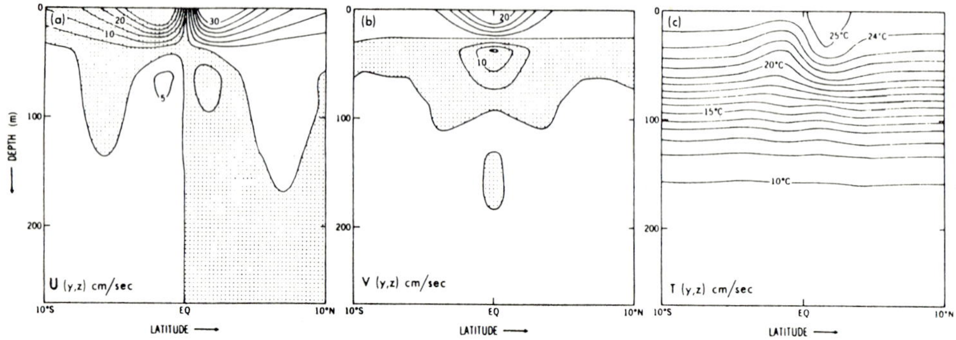

Fig. 2 : A meridional section of the two horizontal velocity components and temperature along a meridian in the center of the basin. The fields were averaged over the 13 day period from day 34 to day 46 in order to eliminate the Rossby-gravity oscillations. (The winds start to blow on day zero.)

Along the western and the eastern boundaries the evolution is different. At mid latitudes, the coastal current along the western boundary accelerates until the arrival of Kelvin waves propagating equatorward, which are excited at the boundaries of the domain (Allen, 1976). This explains why the current becomes steady first at mid latitudes in Figure 3. But in low latitudes, Kelvin waves have different properties as they transfer their energy to equatorial waves. Along the eastern boundary, poleward Kelvin waves are regularly excited by equatorial mixed Rossby gravity waves ; this results into oscillations of the coastal current.

The structure of currents and temperature with depth is presented in Figure 4 and clearly shows the differences between the two boundaries. Along the eastern side, the impact of the eastward jet on the boundary increases the thermocline depression in the northern hemisphere and in the southern hemisphere the upwelling increases along the coast as the westward current leaves the coast. This contributes to the pressure force and tends to reduce the surface current and to create strong southward current just under the surface layer. As the equator is approached from the south, the meridionally integrated Ekman transport offshore grows and the coastal current intensity reverses : there will be a current minimum at the equator. Along the western boundary, the coastal current drains water from the westward current (and there will be downwelling south of the equator along the coast) but north of the equator it leaves the coast to feed the eastward current,where the Ekman transport is maximum,creating a strong coastal upwelling. The pressure gradient force is then northward and contributes to increase the northward current in the surface layer and below the surface. As the meridionally integrated Ekman transport toward the coast increases as it approaches the equator and decreases north of the equator, a current maximum exists at the equator.

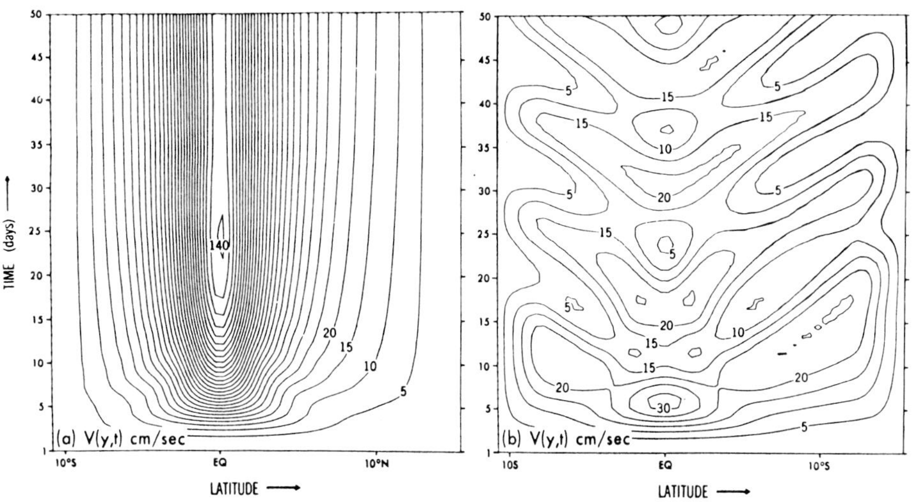

Fig. 3 : The evolution of the meridional velocity comoponents at the surface, 60 km from the eastern and western coasts respectively.

In order to reach an equilibrium, friction must be considered in the equation. Its typical time scale is around two weeks :

$$T = (\beta^2 A_H)^{-1/3}$$

The importance of friction for western boundary adjustment was already noticed by Munk (1950) for mid-latitudes; Hurlburt and Thompson (1976) in a simulation of the Somali current with a two layer model showed its role in equatorial area.

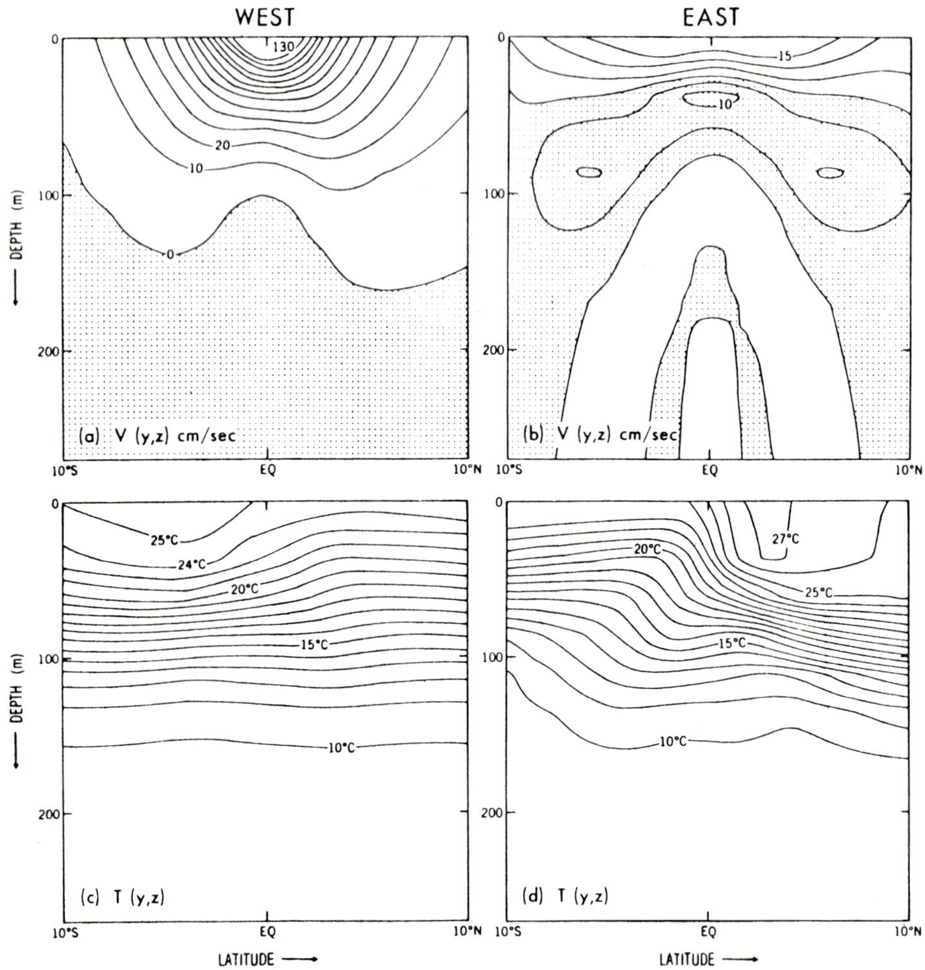

Fig. 4 : Meridional sections of the alongshore velocity component and temperature 60 km from the western ((a) and (c)) and eastern ((b) and (d)) coasts. The fields were averaged over a 13 day period, from day 34 to 46, in order to filter out the dominant oscillations. The flow is southward in shaded regions.

3. NON LINEAR RESPONSE AND EFFECT OF A RELAXATION

The same model, run non-linearly, shows interesting modifications of the current structure. Due to the potential vorticity conservation, a particle moving from the equator to higher latitudes gains eastward momentum ; far from the coast, the eastward current is strongly intensified and the westward current is now slow and broad (Fig. 5). The downwelling area, north of the equator, is strong and narrow ; the subsurface currents are consequently modified : the westward current that was under the eastward surface current moves toward the equator and joints the surface westward current; the subsurface eastward current south of the equator strengthens. These modified circulation patterns may be seen in Figure 6.

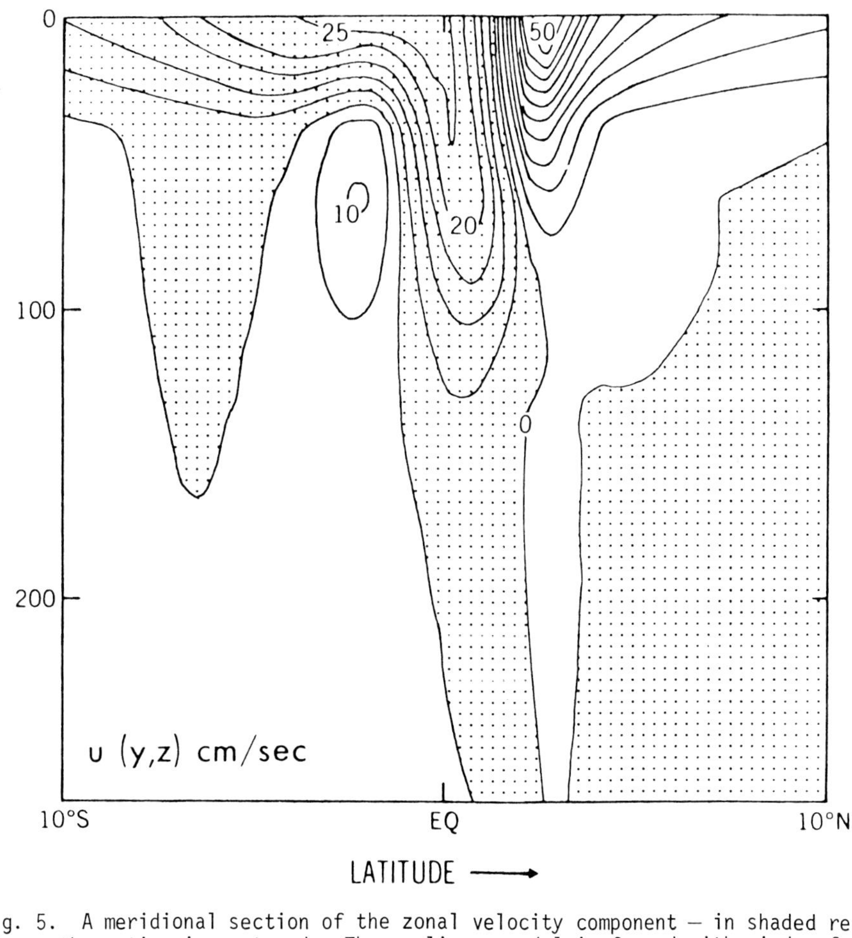

Fig. 5. A meridional section of the zonal velocity component — in shaded regions, the motion is westward — The nonlinear model is forced with winds of intensity 0.5 dynes/cm^2.

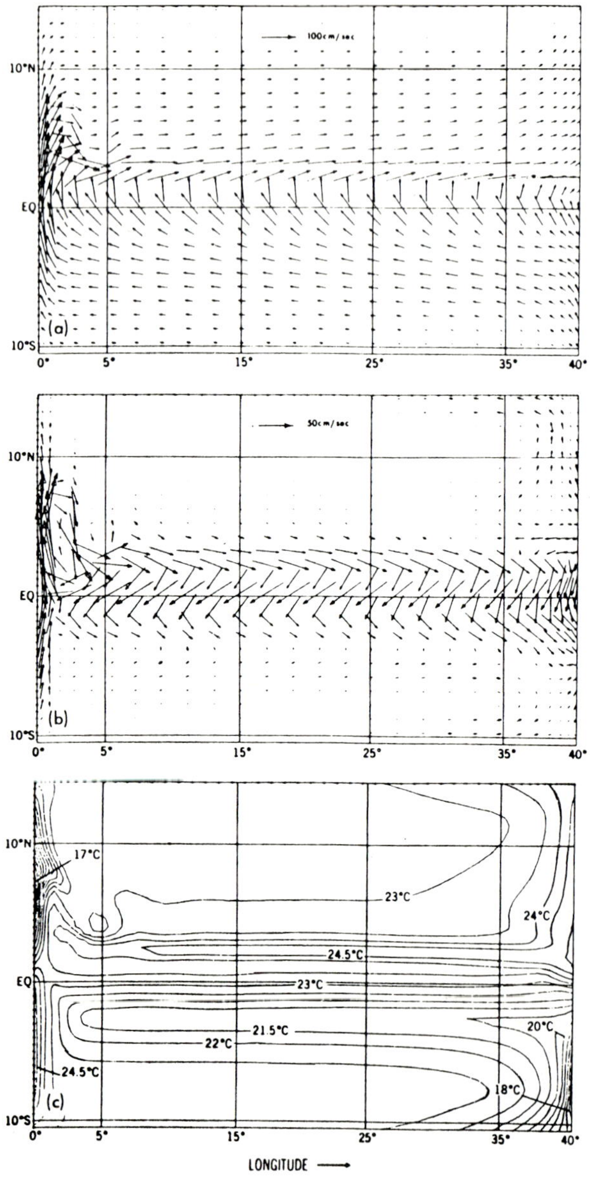

Fig. 6 : The horizontal velocity vectors at depths of (a) 12.5 m and (b) 37.5 m, and the temperature at a depth of 12.5 m, in a nonlinear model 50 days after the wind started to blow.

Along the eastern coast, non linear effects are light ; but along the western coast, there is a strong advection of the coastal current core into the northern hemisphere and the associated upwelling is very well developed (Fig. 7). An anti-cyclonic cell exists now just under the surface, north of the equator, along the coast.

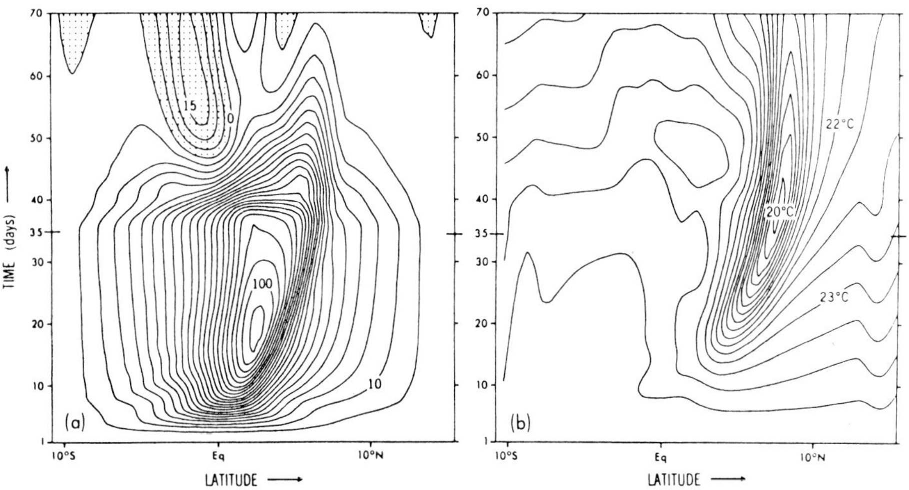

Fig. 7 : The evolution of the alongshore flow and temperature, at the surface, 60 km from the western coast, when the winds start to blow at day zero and suddenly relax at day 35, over a nonlinear model.

After 35 days, the wind-stress was relaxed with a cosine taper of five days in order to avoid the excitation of strong inertia gravity waves. As the wind decays, the water surges following the pressure gradient forces and the surface flow is southward at the equator from the interior to the eastern boundary. Currents in the surface layer and the second layer (Fig. 8) are very similar. Non linear effects allow the currents to persist longer and they decay slowly by friction.

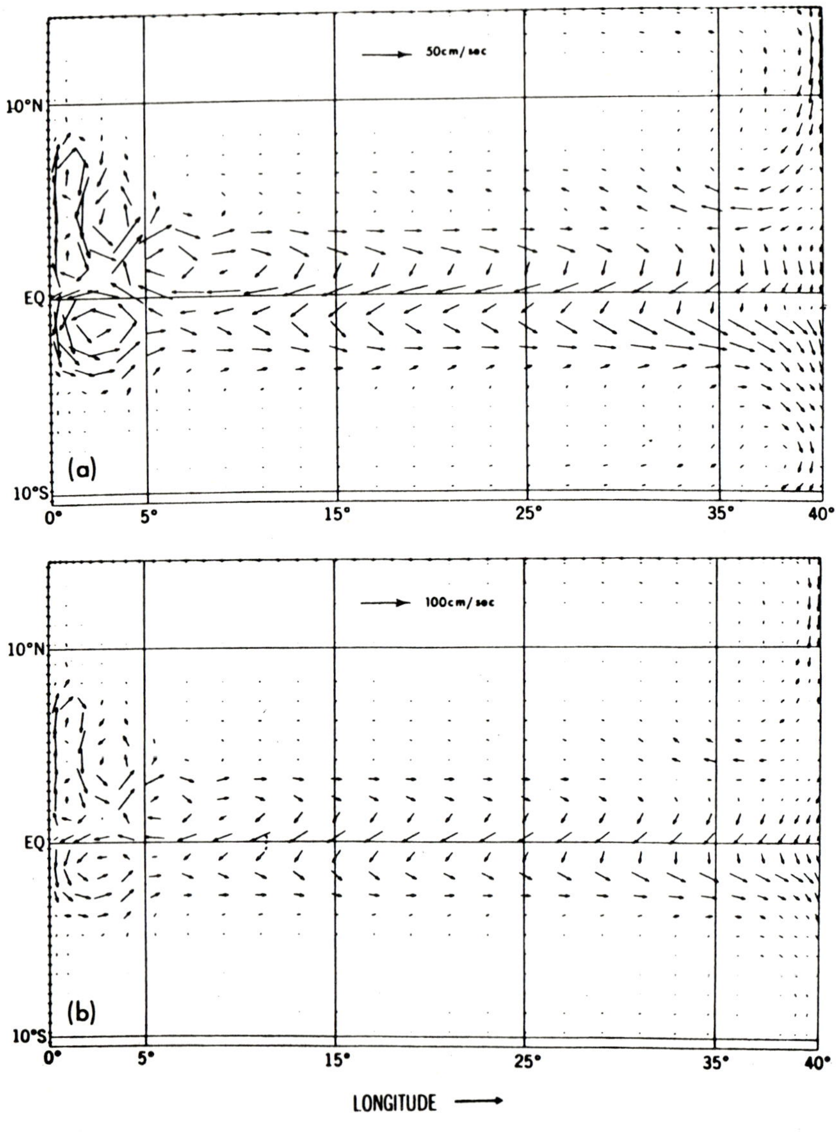

Fig. 8 : The horizontal velocity vectors at depths of (a) 12.5 m and (b) 37.5 m, fifteen days after the wind had stopped blowing. (The wind turned on at day zero and turned off at day 35).

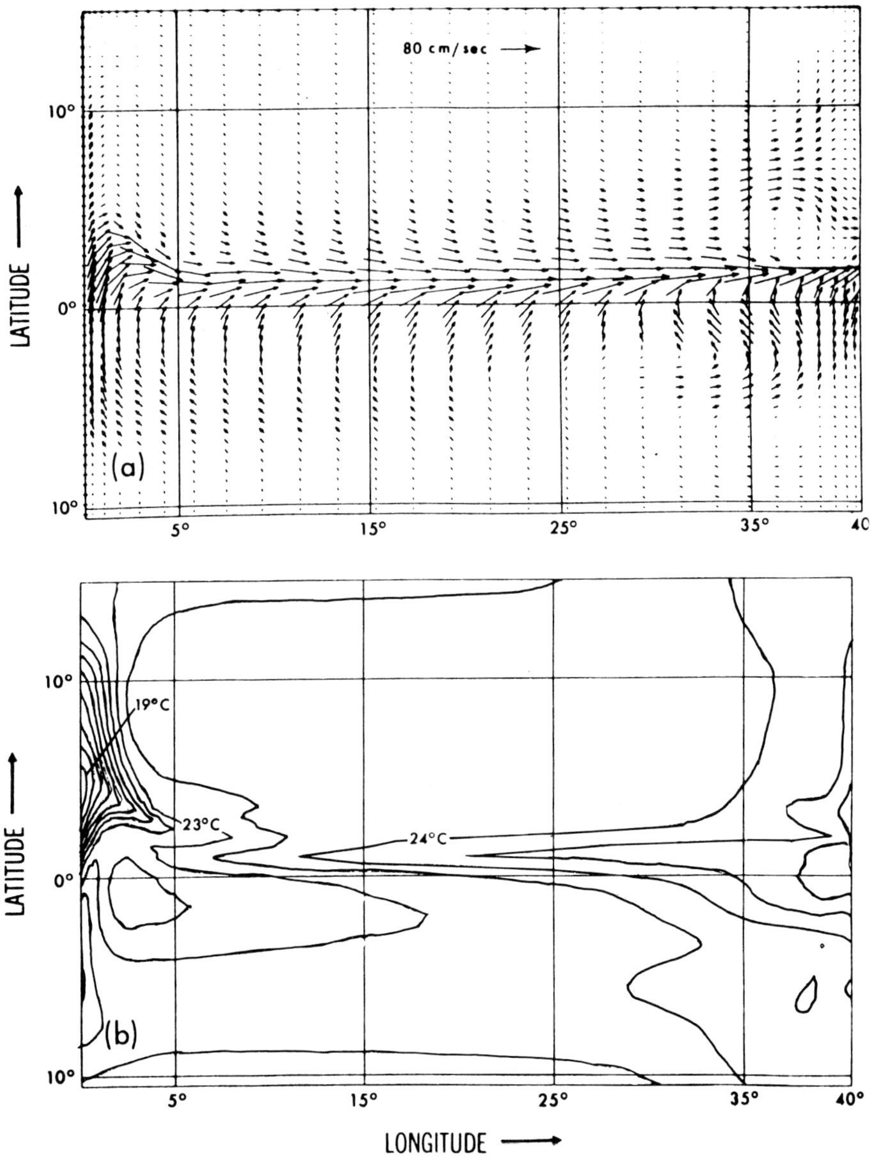

Fig. 9 : The horizontal velocity vectors and temperature at depth of 12.5 m in a non linear model, 50 days after the southwesterly wind started to blow.

4. DISCUSSION

4.1 The Somali current

The important point demonstrated above is that the separation point of the northward current from the coast is determined by the offshore Ekman transport and depends strongly on equatorial dynamics. In a linear model, this point is situated around 4° N. When non linear effects are added, it moves northward as the wind intensifies (it reaches 6° N for a wind of 0.5 dyne/cm2). Southerly winds drive a westward flow at the equator (see Figure 6) ; when westerly winds blow, the current at the equator is an eastward jet (Philander and Pacanowski, 1980). Considering the southwest monsoon, there will be very little zonal flow at the equator and the latitude of the eastward jet is lower than in case of purely southerly wind. Velocity and temperature fields are presented in Figure 9, for a southwest wind stress of 0.5 dyne/cm2 and it is clear that the inclination of the wind tends to inhibit the penetration of the coastal jet into the northern hemisphere (the latitude of separation is around 4° N).

A more realistic simulation of the Somali current is realized by taking into account the coastline inclination. There the separation occurs around 3° N and the upwelling tongue is strongly marked around 4° N (Fig. 10). The characteristics of the southern tongue are well represented (compare Fig. 1 and Fig. 10), considering the crude assumptions required by the numerical model.

An important point is missing in that model : the anticyclonic gyre centered at 8° N. This gyre must be due to the spatial structure of the wind (Anderson, 80). There is, in fact, a strong intensification of the monsoon wind offshore around 12° N. In order to get a complete scenario for the Somali current, a detailed description of the wind field is required (the wind may be highly variable from one year to the other) and a good knowledge of temperature and salinity structure is needed because the Indian ocean is not at rest when the monsoon starts and it evolves from a given density state.

4.2 The eastern boundaries

Very few measurements have been made along the equatorial eastern coast and no strong currents have been reported there. Eastern boundary currents are slow and broad. In the eastern part of the Gulf of Guinea, the winds are mostly southerly and they relax during the months of February-April. In the numerical model, currents are weak along the eastern boundary and oscillate around a mean value which is northward at the surface and southward in subsurface. When the wind relaxes, the flow reverses at the surface and goes southward during a few

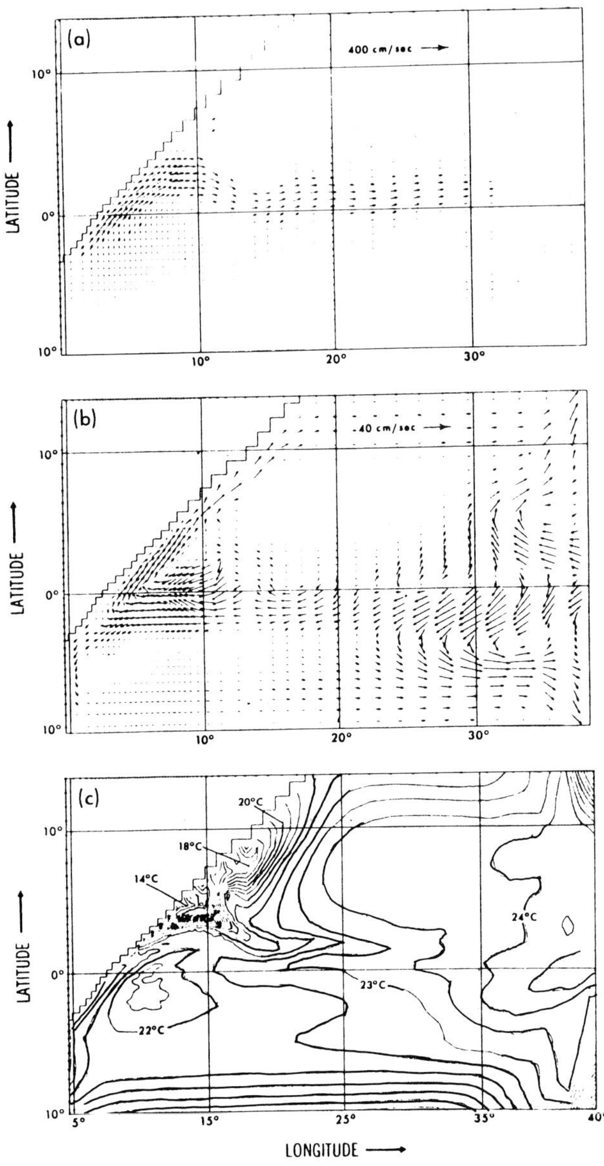

Fig. 10 : The horizontal velocity vectors at depths of (a) 12.5 m and (b) 11.5 m, 50 days after the onset of southwesterly winds of intensity 1.0 dyne/cm2 over a basin with a western coast that is inclined to meridians.

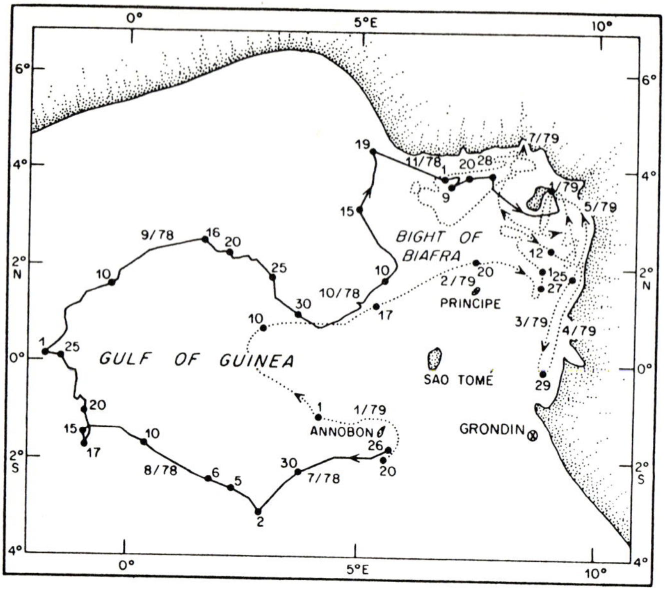

Fig. 11. Trajectories of MARISONDE buoys B 10 (solid line) from 26 July 1978 to 1 February 1979 and B 37 (dashed line) from 20 January to 10 July 1979 in the Gulf of Guinea.

weeks . This scheme is coherent with the observations in the Gulf of Guinea : during most of the year, there is stagnation in the eastern part (Hisard's personal communication) and some drifting buoys trajectories (Fig. 11) observed during the fall and winter 78-79 confirm this circulation (Piton, 1982). It is interesting to point out that, in March 79, one of the buoy, along the eastern coast starts to move southward during about a week and then, it reverses.

This wind relaxation happens also in the Pacific ocean and may cause the warm southward current that appears during a few weeks along the coast of Ecuador and Peru at the beginning of Spring.

Currents along the eastern coasts cannot persist longer than a few weeks because they are dispersed by long westward propagating Rossby waves. Thus variability is likely to be high in these areas. Moreover, remote effects may increase this variability. In this period range (a week - a month), Kelvin waves and mixed Rossby gravity waves are easily excited and they both contribute to transport the energy eastward. They may greatly modify the coastal phenomena. Further long term measurements are required to understand the dynamics of these boundaries.

Appendix : The numerical model

The equations of motion and temperature (the primitive equations) are solved numerically following the method and schemes described by Bryan (1969). The equation of state is assumed of the form :

$$\rho = \rho_0 (1 - \alpha T)$$

where ρ is the density, T is the temperature, $\alpha = 0.0002/°C$ is the coefficient of thermal expansion, and $\rho_0 = 1\ g/cm^3$. The coefficients of horizontal ν_H and vertical ν_V eddy viscosity, and the thermal diffusivity κ are assigned the values :

$$\nu_V = 10\ cm^2/s \qquad \nu_H = 2 \times 10^7\ cm^2/s$$

$$\kappa_V = 1\ cm^2/s \qquad \kappa_H = 10^7\ cm^2/s$$

except in the neighbourhood of zonal boundaries where the values of ν_H and κ_H are increased by a factor of 20 in order to dampen westward propagating Kelvin waves along those coasts.

The ocean is 3 000 m deep and has a flat bottom, it is represented by 16 grid points in the vertical. Profiles of the initial temperature and stratification and distribution of the upper grid points are depicted in Fig. 12.

The model ocean is a rectangular box which stretches from 11° S to 16° N, and which has a longitudinal extent of 4 600 km. The latitudinal grid spacing is 5.6 km ; the longitudinal grid spacing is 111.2 km in the interior of the basin but this decreases to 22.4 km near the eastern and western coasts. (The horizontal grid spacing is represented on the frame in Figure 12).

No slip conditions are applied along the vertical walls and at the bottom.

$u = v = T_\lambda = 0$ along the meridional boundaries

$u = v = T_\varphi = 0$ along the zonal boundaries.

The ocean is at rest initially and is forced by an uniform northward windstress of intensity 0.5 dyne/cm2. The winds are set up with a cosine taper over a period of five days in order to filter out inertia-gravity waves. In the case where the winds relax after blowing steadily for 35 days, this happens again over a period of five days. The heat flux across all boundaries, including the surface, is zero. When the vertical gradient of density becomes unstable, convective adjustment between adjacent layers restablishes the equilibrium.

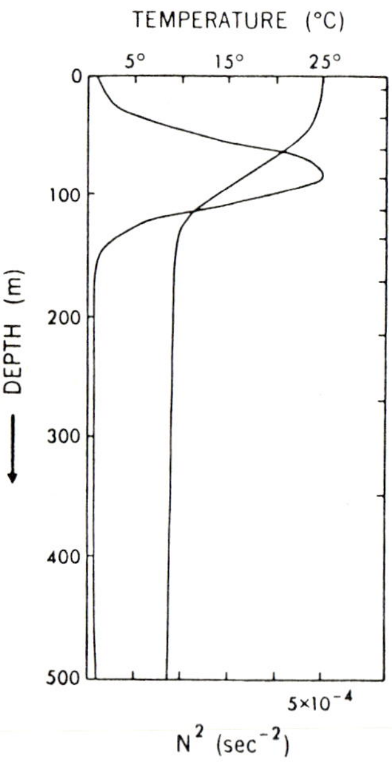

Fig. 12. The initial temperature and Brunt-Väisälä frequency in the upper 500 m of the model. At greater depths the temperature decreases linearly to zero. The vertical spacing of grid points is indicated.

REFERENCES

Allen, J.S. (1976) : Some aspects of the forced wave response of stratified coastal regions. J. Phys. Oceanogr., 6, 113-119.
Anderson, D.L.T., (1980) : The Somali Current (unpublished manuscript)
Brown, D.B., J.G. Bruce and R.H. Evans, (1980) : Evolution of sea surface temperature in the Somali Basin during the southwest monsoon of 1979. Science, 209, 595-597
Bryan, K. (1969) : A numerical method for the study of the world ocean. J. Comp. Phys., 4, 347-376
Cane, M.A. (1979) : The response of an equatorial ocean to simple wind-stress patterns : II. Numerical results. J. Mar. Res., 37, 253-299
Charney, J.G., (1955) : The generation of oceanic currents by winds. J. Mar. Res. 14, 477-498.
Duing, W., R.L. Molinari and J.C. Swallow, (1980) : Somali current : evolution of the surface flow. Science, 209, 588-590
Hurlburt, H.E. and J.D. Thomson, (1976) : A numerical model of the Somali Current J. Phys. Oceanogr., 6, 646-664
Moore, D.W. and S.G.H. Philander, (1977) : Modeling of the tropical oceanic circulation in The Sea, Vol. VI., Wiley Interscience, NY. pp. 319-361
Munk, W.H., (1950) : On the wind-driven ocean circulation. Jour. of Meteo. Vol 7, n° 2.
Philander, S.G.H. and R.C. Pacanowski, (1980) : The generation of equatorial currents. J. Geoph. Res., 85, 1123-1136
Philander, S.G.H. and R.C. Pacanowski, (1981) : The oceanic response to cross-equatorial winds (with application to coastal upwelling in low latitudes). Tellus, 33, 201-210
Philander, S.G.H. and P. Delecluse,(1983): Coastal currents in low latitudes. Deep Sea Research, In press.

REFLECTIONS OF LOW FREQUENCY EQUATORIAL WAVES ON PARTIAL BOUNDARIES

Y. du PENHOAT[1], M.A. CANE[2] and R.J. PATTON[2,3]

[1]Antenne ORSTOM - Centre Océanologique de Bretagne, BP 337, 29273 Brest cédex, France

[2]Department of Meteorology and Physical Oceanography, M.I.T. Cambridge, MA 02139, U.S.A.

[3]Present address : Dynamics Technology, 22939 Hawthorne Blvd, Torrence, CA 90505, U.S.A.

ABSTRACT

We develop a linear theory for the effects of partial boundaries on low frequency waves of the kind thought to be important in the seasonal and interannual variations of the equatorial circulation. The western partial boundary case (e.g. Brazil) differs from the eastern one (e.g., the Gulf of Guinea) by the presence of short Rossby waves trapped along the north-south part of the boundary which form a boundary current accomplishing the required meridional redistribution of the zonal mass flux. There is a discontinuity in the dynamic topography at the corner of the east-west current at this point.

Calculations for the world's equatorial ocean basin shapes are discussed. Calculations carried out for equatorial islands show that propagation of such low frequency waves will not be affected significantly by any island of the real equatorial ocean.

INTRODUCTION

The dynamical effect of the reversal sign of the Coriolis force at the equator makes it a very effective wave guide which supports planetary waves unique to the tropics. Plane waves solution to linearized equations have a faster propagation near the equator than they have at higher latitude. As a consequence, the equatorial ocean exhibits a strong and rapid response to seasonal and interannual variations in the wind stress.

We develop a complete linear theory for the effects of partial boundaries on the low frequency waves which play an important role in these variations. By partial boundaries, we mean a coast which presents discontinuities (or caps) and which is not a straight north-south coast. For example, we may think of the coast of the

Gulf of Guinea and Brazil in the Atlantic ocean, the New Guinea coast in the Pacific and the Somali coast in the Indian ocean. Furthermore, islands in the tropical oceans are a particular case of partial boundaries.

Our interest is in the effect of such boundaries on the wave motions essential to basin wide adjustment rather than on details of the perturbations near the boundaries.

We will solve the linear shallow water equations appropriate to a single baroclinic mode using the usual equatorial scaling, namely, the length scale : $L = (gH/\beta^2)^{1/4}$ (the equatorial radius of deformation) and the timescale $T = (gH\beta^2)^{-1/4}$ with H the equivalent depth of the baroclinic mode, g the acceleration due to gravity and β the meridional derivative of the Coriolis parameter.

Since we are interested in low frequency motion, the frequency is small compared to the equatorial scaling frequency so that :

1 - The mixed gravity wave and short Rossby waves cannot propagate very far into the interior before friction destroys them.

2 - Geostrophic balance holds in the meridional direction.

3 - The westward propagating long Rossby waves are approximately non dispersive.

We choose as our canonical problem the large t asymptotic flow that results when a wave source is switched on at t = 0 and remains steady thereafter ; as discussed in Cane and Sarachik (1976), the solution for a periodic forcing or any forcing can be deduced from the solution to this problem.

Cane and Sarachik (1976) (see also Anderson and Rowlands, 1976) have shown that for large t, the asymptotic motions are of 3 kinds :

(i) Equatorial Kelvin waves, propagating energy eastward with u and h proportional to ψ_0, the zeroth order Hermite function : $\psi_0 = \pi^{-1/4} e^{-y^2/2}$. The large t response is steady for an H(t) time dependence and is independent of x..

(ii) Long Rossby waves, propagating energy westward. The equatorial Kelvin wave has v = 0 and the long Rossby waves have $v \simeq 0$; both satisfy the geostrophic relation

$$yu + \frac{\partial h}{\partial y} = 0 \qquad (1)$$

Again, the large t asymptotic form is independent of t and x.

(iii) Short Rossby waves (including the mixed Rossby-gravity wave) of the form (see Cane and Sarachik, 1977 p 404).

$$(u^s, v^s, h^s) = \left[-\frac{\partial}{\partial y}, \frac{\partial}{\partial x}, y\right] \{J_o(2\sqrt{xt})\ \chi(y)\} \tag{2}$$

Note that as $t\to\infty$, $J_o(2\sqrt{xt}) \longrightarrow \delta(x)$; a sum of such modes is an ever thinning boundary layer trapped at $x = 0$.

The usual boundary conditions prescribed at the eastern and western full boundaries must be carefully applied to the case of partial boundary to ensure conservation of mass. For a full boundary, no normal flow exists at the longitude of the boundary, but if the boundary does not extend accross all latitudes in the basin, the condition of no normal flow can no longer be applied in the open ocean region.

The plan of the remainder of this paper is as follow : in section 2, we solve the problem of a Kelvin wave and Rossby waves separately, for an eastern partial boundary. The case for a western partial boundary will differ from the eastern one by the presence of short Rossby waves trapped along the north-south part of the boundary and is discussed in section 3. In section 4, we will extend the results to a more complex geometry (for example a "zigzag" step coast). In section 5, we summarize our results and consider their applicability to the world ocean.

EASTERN PARTIAL BOUNDARY CASE

Incoming Kelvin wave

We first consider the case of a unit amplitude Kelvin wave impinging on a partial coast at $X = X_B$ extending from the latitude $y = b$ at south to infinity at north (see figure 1). The part of the wave north of $y = b$ will be reflected as a set of long Rossby waves as in the case of a full boundary. South of b and east of X_B, a transmitted Kelvin wave of amplitude T^K and short Rossby waves are allowed to propagate.

Therefore west of X_B, we may write :

$$u^W = u_R + \psi_o\ (y), \qquad h^W = h_R + \psi_o\ (y)\ ; \tag{3}$$

while east of X_B

$$u^E = u_s + T^K\psi_o\ (y), \qquad h^E = h_s + T^K\psi_o\ (y)\ ; \tag{4}$$

where (u_R, h_R) are components of long Rossby waves and (u_s, h_s) are components of short Rossby waves.

At $X = X_B$ and south of the corner, the matching conditions are that u and h be continuous at $X = X_E$, $y<b$ so that $u^E = u^W$ $h^E = h^W$ at $X = X_E$, $y<b$.

Making use of (2),

$$u^W = T^K\psi_o(y) - \frac{\partial X}{\partial y},$$

$$h^W = T^K\psi_o(y) + y\chi. \tag{5}$$

Since (u^W, h^W) satisfies (1), substituting (5) into (3) yields :

$$y\left[T\psi_o(y) - \frac{\partial \chi}{\partial y}\right] + \frac{\partial}{\partial y}\left[T\psi_o(y) + y\chi\right] = 0$$

or, since the Kelvin wave is also geostrophic,

$$- y\frac{\partial \chi}{\partial y} + \frac{\partial}{\partial y}\left[y\chi\right] = 0$$

Therefore $\chi = 0$ for $y<b$; that means there is no reflected short Rossby wave and only Rossby waves and Kelvin wave are reflected and transmitted.

Integrating (1) across the boundary longitude yields :

$$\int_{b-}^{b+} y u dy + h(b+) - h(b-) = 0$$

Therefore there can be no jump in h at $y = b$.

We now summarize conditions that must be satisfy at an eastern partial boundary :

(a) h and u must be continuous in x at the boundary longitude. South of the boundary, the Rossby modes must conspire to cancel the untransmitted part of the Kelvin wave to ensure continuity in x :

$$h_R = u_R = (T^K - 1)\psi_o \qquad \text{for } y<b \tag{6}$$

(b) Above the boundary, the conditions are the same as for a full boundary, namely, no normal flow and the total h a constant :

$$u_R = -\psi_o$$

$$h_R = D^K - \psi_o \qquad \text{for } y>b \tag{7}$$

(c) h is continuous in y at the corner $X = X_B$, $y = b$. Using the expressions (6) and (7) for h this implies

$$\overset{\text{south}}{h_R(b)} = (T^K - 1)\psi_o(b) = D^K - \psi_o(b) = \overset{\text{north}}{h_R(b)}$$

$$\text{or} \quad T^K\psi_o(b) = D^K \tag{8}$$

(d) The Rossby waves generated to the west of the boundary are orthogonal to the Kelvin mode (see Cane and Sarachik, 1979, Appendix A). This condition may be expressed in the form

$$[(\psi_o, \psi_o), (u_R, h_R)] = 0 \tag{9}$$

where the inner product is defined by

$$[(u_a, h_a), (u_b, h_b)] = \int_{-\infty}^{+\infty} (u_a u_b + h_a h_b)\,dy.$$

Substituting (6) and (7) into (9) yields

$$2(T^K - 1)\int_{-\infty}^{b} \psi_o^2 + D^K \int_b^{\infty} \psi_o - 2\int_b^{\infty} \psi_o^2 = 0$$

Finally using (8) and the normalization condition $\left[\int_{-\infty}^{+\infty} \psi_o^2 = 1\right]$,

$$T^K = \frac{2}{2\int_{-\infty}^{b} \psi_o^2\,dy + \psi_o(b)\int_b^{\infty} \psi_o(y)\,dy} \tag{10}$$

Equations (6), (7), (8) and (10) give the entire solution for the problem. We postpone discussion after the incoming Rossby wave case has been solved.

Incident Rossby wave motion

Another possible situation is to have a set of long Rossby waves propagating from the east and encountering the corner at $y = b$. Let $\hat{u}$ and $\hat{h}$ be the components of the incoming Rossby waves ($\hat{v} \simeq 0$). Since it is made up of long Rossby waves, the geostrophic relation (1) still holds so, as before, we may conclude that there are no short Rossby waves generated at the boundary. However, there is the possibility of a Kelvin wave (of amplitude T^R) being reflected eastward from the boundary. Hence for $X > X_B$, $u^E = T^R\psi_o + \hat{u}$ and $h^E = T^R\psi_o + \hat{h}$.

On the other hand there can be no Kelvin wave west of X_B; the solution there must be solely Rossby waves (u_R, h_R).

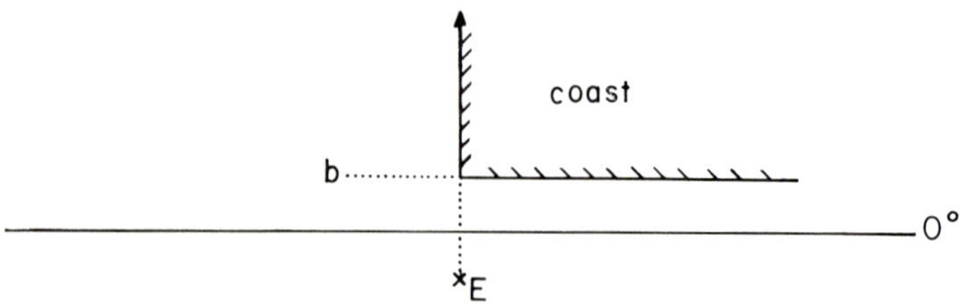

Fig.1: Diagram of partial boundary near equator used in calculating transmission coefficients of reflected planetary waves.

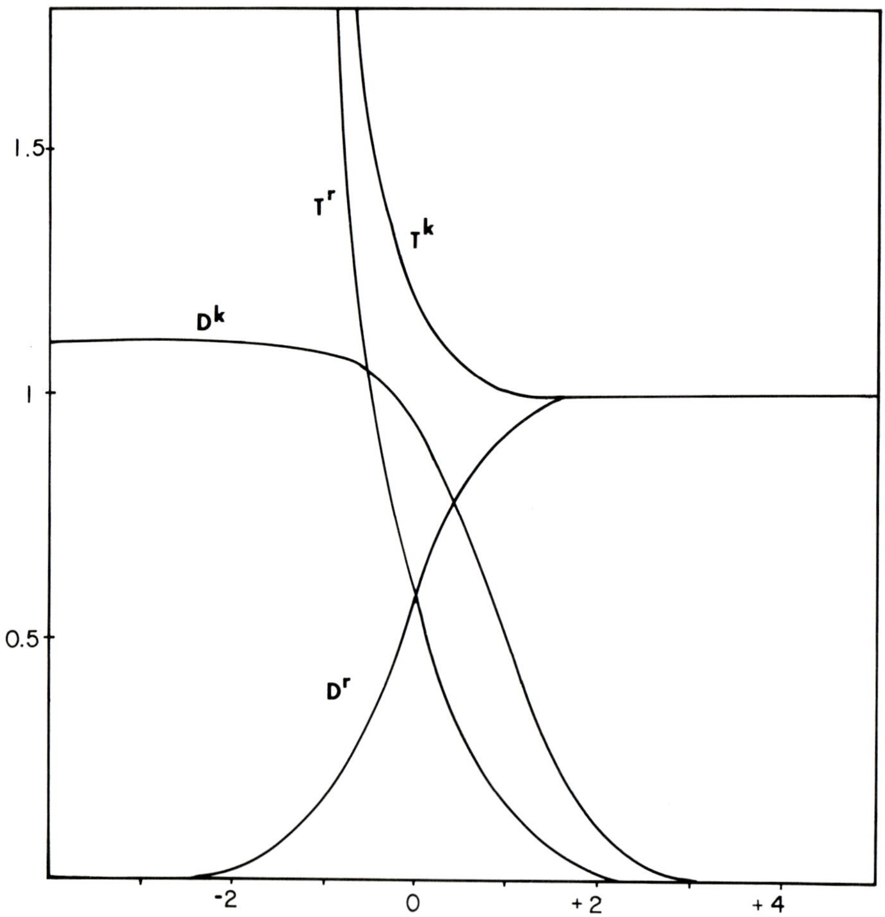

Fig.2: Transmission coefficients of Kelvin mode and height constants along upper wall for partial boundary as functions of distance b from the equator to the zonal coast. T^k transmission coefficient for incident Kelvin waves, T^r transmission coefficient for incident Rossby modes with unit amplitude at corner, D^k height constant set-up for incident Kelvin waves, D^r height constant for incident Rossby waves.

Since there is no incident zonal velocity above the boundary, we have :

$$u_R = 0 \qquad h_R = D^R = \text{constant} \qquad \text{for } y>b \text{ at } X = X_B, \tag{11}$$

while continuity in x below the boundary means :

$$\left.\begin{aligned} u_R &= \hat{u} + T^R\psi_o, \\ h_R &= \hat{h} + T^R\psi_o. \end{aligned}\right\} \quad \text{at } X = X_B \qquad y<y_b \tag{12}$$

As in the previous case h must be continuous in y at the corner:

$$D^R = \hat{h}(b) + T^R\psi_o(b). \tag{13}$$

West of X_B, the Kelvin wave amplitude is zero. Therefore projecting a Kelvin wave onto the solution west of the boundary using (12) and (13) gives :

$$\hat{h}(b)\int_b^\infty \psi_o(y)\,dy + T^R\psi_o(b)\int_b^\infty \psi_o(y)\,dy + 2T^R\int_{-\infty}^b \psi_o^2(y) + $$
$$+ \int_{-\infty}^b (\hat{u}+\hat{h})\psi_o(y) = 0$$

Finally, solving for T^R, we get :

$$T^R = -\frac{\hat{h}(b)\displaystyle\int_b^\infty \psi_o\,dy}{2\displaystyle\int_{-\infty}^b \psi_o^2(y) + \psi_o(b)\int_b^\infty \psi_o(y)\,dy} \tag{14}$$

or $T^R = -\hat{h}(b)T^K \int_b^\infty \psi_o(y)\,dy$

We note that T^R and D^R (and by extension the entire solution for this case) depend only on the value of the height of the Rossby mode at $y = b$, i.e. on $\hat{h}(b)$.

Results

Figure 2 displays the solution coefficients T and D for an incoming Kelvin wave and for an incident Rossby wave motion of unit height amplitude at the corner. For the Kelvin wave case, we obtain the surprising result that the Kelvin wave transmission coefficient is greater than one : the curve asymptotes to one from above as b recedes to infinity to the north (b>0), but T^K goes to infinity as b increases south of the equator (i.e. as the corner is situated south of the equator and $b \to -\infty$). To understand this behavior, we must consider the entire mechanism.

First, the transmitted equatorial Kelvin wave can be considered as a coastal Kelvin wave. To see that the two are the same, we normalize the equatorial Kelvin wave to be one at $y = b$ and define $\eta = y - b$, the distance from the coast. Therefore :

$$u = h = e^{b^2/2}\, e^{-y^2/2} = e^{-b^2/2}\, e^{-(b+\eta)^2/2} = e^{-b\eta}\, e^{-b^2/2}$$

η is small and $e^{-\eta^2/2} \simeq 1$

$1/b$ is liked a (scaled) radius of deformation at $y = b$, so the wave amplitude behaves like the usual f-plane coastal Kelvin wave. The amplitude of this wave must be such as to make its height at the corner match the constant height set up among the boundary as given by (8).

In the other hand, as b goes south ($b \to -\infty$), the reflected Rossby waves have the same structure as they would if the barrier were infinite. In particular, the height sets up to have the same value as for an infinite coastline case. The process that brings this set up has been discussed by Anderson and Rowlands (1976) and Cane and Sarachik (1977). Mass is carried towards the poles in a meridional current that is like a coastal Kelvin wave and is geostrophically balanced. The transport in this current at $y = b$ is given by :

$$V\,(y=b) = \int_{-\infty}^{X_E} v\,(y=b)\,dx = \frac{D}{b}$$

If we now calculate the transport under the boundary carried by the transmitted Kelvin, we find by asymptotic expansions as $b \to \infty$, that :

$$T^K \int_{-\infty}^{b} \psi_o \approx \frac{D}{b} = \sqrt{2}\,\pi^{1/4} \qquad \text{as } b \to -\infty$$

The mass transport around the corner is just what the coastal Kelvin wave under the boundary is able to supply.

As b goes north ($b \to +\infty$), the amplitude of the transmitted Kelvin wave goes to one and the height along the north-south coast goes to zero. There is no transmitted Rossby wave.

For an incoming Rossby wave motion under the corner, away from the equator to the north, the height along the wall is simply the height of the incident Rossby modes at the corner and no Kelvin wave is reflected. This is to be expected since the height along the wall will have an increasing smaller projection on the equatorial Kelvin mode as the boundary recedes from the equator.

THE WESTERN PARTIAL BOUNDARY CASE

Incoming Kelvin wave

We now deal with a western partial boundary extending from $y = -\infty$ to $y = b$ at a longitude $X = X_W$ (figure 3). The only possible motion generated at the corner are long Rossby waves to the west (above the east-west coast), and Kelvin waves plus boundary-trapped short Rossby waves to the east of the corner.

West of the longitude X_W, the solution is made up of the incoming Kelvin wave (assumed to have unit amplitude), and a set of long Rossby waves reflected at the boundary at X_W. The part of the Kelvin wave not reflected in Rossby waves is transmitted with an amplitude T^K. Equations (3) and (4) still hold. Both u and h are continuous in x, so that $u^E = u^W$ and $h^E = h^W$ for $y>b$. Since u and h are in geostrophic balance at $X = X_W$ $y>b$, we can say as in section 2 that there is no disturbance created by short Rossby waves and

$$h^R = u^R = (T^K - 1)\psi_o(y)$$

At $X = X_W$ and $y<b$, the boundary condition is $u^E = 0$; hence

$$\left.\begin{aligned} u^E &= T^K\psi_o - \chi_y = 0 \;; \\ h^E &= T^K\psi_o + y\chi \end{aligned}\right\} \quad X = X_W \qquad y<b$$

This condition leads to :

$$\chi(y) = T^K\int_{-\infty}^{y} \psi_o(y) + C, \qquad (15)$$

$$C = \chi(-\infty) = 0 \;;$$

and $h^E = T^K\psi_o(y) + y\, T^K\int_{-\infty}^{y} \psi_o(y) \qquad \text{at } X = X_W \quad y<b \qquad (16)$

So at $X = X_W$, $y = b_+$, $h^E = T^K\psi_o(b)$,

$y = b_-$, $h^E = T^K\psi_o(b) + bT^K\int_{-\infty}^{b} \psi_o(y)$.

Discontinuities in h are thus possible at $y = b$. Since west of the boundary u and h are in geostrophic balance for all y, we must allow for the possibility on an infinite zonal velocity at $y = b$

to balance the jump in h : i.e. $u = A\delta(y - b)$ at $y = b$. Hence west of X_B, we have

$$u^W = T^K\psi_o(y) + A\delta(y - b) \quad ; \quad h^W = T^K\psi_o(y) \tag{17}$$

Using the orthogonality and completness properties of the eigen functions of the shallow water equations, we may project the Kelvin wave on the solution east of the barrier :

$$2T\int_{-\infty}^{+\infty}\psi_o^2 = 2T\int_b^{\infty}\psi_o^2 + A\psi_o(b) + T\int_{-\infty}^{b}\{\psi_o(y) + y\int_{-\infty}^{y}\psi_o(y')dy'\}\psi_o(y)dy \;;$$

using the fact that $y\psi_o = -\psi_o'$ and integrating the last term by parts leads to :

$$A = T\int_{-\infty}^{b}\psi_o dy \tag{18}$$

Note that it follows from (2), (15) and (18) that

$$\int_{X_W}^{-\infty} V(y{=}b)dx = -A$$

That is, the eastward transport A in the boundary current at the corner is all carried southward by the boundary current along the north-south coast. There is no net flux in this boundary layer (see Cane and Sarachik, 1977) ; its only role is to redistribute zonally the mass flux so that the Kelvin wave may carry it off.

Projecting the Kelvin wave west of the boundary where the Kelvin wave amplitude is known to be one leads to :

$$T^K = \frac{2\int_b^{\infty}\psi_o^2}{2\int_b^{\infty}\psi_o^2 + \psi_o(b)\int_{-\infty}^{b}\psi_o(y)dy} \tag{19}$$

Incident Rossby wave motions

In this section, we are looking at the effects of the western corner on an incoming Rossby wave.

Again, the only possible motions generated at the corner are a set of long Rossby waves to the west, reflected Kelvin wave and boundary-trapped short Rossby waves to the east. As in the previous section, there is no perturbation created by short Rossby waves above the latitude of the corner ($\chi = 0$ for $y>b$). We write the solution as

$$\left.\begin{aligned} u^r &= \hat{u} + T^R\psi_o \\ h^r &= \hat{h} + R^R\psi_o \end{aligned}\right\} \quad \text{at } X = X_W \qquad y>b \tag{20}$$

$$\left.\begin{aligned} u^E &= 0 = \hat{u} + T^R\psi_o - \frac{\partial\chi}{\partial y} \\ h^E &= \hat{h} + T^R\psi_o + y\chi \end{aligned}\right\} \quad \text{at } X = X_W \qquad y<b \tag{21}$$

Therefore, $u^E = 0$ on the wall implies

$$\chi(y) = \int_{-\infty}^{y} (\hat{u} + T^R\psi_o) \qquad \text{for } y<b \tag{22}$$

As in the previous section, , there is a discontinuity for h at y = b and an infinitesimally thin boundary current must be allowed for to balance it :

$$\begin{aligned} u^R &= \hat{u} + T^R\psi_o(y) + B\delta(y - b) \\ h^R &= \hat{h} + T^R\psi_o(y) \\ h^E &= \hat{h} + T^R\psi_o(y) + y\int_{-\infty}^{b} (\hat{u} + T^R\psi_o)dy \end{aligned} \tag{23}$$

Projecting the Kelvin wave on the solution east of the boundary leads to :

$$B = \int_{-\infty}^{b} (\hat{u} + T^R\psi_o)dy \tag{24}$$

The zero projection of the Kelvin wave on the Rossby waves west of the boundary leads to :

$$\int_{-\infty}^{+\infty} (u^R, h^R)(\psi_o, \psi_o)dy = 0 = \int_b^{\infty} (\hat{u} + \hat{h})\psi_o dy + 2T^R\int_b^{\infty} (\psi_o^2)dy + B\psi_o(b)$$

and with equation (24) :

$$T^R = -\frac{\int_b^{\infty} (\hat{u} + \hat{h})\psi_o(y)dy + \psi_o(b)\int_{-\infty}^{b} \hat{u}dy}{2\int_b^{\infty} \psi_o^2(y) + \psi_o(b)\int_{-\infty}^{b} \psi_o(y)dy} \tag{25}$$

Results

Figure 3 shows the results obtained for the western boundary case. For an incoming Kelvin wave, A^K the amplitude of the boundary

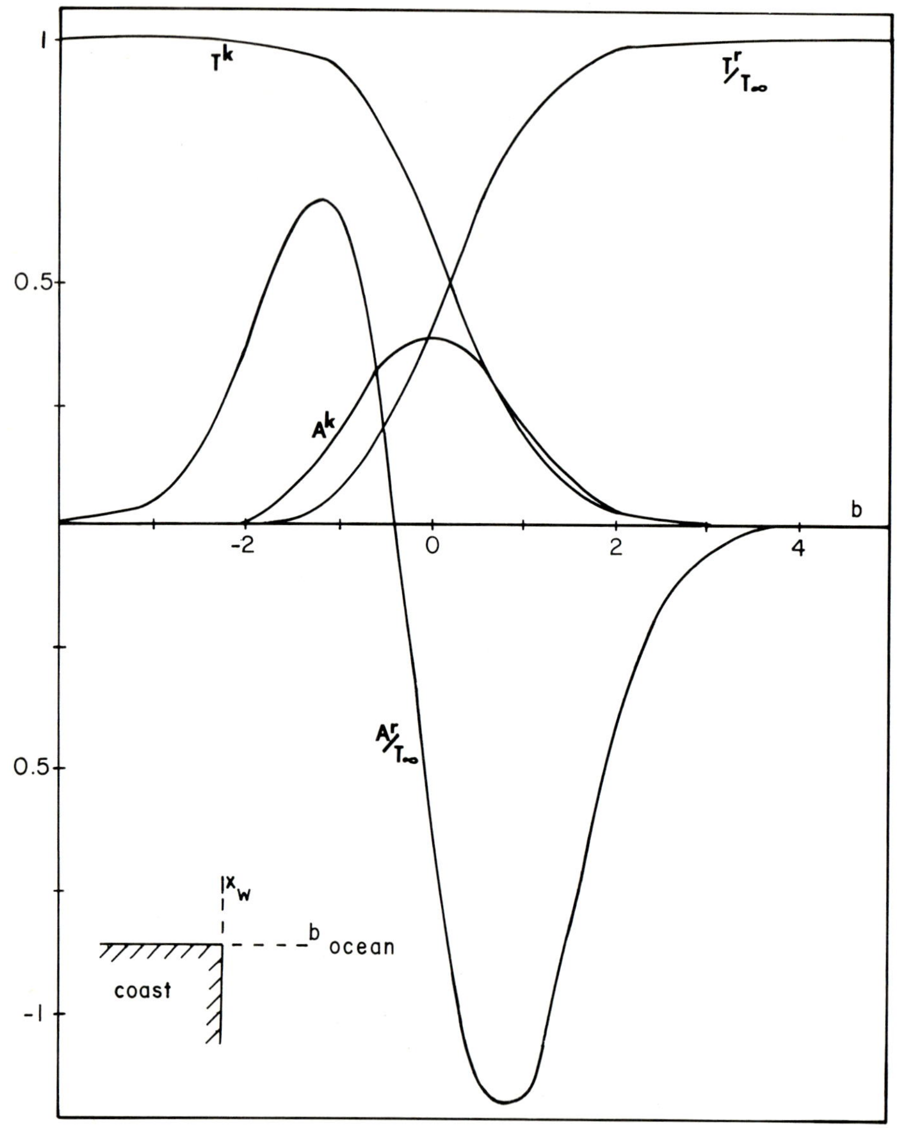

Fig.3: Transmission coefficient (T) and amplitude of the zonal current (A) at the latitude of the western corner as function of the latitude of the zonal coast. Subscript (k) applies to an incident Kelvin wave and subscript (r) for an n=1 incident Rossby wave. T_∞ is the value of T^k as b → +∞

current at the corner is maximum when the corner is situated at the equator and becomes negligible when the position of the corner is of the order of two radii of deformation : for $b \lesssim -2$, the equatorial Kelvin wave does not feel the corner and propagates further far without significant change in amplitude ($T^K \approx 1$). As b increases northward from the equator, the amplitude of the boundary current and the transmission coefficient both decrease very rapidly at first becoming negligible by $b \approx 2$. By this latitude, there is almost no equatorial Kelvin wave any more.

The results in figure 3 are for the $n = 1$ Rossby wave and have been normalized to the value of T^R at infinity, i.e. the value of T^R when the boundary extends from $-\infty$ to $+\infty$. As b goes to $+\infty$, $T^R_\infty \longrightarrow 2^{-1/2}\,\pi^{-1/4}\int_{-\infty}^{+\infty} \hat{u}dy$: as the barrier becomes infinite all of the incident mass flux is returned in the reflected Kelvin wave (see Cane and Sarachick, 1977). The north-south boundary current redistributes meridionnally the incoming mass flux to make this possible ; it has no net mass flux. In contrast to the eastern boundary case, where the amplitude of the reflected Kelvin wave depends only on the value of height at the corner, the western boundary case is quite complicated. As noted above, as the barrier becomes infinite, the answer depends on the total zonal mass flux. For a partial barrier, equation (25) shows that the response depends on the structure of the incident motions as well as on the zonal mass flux incident on the barrier. Note that the first term in the numerator is small for $b \gg 1$ since ψ_o is then small. It is also small for $b \ll -1$ because the orthogonality of the Kelvin and Rossby modes then implies that the integral is small.

To know how well the nth mode is transmitted past the corner, we also compute the transmission factor γ that is the ratio of the energy in the nth mode west of the corner to the incident energy. $(T^K)^2$ is the comparable measure for the Kelvin wave case. Projecting the nth Rossby mode on the solution west of the corner, we obtain :

$$\gamma = \frac{\int_b^{\infty}(\hat{u}^2 + \hat{h}^2)dy + B\hat{u}(b) + T^R\int_{+b}^{+\infty}\psi_o(\hat{u} + \hat{h})}{\int_{-\infty}^{+\infty}(\hat{u}^2 + \hat{h}^2)dy}$$

Figure 4 shows the transmission coefficient for the first 6 Rossby modes. For the nth mode, the value approaches zero as $b \sim (2n + 1)^{1/2}$ which is the turning latitude. At that latitude,

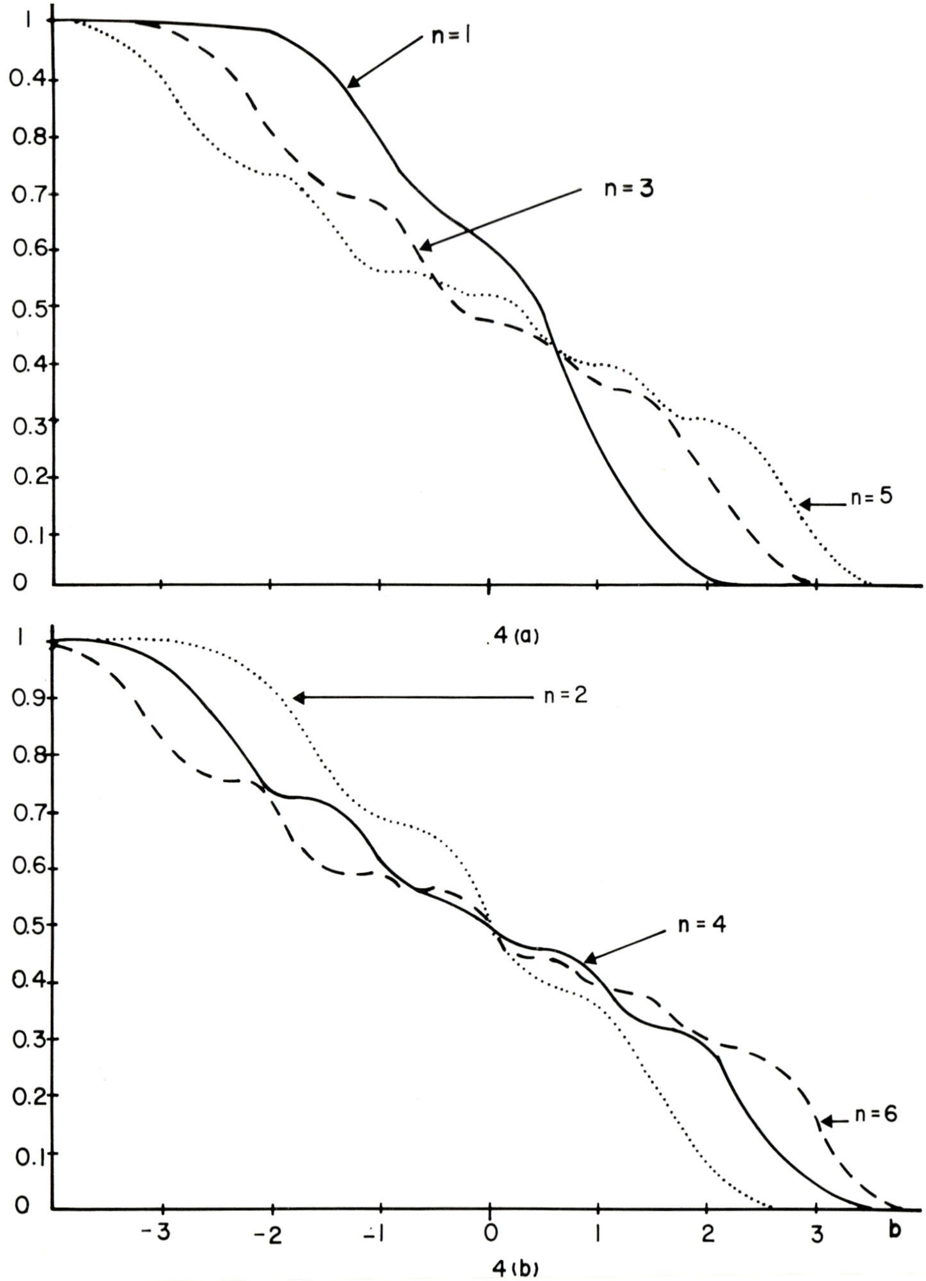

Fig. 4(a)_ 4(b): The transmission factor γ for the first 6 Rosby waves in the case of western partial boundary in function of the latitude b of the corner. (a) symetric modes, n odd
(b) antisymetric modes, n even

the barrier is extended enough to block the incident motion. The bumps in the curves are due to the oscillatory nature of the u component.

GENERALIZATION

The configuration of oceanic coastline (e.g. the South America coastline) cannot be drawn schematically with only one step but, more realistically, with several steps. The case of low frequency waves impinging on a north-south zigzag coast can be deduced in a straightforward way from the results of the previous sections.

If a partial westward boundary is constituted of N steps of north-south extension $a_{i+1} - a_i$, $i = 1, N$, we get for an incoming Kelvin wave the expression :

$$T_i^K = \frac{2^i \prod_{j=1}^{i} \int_{a_j}^{\infty} \psi_o^2(y)\,dy}{\prod_{j=1}^{i} \left\{ 2\int_{a_j}^{\infty} \psi_o^2(y)\,dy + \psi_o(a_j)\int_{a_{j+1}}^{a_j} \psi_o(y)\,dy \right\}}$$

with T_i^K the amplitude of the transmitted Kelvin wave and i the number of steps, $i = 1, N$.

The transport in the west-east boundary current is given by :

$$A_i = T_i^K \int_{a_{i+1}}^{a_i} \psi_o(y)\,dy$$

In the case of an incoming Rossby wave, we get for the amplitude of the reflected Kelvin wave :

$$T_i^R = -\frac{\int_{b_i}^{\infty} (\hat{u} + \hat{h})\psi_o(y)\,dy + (\sum_{j=1}^{i-1} T_{j-1})(2\int_{a_i}^{\infty} \psi_o^2(y)\,dy + \psi_o(a_i)\int_{a_{i-1}}^{a_i} \psi_o(y)\,dy)}{2\int_{a_i}^{\infty} \psi_o^2(y)\,dy + \psi_o(a_i)\int_{a_{i-1}}^{a_i} \psi_o(y)\,dy}$$

$$+ \frac{\psi_o(a_i)B_{i-1} + \psi_o(a_i)\int_{a_{i-1}}^{a_i} \hat{u}\,dy}{2\int_{a_i}^{\infty} \psi_o^2(y)\,dy + \psi_o(a_i)\int_{a_{i-1}}^{a_i} \psi_o(y)\,dy}$$

(in this case, the subscript i counts the number of steps from east to west ; see figure 5).

and the east-west transport at the corner i :

$$B_i = B_{i-1} + \int_{a_{i-1}}^{a_i} \hat{u}dy + (\sum_{j=1}^{i} T_j) \int_{a_{i-1}}^{a_i} \psi_o(y)dy$$

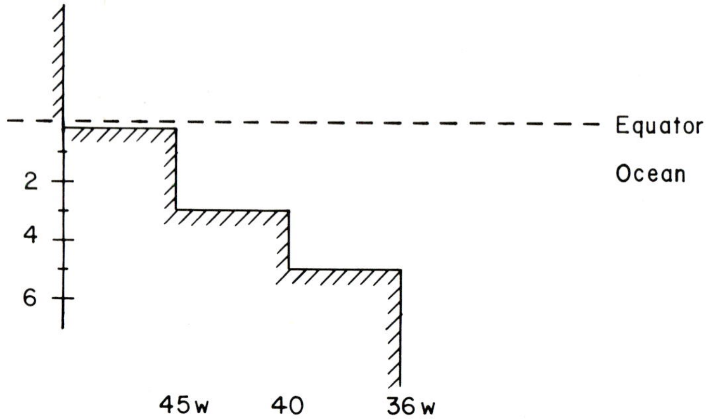

Fig.5 Step structure used to schematize the Brasil coast.

Similar arguments have been used to compute the solution of low frequency waves impinging on a thin island or chain of islands (see Cane and du Penhoat, 1982). We have found a constant depth variation east of the island, two boundary currents on the eastern side of the islands. The solution for an incoming Kelvin wave is sketched in figure 6.

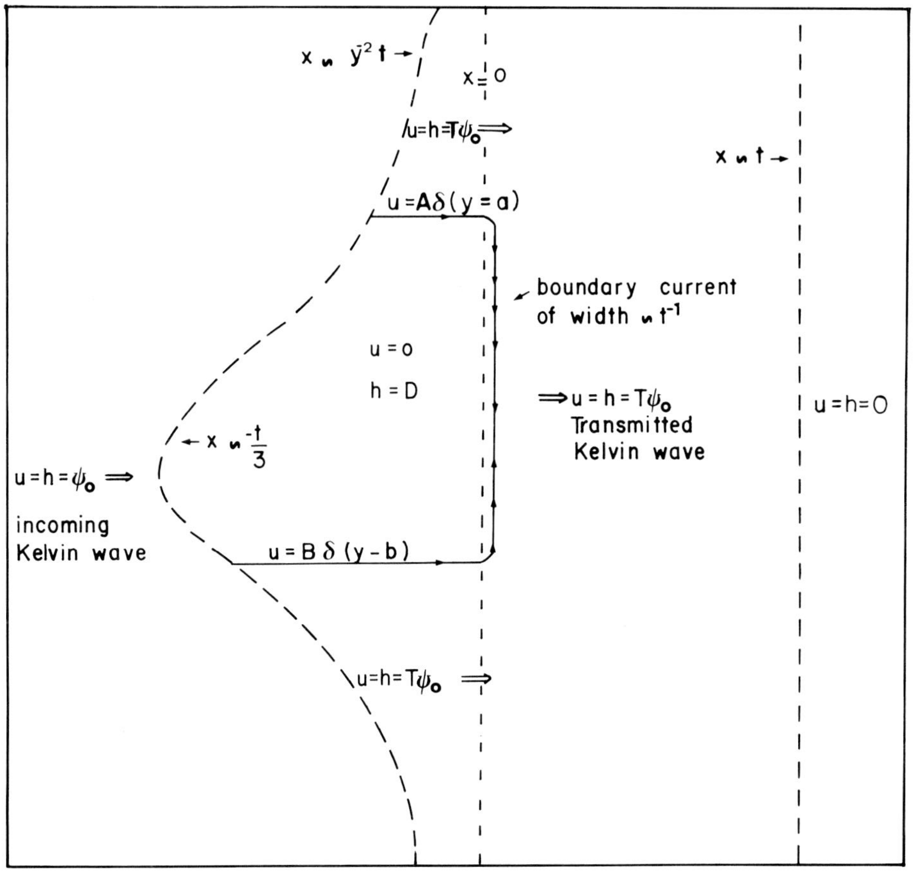

Fig. 6 Sketch of the solution for an incoming Kelvin wave on a thin island. For $t>>1$, the asymptotic solution holds out to a longitude $x \sim t$, the farthest eastward the Kelvin wave travels in time. To the west, it applies behind a front with a form $x \sim -1/3t$ near the equator and $x \sim y^{-2}t$ for high latitude.

If an island is small (i.e., $a - b << 1$ radius of deformation, a and b latitude of the north and south tip of the island), then low frequency waves pass it almost undisturbed with the mass flux incident on the upstream side flowing around it about equally to the north and south and continuing on downstream in the lee of the island. If the island is large ($|a|$, $|b| \gtrsim 2$ radii of deformation), then the principal response is organized as it would be if the island barrier were meridionally infinite.

DISCUSSION

We now apply our results to the real ocean. In the Atlantic the effect of the coast of the Gulf of Guinea (roughly at 5° latitude north) is very small for either low frequency Rossby waves or the Kelvin wave. (see fig. 2 for n = 1, a ≃ 2 in non dimensional units, and for higher baroclinic modes a ⩾ 2). A constant height is set up on the north-south coast and if we add friction, there is a boundary current along the east-west coast, which widens westward.

In the case of the South America coast, in the Atlantic ocean, represented schematically by 3 steps (fig. 5) : the first corner between 0.20 S and 3 S affects the long equatorial Kelvin wave with an intense westward current at this point (see table 1). For the two other steps, the amplitude of the transmitted Kelvin wave decreases only slightly and results will not change by adding more steps : low frequency waves will not feel a more detailed coast very much. For the n = 1 Rossby mode, only approximately 63 % of the energy in the first horizontal mode is transmitted through the 3 steps.

TABLE 1

Parameter values for the coast of South America (a) for an incident Kelvin wave, (b) for n = 1 Rossby wave. The equatorial length scale R_1 et R_2 are from Cane and Moore (1981).

(a)

	y	T^K	B^K
first baroclinic mode	-0.113	0.727	0.412
Equatorial length scale	-1.022	0.693	0.142
R_1 = 326 km	-1.704	0.687	0.057
Second baroclinic mode	-0.149	0.704	0.465
Equatorial length scale	-1.344	0.688	0.099
R_2 = 248 km	-2.240	0.687	0.016

(b)

n = 1 Rossby wave	y	T^R	B^R	γ
First baroclinic mode	-1.704	0.001	0.09	0.95
R_1 = 326 km	-1.022	0.01	0.111	0.79
	-0.1136	0.06	-0.078	0.63
Second baroclinic mode	-2.24	0.0001	0.043	0.99
R_2 = 248 km	-1.344	0.004	0.114	0.88
	-0.149	0.061	-0.064	0.63

We have computed the different coefficients for the coast of New Guinea (Pacific ocean) schematically represented with 6 steps (table 2). For an incident Kelvin wave, the amplitude of the transmitted wave is greater than 0.73 for the first and second baroclinic modes and more steps will not change it significantly. For the n = 1 Rossby wave, 64 % of the energy is transmitted in this mode pass the last step. For higher horizontal mode (not shown), this value decreases.

TABLE 2

Parameter values for the coast of New Guinea (a) for an incident Kelvin wave, (b) for n = 1 Rossby wave.

(a)

	y	T^K	B^K
First baroclinic mode	-0.155	0.877	0.193
	-0.465	0.780	0.214
R_1 = 357 km	-0.933	0.754	0.097
	-1.245	0.735	0.104
	-1.867	0.733	0.041
	-3.112	0.733	0.001

(b)

n = 1 Rossby wave	y	T^R	B^R	γ
First baroclinic mode	-3.112	∿0	0.005	0.999
	-1.867	$7x10^{-4}$	0.076	0.971
R_1 = 357 km	-1.245	0.006	0.117	0.860
	-0.933	0.009	0.104	0.773
	-0.465	0.027	0.018	0.671
	-0.155	0.031	0.075	0.643

New Guinea is situated at the connection between the Indian and Pacific oceans and the flow through this connection could turn out to have considerable significance in the exhange between the two oceans and its possible relations to El Niño. A crude calculation, using Cane and du Penhoat's (1982) results and a rough schematization of the area shows that for a n = 1 incident Rossby wave, only less than 9 % of the incoming energy is transmitted pass Borneo island and that Java-Sumatra islands act almost as an infinite barrier for low frequency equatorial Rossby waves.

Calculations carried out for equatorial islands show that no island, in the world ocean, influences low frequency equatorial waves very much, because their north-south extension is too small compared to the equatorial radius of deformation (see Cane and du Penhoat, 1982). West of the island, the sea level signal is slightly enhanced (and constant at the coast) and the thermocline is thickened (assuming it to be described by the second baroclinic mode). East of the island, the thickness decreases toward the equator due to the presence of short Rossby waves. Even for the Galapagos archipelago, our theory predicts a transmission coefficient over 0.98 for the first and second baroclinic mode Kelvin waves, so that the islands do not affect the propagation of these waves. This result agrees with Yoon's (1981) numerical calculations. An incident Rossby wave is transmitted without major loss of energy and there is only a weak reflected Kelvin wave. In fact, its propagation will be more severely influenced by the mean current system (Philander, 1978). Calculations carried out for the Maldives islands show that they do not act as a significant barrier, because, although they have a greater latitudinal extension, the islands close to the equator are small.

We conclude that islands in the real equatorial oceans will not affect the propagation of low frequency waves significantly and that perturbations will occur only in their vicinity. The irregularities in the Brazilian coast have a noticeable effect but little is gained representing them by more than a single step. The Gulf of Guinea has little effect on incoming Kelvin waves (though the response along its coast is of course of interest for its own sake) and the complex, ragged boundary in the western Pacific is an effective boundary for such low frequency waves.

REFERENCES

Abramowitz and Stegun, 1965. Handbook of mathematical functions. Dover, New York, 1046 pp.

Anderson, D.L.T. and Rowlands, P.B., 1976. The role of inertia-gravity and planetary waves in the response of a tropical ocean to the incidence of an equatorial Kelvin wave on a meridional boundary. J. Mar. Res., 34: 295-312.

Bye, J.A.T. and Gordon, A.H., 1982. Speculated cause of interhemispheric oscillation. Nature, 296: 52-54.

Cane, M.A. and Moore, D., 1981. A note on low frequency equatorial basin modes. J. Phys. Oceanogr., 11: 1578-1584.

Cane, M.A. and du Penhoat, Y., 1982. The effects of islands on low frequency equatorial motions. J. Mar. Res. (in press).

Cane, M.A. and Sarachik, E.S., 1976. Forced baroclinic ocean motions : I. The linear equatorial unbounded case. J. Mar. Res., 34: 629-665.

Cane, M.A. and Sarachik, E.S., 1977. Forced baroclinic ocean motions : II. The linear equatorial bounded case. J. Mar. Res., 35: 395-432.

Cane, M.A. and Sarachik, E.S., 1979. Forced baroclinic ocean motions : III. The linear equatorial basin case. J. Mar. Res., 37: 366-398.

Cane, M.A. and Sarachik, E.S., 1981. The response of a linear baroclinic equatorial ocean to periodic forcing. J. Mar. Res., 39: 652-693.

Gill, A.E., 1975. Model of equatorial currents. Symposium on numerical models of ocean circulation. Nat. Acad. Sci., Durham, N.H., U.S.A., Oct. 17-20-72.

Patton, R.J., 1981. A numerical model of equatorial waves with application to the seasonal upwelling in the Gulf of Guinea. MS thesis, M.I.T., 120 pp.

Pedlosky, J., 1965. A note on the western intensification of the oceanic circulation. J. Mar. Res., 23: 207-210.

Philander, S.G.H., 1979. Equatorial waves in the presence of the equatorial undercurrent. J. Phys. Oceanogr., 9: 254-262.

Philander, S.G.H. and Pacanowski, 1980. The generation of equatorial currents. J. Geophys. Res., 85 (C2): 1123-1136.

Ripa, P. and Hayes, S.P., 1981. Evidence of equatorial trapped waves at the Galapagos Islands. J. Geophys. Res., 86: 6509-6516.

Rowlands, P.G., 1981. The flow of equatorial Kelvin waves and the equatorial undercurrent around an island. J. Mar. Res. in press.

Wirtky, K., 1961. Physical oceanography of the Southeast Asian waters. Naga report, volume 2 - The University of California Scriffs Institutions of oceanography, La Jolla, California.
Yoon, 1981. Effects of islands on equatorial waves. J. Geophys. Res., 86: 10913-10920.

COASTAL EFFECTS ON UPWELLING.

Pascale DELECLUSE

Laboratoire d'Océanographie physique/MNHN, Paris, France

Abstract

Steady wind parallel to the coast generates upwelling and the influence of the spatial structure of the wind or the coastal indentattions must be taken into account in the determination of the lifting of the thermocline. A shallow water model is used to display the response of the ocean around a cape. The response to steady wind forcing is described in (2). A spatially uniform wind, varying with time according to the real wind records in Dakar, is then used to force the ocean and the response is examined off-shore and along the coast for different frequency bands (3). Ekman pumping dominates offshore for intermediate range frequencies (20 - 3 days) and the coastal dynamics is driven by Kelvin waves except for very high and very low frequencies. The relevance of these results for Dakar is discussed in (4).

1. INTRODUCTION

Recently satellite pictures have shown synoptic maps of the SST (Sea Surface Temperature) in the area of Dakar and upwelling was clearly visible north and south of the cape (Citeau, Guillot and Roy, personal communication). The purpose of this paper is to demonstrate that the presence of a cape may affect the local dynamics. The Dakar area was chosen because the coast is mainly north-south with a small cape at Dakar. During a long part of the year (except the summer months), Dakar is situated in the trade winds and upwelling appears. The area is well documented by the work done by ORSTOM (Rossignol, 1973; Merle, 1973; Portolano, 1981; ...). Dakar is located at 15° N, very close to the equatorial band and Rossby waves can evacuate efficiently the energy away from the eastern boundary.

In part 2, a review of analytical results for upwelling is given and the academic problem of the upwelling driven by an uniform wind stress in presence of a cape is solved numerically. Then, in part 3, the wind recorded at the Dakar airport is used to drive the model during one year. A description of the oceanic response is given, considering the different frequencies of the wind forcing. Finally part 4 contains the discussion of the results and tries to estimate their relevance to the real oceanic response.

2. UPWELLING IN PRESENCE OF A CAPE FOR AN UNIFORM WIND FORCING

2.1 Analytical results

Despite the fact that the upwelling event is a 3-dimensional phenomenon (Halpern, 1976) , a reasonably good description of the physical processes can be obtained with a two-layer model. Many authors have studied analytically and numerically the coastal response of a two-layer ocean to simple wind-stress patterns (Allen, 1976; Charney, 1955; Csanady, 1975; Crépon and Richez, 1982; ...) and have shown that the contribution of the baroclinic terms to the thermocline elevation was essential.

In order to describe the oceanic circulation around a cape, it is useful to give a brief review of the analytical results obtained by Crépon and Richez (1982). Let h denote the interface elevation from its initial position; $\mathbf{T}$ (T_x, T_y), the applied wind-stress; f, the Coriolis parameter (constant for this analysis); $\Delta\rho$, the density difference between the surface layer and the deep layer and ρ the reference density; the internal radius of deformation is

$$R = \frac{c}{f}$$

where c is the propagation speed of the internal mode :

$$c = \left(\frac{\Delta\rho}{\rho}\, g\, H\right)^{1/2}$$

The surface layer thickness H is considered negligible compared to the deep layer thickness. With the Boussinesq and the hydrostatic assumptions, the linearized equations that govern the dynamics are the shallow water equations :

$$\frac{\partial u}{\partial t} - fv = -g\,\frac{\Delta\rho}{\rho}\,\frac{\partial h}{\partial x} + \frac{T_x}{\rho H}$$

$$\frac{\partial v}{\partial t} + fu = -g\,\frac{\Delta\rho}{\rho}\,\frac{\partial h}{\partial y} + \frac{T_y}{\rho H}$$

$$\frac{\partial h}{\partial t} + H\left(\frac{\partial u}{\partial x} + \frac{\partial v}{\partial y}\right) = 0$$

A propagation equation can be deduced for h :

$$\frac{\partial}{\partial t}\left[c^2\left(\frac{\partial^2}{\partial x^2} + \frac{\partial^2}{\partial y^2}\right)h - \left(\frac{\partial^2}{\partial t^2} + f^2\right)h\right] = \frac{1}{\rho}\left[\frac{\partial}{\partial t}\left(\frac{\partial T_x}{\partial x} + \frac{\partial T_y}{\partial y}\right) + f\left(\frac{\partial T_y}{\partial x} - \frac{\partial T_x}{\partial y}\right)\right]$$

The equation for h is solved by Fourier transform in space and Laplace transform in time and asymptotic solutions are calculated for specific wind forcing.

Let us consider the case of an uniform wind blowing over an ocean limited by two solid boundaries at right angles at the eastern edge and at the southern edge. Two different cases must be considered. First, let us assume that the wind blows southward (fig. 1) [this case will be called Bay A].

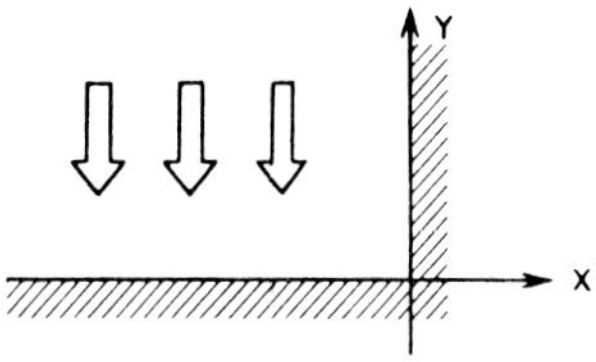

Fig. 1. Bay A.

Away from the corner, along the eastern boundary (x = 0) the wind creates a boundary coastal jet growing linearly with time. Upwelling compensates for the Ekman drift away from the coast. Along the southern boundary (y = 0), a small downwelling due to the wind normal to the coast, bounded with time, appears. A Kelvin front is created in the corner for the current to match the boundary conditions (no flux through boundaries). This front propagates poleward along the eastern boundary and compensates exactly the acceleration due to the wind stress. After its passage, the flow is steady and the upwelling is stabilized.

The analytical solution along the eastern coast that summarizes this mechanism is :

$$h = \frac{T_y}{\rho c f} \exp\left(\frac{x}{R}\right) \{ \underbrace{[ft - (ft + \frac{y}{R})\, \mathcal{H}(ct - y)]}_{①} - \underbrace{(1 - \frac{1}{1 + |ct - y| \frac{f}{c}})\, \mathcal{H}(ct - y)}_{②} \}$$

where $\mathcal{H}$ is the Heaviside function.

The term ① represents the accelerating flow and the Kelvin front. The term ② shows the effect of the small downwelling and its propagation northward as a Kelvin front. The long term aspect of the solution is a constant pressure gradient along the eastern coast in equilibrium with the wind forcing.

Let us now consider the same wind but assume a different geometry. The ocean is limited by two solid boundaries at its northern edge and its eastern edge (Bay B, fig. 2).

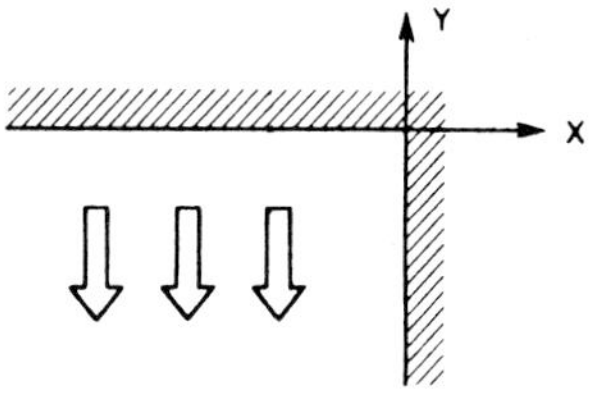

Fig. 2. Bay B.

After the wind's onset, there is an upwelling along the eastern coast and no mechanism will prevent it from growing. A small bounded upwelling appears along the northern coast (due to the effect of wind forcing normal to the coastline) but this signal is quickly masked by the arrival of a Kelvin front, generated in the corner, that propagates the flow acceleration westward in order to feed the increasing eastern transport. A pressure gradient is slowly established along the coast (in balance with the coastal current). This system is continuously growing with time. The mathematical expression of the solution along the northern coast is the following :

$$h = \frac{T_y}{\rho c f} \exp\left(\frac{y}{R}\right) \left[\underbrace{\left(ft + \frac{x}{R}\right) \mathcal{H}(ct + x)}_{①} + \underbrace{\left(1 - \frac{1}{1 + |ct + x| \frac{f}{c}}\right) \mathcal{H}(ct + x)}_{②} \right]$$

The term ① represents the Kelvin front that accelerates the flow and the term ② is the small upwelling due to the direct wind effect along this coast. Symmetrical solutions can be deduced for different coast directions and other wind directions. The same approach is now used to observe the effect of capes. As previously, there are two different classes.

For the case Cape A (fig. 3) which is similar to Bay A, a small bounded upwelling appears along the southern coast and the linearly growing upwelling, along the eastern coast, is stabilized by a Kelvin front generated in the corner.

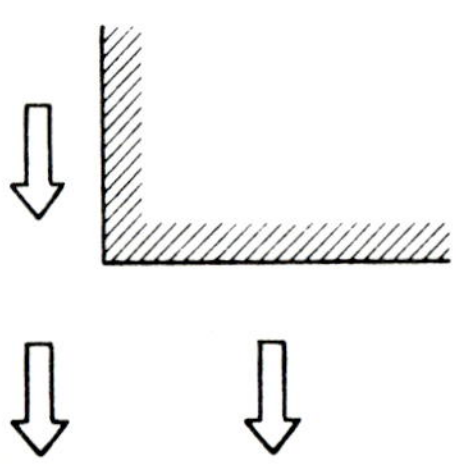

fig. 3. Cape A.

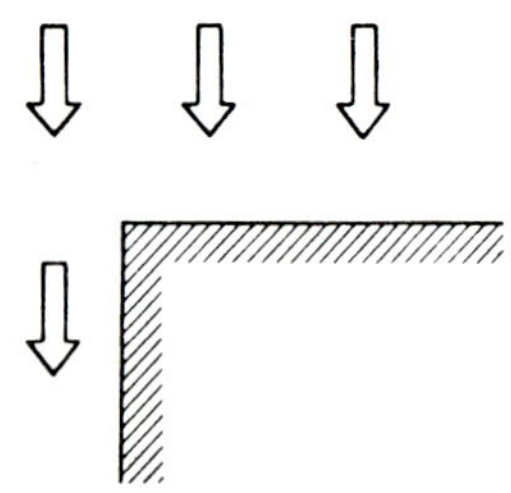

fig. 4. Cape B.

For the case Cape B (fig. 4), similar to Bay B, there is continuously increasing upwelling along the eastern coast and the flux is deviated as a Kelvin front around the corner. This front cancels the weak downwelling and creates an upwelling later along the northern coast where a pressure gradient will balance the coastal current.

All the previous results can be summarized to describe the response of the ocean around a cape (fig. 5).

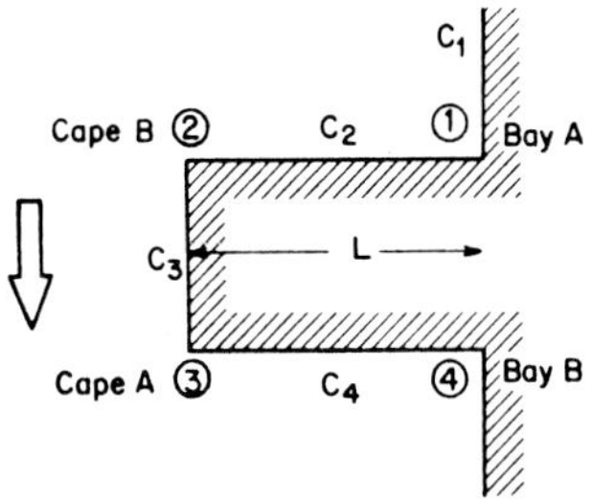

Fig. 5.

The position ① is an example of Bay A. A Kelvin front is generated in ① and creates a pressure gradient along the coast C_1, that balances the wind-stress forcing and arrests current acceleration. In ②, results of Cape B can be applied. A Kelvin front will raise the thermocline along C_2 and restore the acceleration along C_1. The cape ③ will create a stabilizing Kelvin front but the bay ④ will finally re-establish the acceleration all along the coastline. This is a brief summary of the mechanism; secondary effects appear (bounded motions due to the wind normal to the coast, coupling with inertial motions) that are discussed in the paper of Biju-Duval and Chartier (1982).

The important result is that, after a time long enough for a Kelvin front to reach the coast C_1, coming from ④, the upwelling exists everywhere but is slightly weaker on the northern coast because it has been stopped twice during the time it takes for a Kelvin wave to propagate from ② to ①.

2.2 The numerical approach

Finite difference schemes (in space and time) were used to write numerically the generalized shallow water equations on a sphere, with non-linear terms. The code, described in the Appendix, can serve to solve many different problems. The present one is a spin-up problem and only a few days of simulation are required to give the solution patterns. For such a small time integration, currents do not reach very high speed and it is not necessary to keep the non-linear terms. The spherical terms and especially the beta effect will not be included in the computation for the same reason.

A southward wind-stress of $0.5\ \text{dyn/cm}^2$ is applied uniformly over the ocean, starting from rest. The wind is slowly increasing from zero to its final value with a cosine taper of 100 timesteps. The interface is presented during four following days in fig. 6. It is rapidly rising along the coastlines and the Kelvin fronts, generated in each corner at $t = 0$, are clearly visible. They all propagate at the same speed :

$$c = \left(\frac{\Delta\rho}{\rho}\, g\, H\right)^{1/2} = 1\ \text{m/s}$$

with

$\frac{\Delta\rho}{\rho} = 2 \cdot 10^{-3}$ and $H = 50$ m

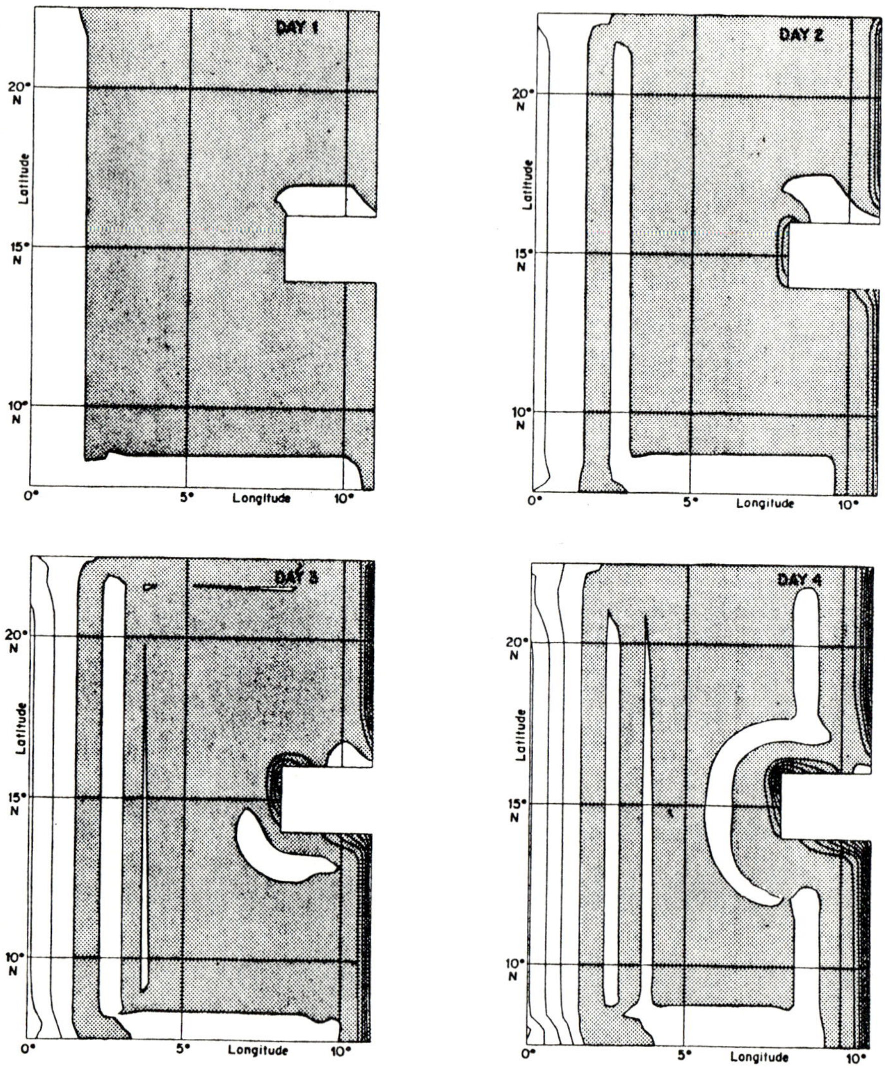

Fig. 6. Interface depth during four following days. The contour interval is 150 m and stippled area indicates upwelling.

Even the secondary downwelling effect along the coast C_2 can be exhibited. The motions are confined along the coast within the radius of deformation :

$$R = \frac{c}{f} = 26 \text{ km}$$

The exponential profile of the velocity is noticeable in fig. 7.

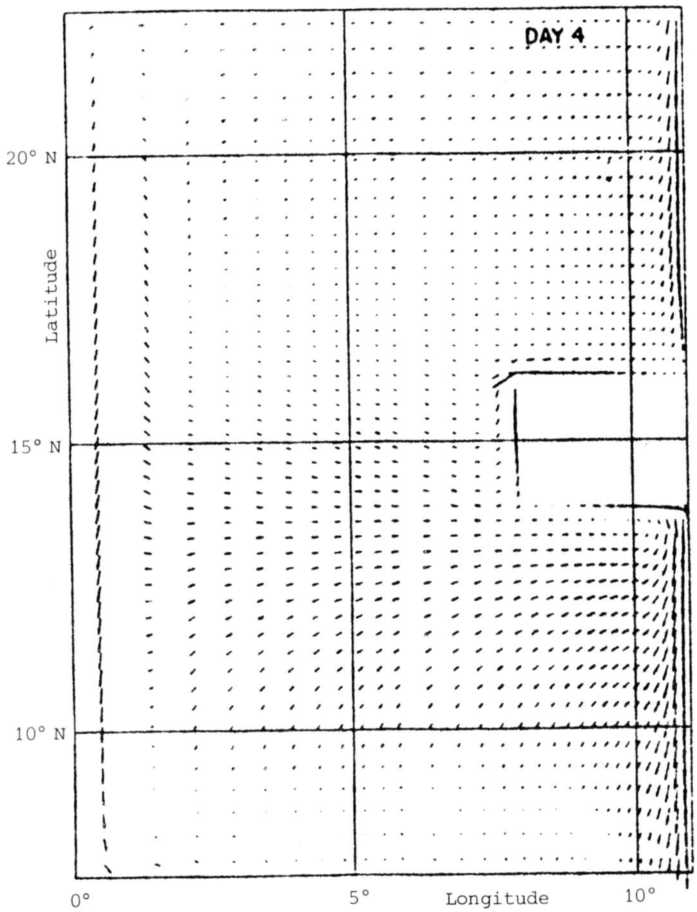

Fig. 7. Velocity vectors for the same experiment after four days.

The best way to outline the front propagation is to present the contours of h as functions of space and time along the eastern coast. This is done in fig. 8. Lines parallel to the space axis mean growing upwelling (when they are equidistant, the growth is linear) and lines parallel to the time axis represent stable situation (constant values). The four Kelvin fronts are represented with dashed lines in the northern part.

Other configurations of capes have been tested with the finite element model written by Lien Hua et al. This model allows a good description of irregular coasts. Trapezoidal, triangular and even rounded capes were put in the model with about the same dimensions and the results were the following :

- for the rectangualr cape, both numerical models (the finite difference model and the finite element model) give the same results in perfect agreement with the analytical approach;

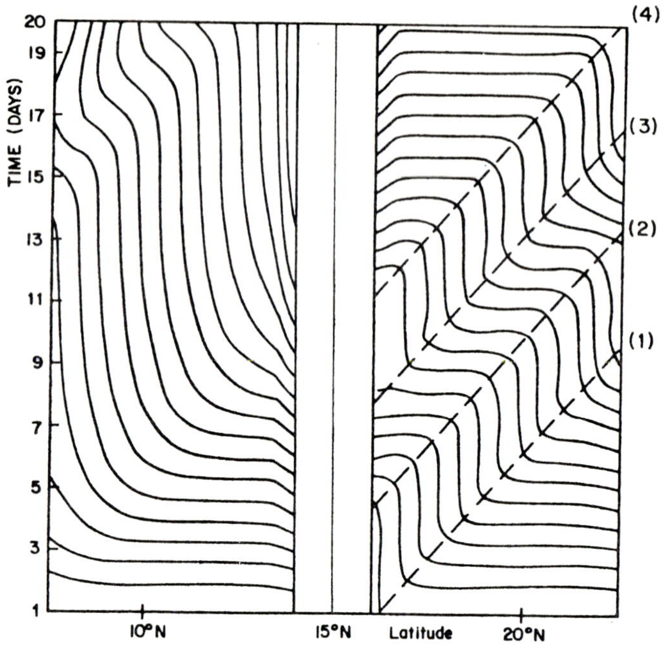

Fig. 8. Space-time diagram of the interface depth along the eastern coast. The contour interval is 2.50 m and the central strip represents the cape. Dashed lines depict the Kelvin front propagation.

• for other shapes, it is more and more difficult to distinguish the different fronts and where exactly they are generated; but however the shape, the final result is the same : the upwelling exists everywhere after a while but is weaker along the northern part because of the delay due to Kelvin front propagation along the faces of the cape normal to the wind. (Compare fig. 6 and fig. 9 where the solution for a triangular cape with the finite element is presented.)

In the next section, simulations will be carried on with a wind-stress uniform in space but having a real time variability. The finite difference model, with a rectangular cape, will be used for the following reason :

• it is much cheaper than the finite element model;

• it has been shown that the events along the northern and southern coasts were similar, however the shape of the cape. (The Kelvin fronts generated in each corner, cancel each other two by two.);

• the finite difference model will keep the spherical term because, for a long run, Rossby waves are required to transport the energy westward.

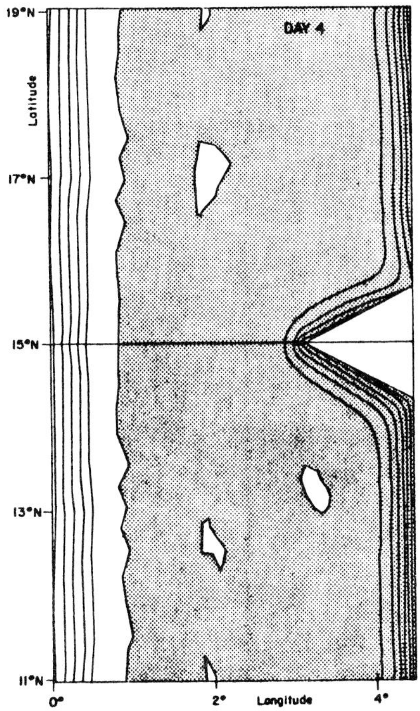

Fig. 9. Interface depth at day 4 before a triangular cape. The contour interval is 2.00 m .

3. SIMULATION WITH A TIME VARYING WIND

3.1 Description of the wind field

The wind used in the simulation has been recorded by the ASECNA (Agence pour la SECurité de la Navigation Aérienne) at the Dakar airport. Direction and amplitude are estimated each three hours and the series were nearly complete. The wind-stress acting on the ocean was estimated with the bulk formula :

$$T_x = - \rho_a C_D U^2 \sin D$$

$$T_y = - \rho_a C_D U^2 \cos D$$

where ρ_a is the air density, C_D the drag coefficient, U the wind amplitude, D the wind direction (given with the meteorological convention).

Careful analyses of the wind in the Dakar area have been done for space and time variability by Portolano (1981). The year chosen (1973) has the main characteristics of the mean climatic year and was well sampled for the wind (2 % of missing values).

Compared to the other coastal stations, Dakar seems to be the most representative for the wind over the ocean. No other measurement (neither on a pier nor a lighthouse ...) was available. Synoptic maps did not exist for long periods. To obtain a spatial coverage of the area, satellite methods combined with assimilation technics in GCM (Global Circulation Model) are required. Such winds are available for the FGGE year 1979 and will be soon introduced in the model. Nevertheless, their relevance for coastal dynamics is questionable.

The wind pattern in Dakar is closely related to the position of the ITCZ (Inter Tropical Convergence Zone) which moves meridionally back and forth during the year. Two seasons appear :

• the boreal winter when the ITCZ is far south. Dakar is situated in the trade winds (northerly to north-easterly winds). Their amplitude is important (5 m/s) and they are relatively regular. These winds can easily induce coastal upwelling;

• the boreal summer when the ITCZ is north of Dakar. Its position is not very well defined. The wind is weak and variable, mainly southerly, south-easterly.

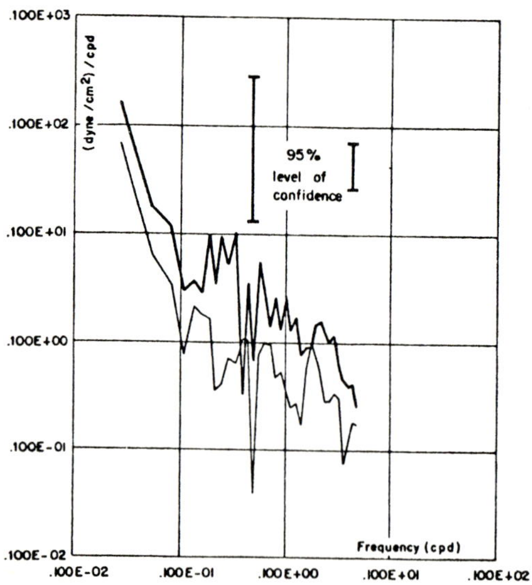

Fig. 10. Spectra of the wind stress component. The thin line represents the zonal component and the dark line represents the meridional component.

In both seasons, a strong diurnal cycle is superimposed to the wind. In fig. 10, spectral power for T_x and T_y are presented (for periods longer than two days). The most important point to notice is the energy excess of T_y on T_x at all frequencies. There is no well defined peak in both spectra, just a slightly higher amount of energy for T_y between 40 and 50 days.

3.2 The oceanic response

The ocean model described in part 2 was run linearly on a sphere during a year (with 72 timesteps per day) with a continuously varying wind stress (the wind input was modified each three hours). The eastern coast of the model had the same orientation than the Senegal coast (mainly north-south) and the rectangular cape emphasized the cape of Dakar. Mean kinetic energy (KE) and mean available potential energy for the ocean (APE) are rising until the beginning of May where after 150 days of run (the run starts on the 1st of January 1973) a peak exists (for APE and KE). There seems to be a delay (about 25 days) between the maximum of the energy response in the ocean and the wind input (fig. 11).

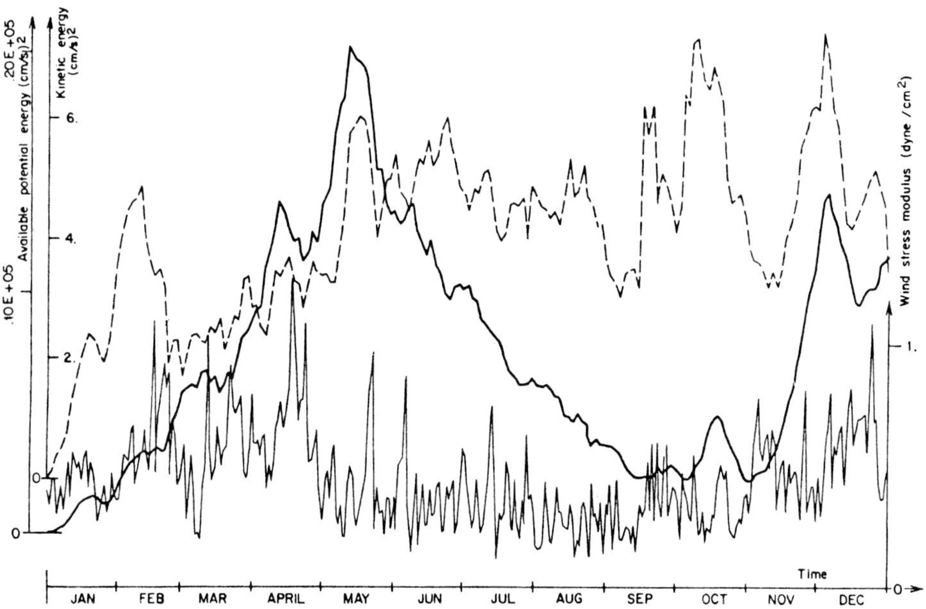

Fig. 11. Wind stress modulus (thin line), mean kinetic energy (dashed line) and mean available potential energy (dark line) in function of time during the one-year run.

To justify this point, the following argument is proposed : the wind drives currents in the ocean but two kinds of ocean responses exist. The first one is the local response (for instance, coastal currents will reach their maxima simultaneously with the local wind); the second one is the remote response due to waves (coastal Kelvin waves are excited and propagate the energy poleward; Rossby waves carry the energy westward, away from the eastern boundary). The presence of these waves may introduce a delay in the oceanic response. For KE, another peak is reached in September and then in November. The APE is decaying from April to September and rises as the trades blow again in the fall. The annual cycle is much more noteworthy on the APE than on the KE.

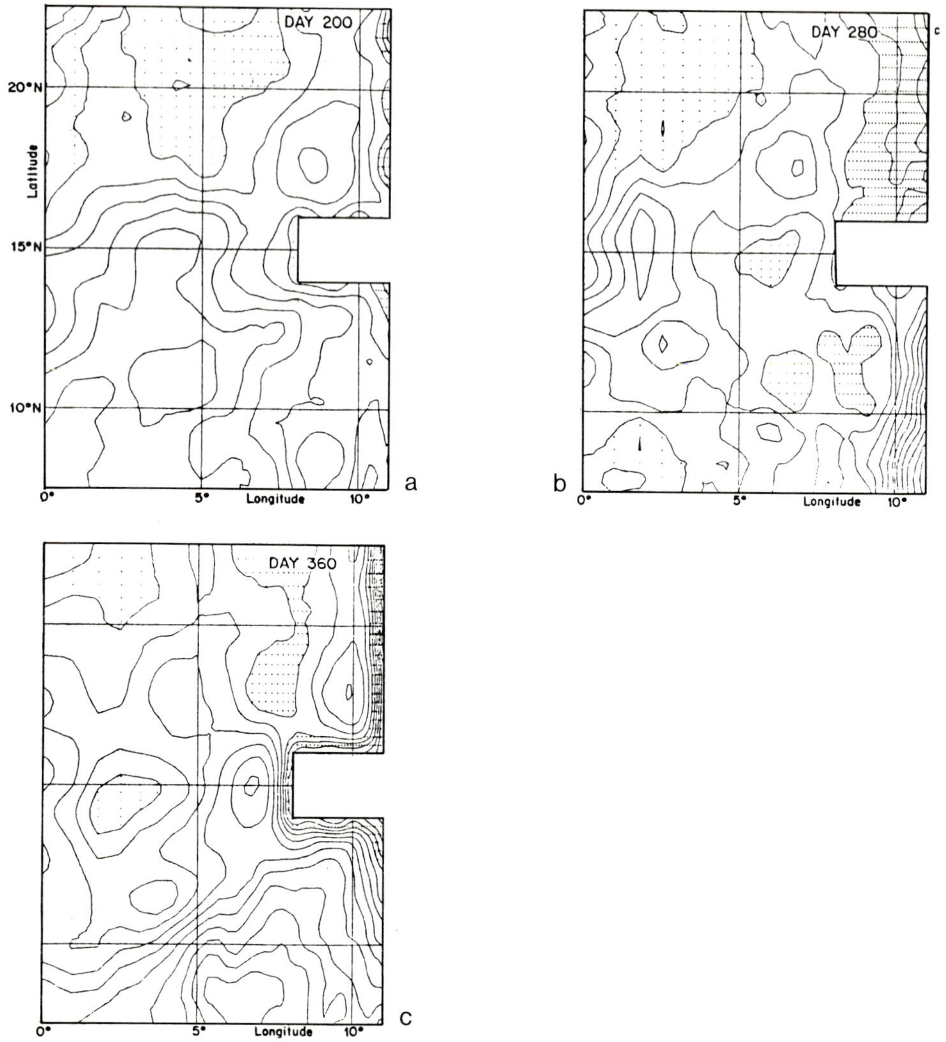

Fig. 12. Interface depth (the contour interval is 2.00 m) for three different days. Dashed areas are upwelling areas.

Figure 12 presents the thermocline depth at three different times (200 days, 280 days and 360 days). The spatial variability is very high but Rossby waves propagation is noticeable across the domain (inclined fronts of equal h value appear because the wave speed depends on the latitude and is larger along the southern boundary). Two space-time diagrams have been selected in order to outline the waves importance in this model (fig. 13). The first one presents h along the eastern coast as a function of latitude and time and shows the succession of coastal Kelvin fronts propagating polewards. The second picture is a

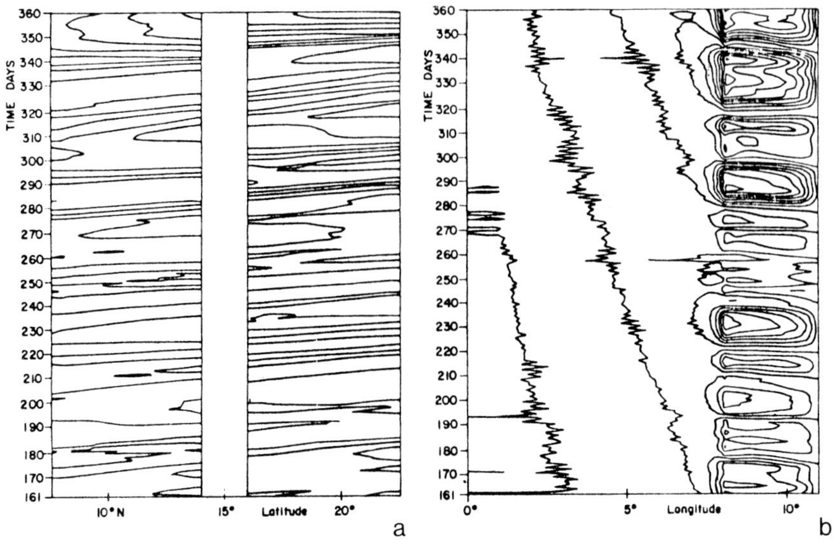

Fig. 13. Space-time diagrams.
a) Interface depth along the eastern boundary.
b) Zonal velocity in function of longitude just south of the cape.

section taken across the basin just south of the cape. Kelvin waves appear along the cape but the propagation across the oceanic interior is done through Rossby waves. From this picture, their speed is estimated around 3 cm/s which is a reasonable value for Rossby front at this latitude.

The last analysis performed with these data is the spectral analysis. Auto spectra for the zonal velocity u and the meridional velocity v in the middle of the southern half of the basin show that for periods between 3 days and 40 days there is an excess in energy for the zonal component compared to the meridional component (fig. 14) but for lower and higher frequencies, their levels are similar. This can be explained, at high frequencies (around two days), by the existence of a strong inertial peak; at low frequencies, the energy equipartition may be justified by the presence of Rossby waves which contribute efficiently to increase the energy level in that range. For the medium range, it is easy to verify that the oceanic transport is directed to the right of the wind, in agreement with the Ekman theory. The coherence between v and T_y is above the 95 % level of confidence (fig. 15) and the phase lag is zero. A smaller coherence is found between u and T_x (this wind component is smaller than the meridional component) but the phase lag is stable around 180°.

Along the coast, spectra are different (fig. 14). In the range 3 - 100 days, kinetic energy for the meridional component is two decades higher along the coast than in the middle of the basin. All the coastal points show a similar spectrum. The direct effect of the wind stress and of the propagation of coastal

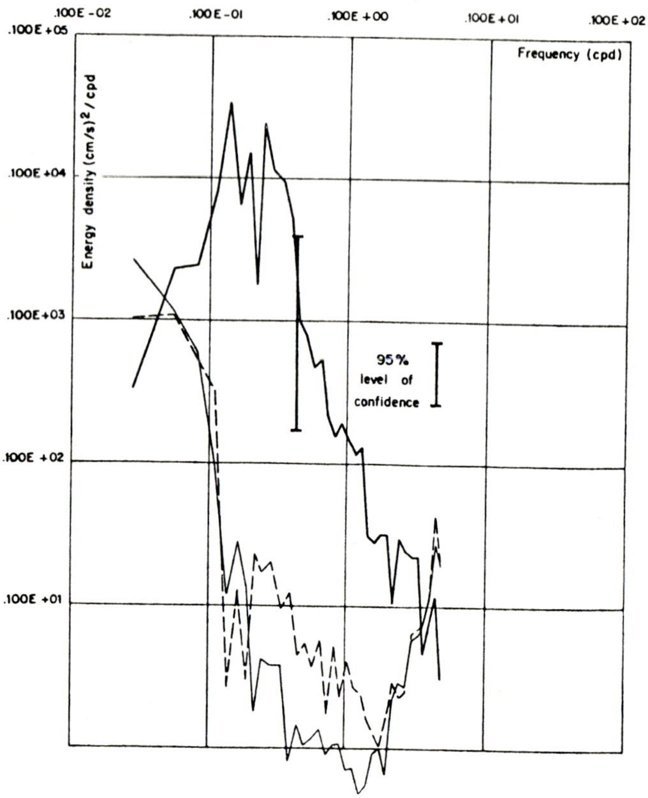

Fig. 14. Kinetic energy density spectra for the meridional velocity along the coast at 12° N (dark line), for the meridional velocity (thin line) and the zonal velocity (dashed line) at 12.12° N and 5° W.

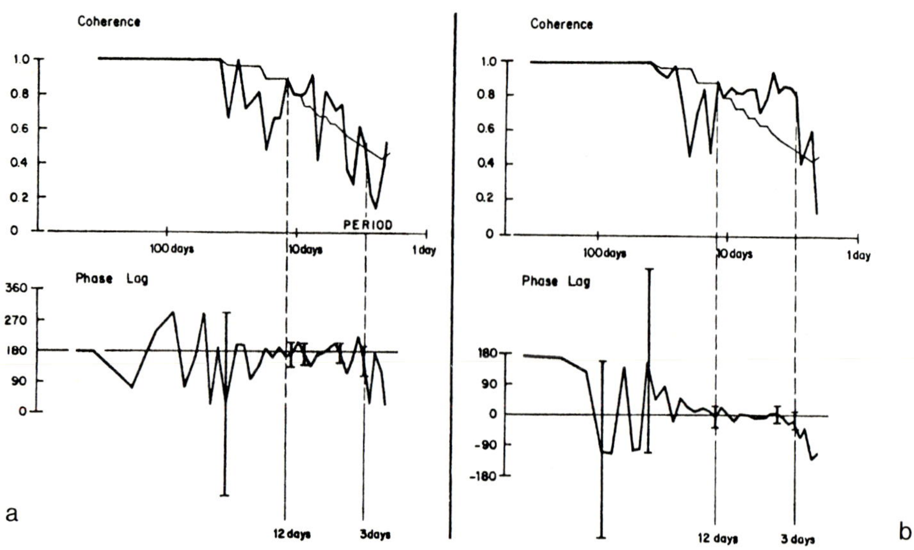

Fig. 15. Coherence and phase lag at 12.12° N, 5° W, between : a) v and T_y ; b) u and T_x.

waves is clearly visible. In order to prove the wave existence, the coherence was calculated between different points along the coast (9.56° N, 12.12° N, 13.73° N, 16.27° N, 17.88° N). The level of coherence was very high at all frequencies and the phase lag analysis revealed that the coastal Kelvin waves propagation was responsible for the high coherence level. In figure 16, the phase lag between different points is presented as a function of frequency. The dashed line is the theoretical straight line obtained by assuming that two points, separated by a distance L, are correlated by Kelvin waves propagating at the speed c (2 m/s for this run). The phase lag can be expressed as a function of frequency w :

$$\varphi \text{ (radians)} = (2\pi \frac{L}{c}) w$$

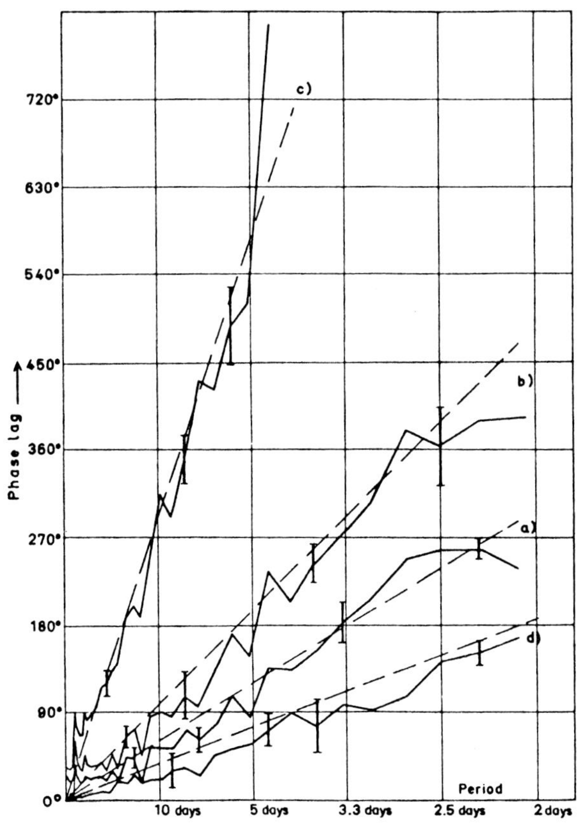

Fig. 16. Phase lag in function of period for the meridional velocity at different points along the coast :
a) 9.56° N and 12.12° N;
b) 9.56° N and 13.73° N;
c) 9.56° N and 16.27° N;
d) 16.27° N and 17.88° N.
The dashed lines represent the theoretical curves.

The agreement is very good in the medium range between points situated either in the northern half of the domain or in the southern half. It is still valid between one point north of the cape and one south of it but with a smaller level of confidence.

4. DISCUSSION

It was shown, in the previous part, that for a large band period (3 days to 300 days) which corresponds to a wavelength band (500 km to 5000 km), the presence of a cape, the dimensions of which are bigger than the radius of deformation, allows the propagation of coastal Kelvin waves without affecting their properties. Secondary waves are generated by the cape itself but they cancel one another. For smaller period (and smaller wavelengths) the coherence level is not high enough for the phase lag to be significant; coastal phenomena appear uncorrelated for periods smaller than the inertial period. The coherence level is not determined either for very long periods (superior to 30 days). Away from the coast, the Ekman theory is well verified for periods between 3 days and a few weeks. But, outside this band, wave dynamics dominates.

With a multi-level numerical model, vertical structure would have been obtained. Such a simulation displays the coastal and the vertical trapping as shown by Yoon and Philander (1982). But the basic adjustment, through coastal Kelvin waves, is similar to the one obtained with a shallow water model; considering the stratification offshore of Dakar, it is reasonable to assume that the first baroclinic mode is responsible for the largest part of the adjustment and that the reduced gravity model reproduces the main patterns of the mixed layer. A stratified model will also be able to present topographic effects due to the continental margin. Coastal trapped waves are generated with speed comparable to Kelvin waves speed and they propagate poleward (see Suginohara, 1982). In the ocean, all the coast indentations, cape or bay, larger than the internal radius of deformation, are able to create series of trapped waves that may cancel one another. Arthur (1965) has considered a different problem; in his approach, there exists a coastal boundary layer where the velocity goes to zero and the dimensions of the cape are smaller than the radius of deformation. Thus, the variation of vorticity around the cape is not negligible. But it does not apply in the present model where the cape is large and the velocity high along the coast. Long Kelvin waves were present in the model though open boundary conditions were applied; they may represent global phenomena coming from equatorial region. The salinity cannot be introduced in the reduced gravity model and its variation may play a role in this area and create important local variability.

The phase lags calculated in the model are not directly comparable with measurements because coastal topography and stratification will alter the wave speed and the physical characteristics (the trapping scale for instance) but with this simple numerical approach, basic features were outlined. The next step will require the input of spatially varying wind-stress that will excite a large wave spectrum in the ocean and add some remote effects. The coast will not be the only source of vorticity. Then, the non-linear problem will be studied. The addition of non-linear terms will modify slightly wave properties but the currents are not very strong and this will not introduce completely different results.

REFERENCES

Allen, J.S., 1976. Some aspects of the forced wave response of stratified coastal regions. J. Phys. Oceanogr., 6, 1: 113-119.

Arthur, R.S., 1965. On the calculation of vertical motions in the eastern boundary currents from determination of horizontal motion. J. of Geoph. Res., vol. 70, n° 12: 2799-2803.

Biju-Duval, N. and Chartier, M., 1982. Effets d'un cap sur les upwellings. Solution numérique à partir d'une méthode aux éléments finis. Projet n° 127, Ecole Nationale Supérieure de Techniques Avancées.

Charney, J.G., 1955. The generation of ocean currents by wind. J. Mar. Res., 14, 4: 477-498.

Camerlengo, A.L. and O'Brien, J.J., 1980. Open boundary conditions in rotating fluids. Jour. of Computational Physics, vol. 35, n° 1: 12-35.

Crépon, M. and Richez, C., 1982. Transient upwelling generated by two dimensional atmospheric forcing and variability in the coastline (submitted to J. Phys. Oceanogr.).

Csanady, G.T., 1975. The coastal jet conceptual model in the dynamics of shallow seas. In : Goldberg, Mc Cave, O'Brien and Steele (Eds.), The Sea, vol. 6, Marine Modelling, pp. 117-144. Wiley Interscience Publ., New York.

Halpern, D., 1976. Structure of a coastal upwelling event observed off Oregon during July 1973. Deep-Sea Res., 23, 6: 495-508.

Hua, B.L., 1981. Modélisation numérique d'upwellings côtiers à l'aide d'une méthode d'éléments finis. Application au golfe du Lion. Thèse de doctorat d'Etat ès Sciences physiques, Université Pierre et Marie Curie et Muséum National d'Histoire Naturelle, Paris, 18 mai 1981.

Merle, J., 1973. Hydrologie saisonnière dans la région de Dakar (unpublished document).

Portolano, P., 1981. Contribution à l'étude de l'hydroclimat des côtes sénégalaises, CRODT - ORSTOM.

Rossignol, M. and Meyrueis, A.M., 1962. Campagne océanographique du "Gérard Tréca", CRODT - ORSTOM.

Rossignol, M., 1973. Contribution à l'étude du complexe guinéen, ORSTOM.

Suginohara, N., 1982. Coastal upwelling : onshore-offshore circulation, equatorward coastal jet and poleward undercurrent over a continental shelf slope. J. of Phys. Oceanogr., vol. 12, n° 3: 272-284.

Yoon, J.H. and Philander, S.G.H., 1982. Formation of the coastal undercurrent. Submitted to the J. Oceanogr. Soc. Japan.

APPENDIX — DESCRIPTION OF THE NUMERICAL MODEL

The equations used to study the dynamics of the upwelling event in the numerical model are the reduced-gravity equations in the shallow water approximation :

$$\frac{\partial \mathbf{u}}{\partial t} + (\mathbf{u}.\nabla)\mathbf{u} + f\,\mathbf{k} \wedge \mathbf{u} = - g'\,\nabla h + \mathbf{D}$$

$$\frac{\partial h}{\partial t} + \nabla(h\mathbf{u}) = 0$$

$$g' = g\,\frac{\Delta\rho}{\rho}$$

where $\mathbf{u}$ represents the horizontal velocity vector, h the thickness of the layer, f the Coriolis parameter, g the acceleration of gravity and $\mathbf{k}$ the vertical normalized vector. $\Delta\rho$ is the density difference between the mixed-layer and the deep motionless layer and $\mathbf{D}$ represents the external forces (wind stress forcing, friction, ...).

For some experiments, a linear version of the above system of equations will be used :

$$\frac{\partial \mathbf{u}}{\partial t} + f\,\mathbf{k} \wedge \mathbf{u} = -\,g'\,\nabla h + \mathbf{D'}$$

$$\frac{\partial h}{\partial t} + H\,\nabla\mathbf{u} = 0$$

H is the initial thickness of the layer. Considering this system in terms of vertical modes, a equivalent depth can be estimated :

$$he = \frac{\Delta\rho}{\rho}\,H = 10\ \text{cm}$$

when $\Delta\rho/\rho$ is equal to $2 \,.\, 10^{-3}$ and H to 50 m. The horizontal phase speed related to this baroclinic mode is equal to

$$c = (g\ he)^{1/2} = 1\ \text{m/s}$$

Without forcing, the non-linear momentum equations will be written with vectorial operators :

$$\frac{\partial \mathbf{u}}{\partial t} + (\nabla \wedge \mathbf{u} + f\mathbf{k}) \wedge \mathbf{u} = -\,\nabla\left(g'h + \frac{\mathbf{u}\cdot\mathbf{u}}{2}\right)$$

The vectorial operators can be expressed in a tensorial form, invariant in any curvilinear orthogonal base transformation.

Let φ be the latitude, λ the longitude and r the distance from the center of the earth. Let i and j describe the horizontal discrete mesh and the vertical axis give the third direction. φ is a function of j, and λ of i.

Let x, y and z denote the distances in a frame referred to the center of the earth.

$$x = r\cos\varphi(j)\ \cos\lambda(i)$$

$$y = r\cos\varphi(j)\ \sin\lambda(i)$$

$$z = r\sin\varphi(j)$$

Assuming a to be the radius of the earth, a new frame (x',y',z') is defined :

$dx' = a\ di$

$dy' = a\ dj$

$dz' = dr$

The infinitesimal displacements in both frames (x,y,z) and (x',y',z') are equal :

$$(ds)^2 = (dx)^2 + (dy)^2 + (dz)^2 = (e_{x'}\ dx')^2 + (e_{y'}\ dy')^2 + (e_{z'}\ dz')^2$$

The coefficients $e_{x'}$, $e_{y'}$ and $e_{z'}$ are easily calculated :

$$e_{x'} = [(\frac{\partial x}{\partial x'})^2 + (\frac{\partial y}{\partial x'})^2 + (\frac{\partial z}{\partial x'})^2]^{\frac{1}{2}} = \frac{r}{a} \cos\varphi(j)\ \frac{d\lambda}{di}$$

$$e_{y'} = [(\frac{\partial x}{\partial y'})^2 + (\frac{\partial y}{\partial y'})^2 + (\frac{\partial z}{\partial y'})^2]^{\frac{1}{2}} = \frac{r}{a}\ \frac{\partial\varphi}{dj}$$

$$e_{z'} = [(\frac{\partial x}{\partial z'})^2 + (\frac{\partial y}{\partial z'})^2 + (\frac{\partial z}{\partial z'})^2]^{\frac{1}{2}} = 1$$

These are the important parameters, typical of the transformation, that appear in the expression of the covariant and contravariant coordinates of a vector.

Let $\mathbf{V}$ $(V_{x'}, V_{y'}, V_{z'})$ be a vector. The covariant components are defined by

$$\tilde{V}_K = e_K\ V_K \qquad (K = x'\ ,\ y'\ \text{ or }\ z')$$

and the contravariant component

$$\tilde{\tilde{V}}_K = \frac{V_K}{e_K} \qquad (K = x'\ ,\ y'\ \text{ or }\ z')$$

The vectorial operators have an invariant expression in function of these coefficients.

Let b be the product $e_{x'} \times e_{y'} \times e_{z'}$ and A a scalar quantity

$$(\nabla A)_K = \frac{1}{e_K}\ \frac{\partial A}{\partial K}$$

$$\nabla . \mathbf{V} = \frac{1}{b} \sum_{K=x',y',z'} \frac{\partial(b\tilde{\tilde{V}}_K)}{\partial K}$$

$$(\nabla \wedge \mathbf{V})_{x'} = \frac{e_{x'}}{b}\ (\frac{\partial\tilde{V}_{z'}}{\partial y'} - \frac{\partial\tilde{V}_{y'}}{\partial z'})$$

We now assume that r is equal to a and the indices $'$ will be dropped to simplify the expressions.

The continuity equation gives :

$$\frac{\partial}{\partial t}(hb) + \frac{\partial}{\partial x}(h\,b\,\tilde{\tilde{u}}) + \frac{\partial}{\partial y}(h\,b\,\tilde{\tilde{v}}) = 0$$

and, setting $P = hb$,

$$\frac{\partial}{\partial t} P + \frac{\partial}{\partial x} P\,\tilde{\tilde{u}} + \frac{\partial}{\partial y} P\,\tilde{\tilde{v}} = 0$$

$P\tilde{\tilde{\mathbf{u}}}$ is the mass flux.

For the momentum equation, let

$$\Phi = g'h + \frac{\mathbf{u}\cdot\mathbf{u}}{2}$$

$$\zeta = \frac{\partial}{\partial x}\tilde{v} - \frac{\partial}{\partial y}\tilde{u}$$

With these notations, the equations for the meridional and zonal velocity components are easily deduced :

$$\frac{\partial \tilde{u}}{\partial t} - \frac{\zeta + bf}{P} P\tilde{\tilde{v}} = -\frac{\partial}{\partial x}\Phi$$

$$\frac{\partial \tilde{v}}{\partial t} + \frac{\zeta + bf}{P} P\tilde{\tilde{u}} = -\frac{\partial}{\partial y}\Phi$$

The shape taken by these equations is very interesting because they are intrinsivally consistent for any curvilinear orthogonal transformation. They allow the use of variable grid-spacings.

The equations will be discretized in a staggered array

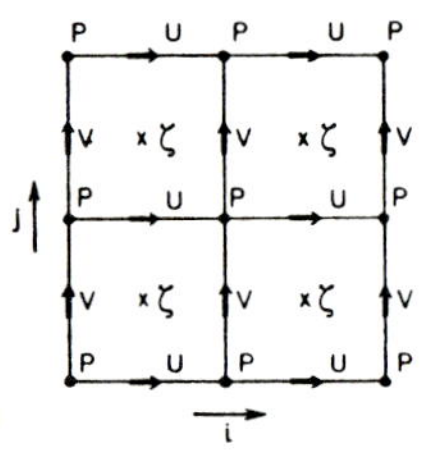

Let

$$\delta_K q = \frac{q(K+\frac{1}{2}) - q(K-\frac{1}{2})}{a\,\Delta K} \qquad K = i,j$$

$$\overline{q}^K = \frac{1}{2}\,[q(K+\frac{1}{2}) + q(K-\frac{1}{2})]$$

Assuming

$$\zeta = \delta_x\tilde{v} - \delta_y\tilde{u}$$

and

$$\Phi = g'h + \frac{1}{2}\,(\overline{u^2}^x + \overline{v^2}^y)$$

the discrete set of equation is

$$\frac{\partial}{\partial t}P + \delta_x(\overline{P}^x\,\tilde{\tilde{u}}) + \delta_y(\overline{P}^y\,\tilde{\tilde{v}}) = 0$$

$$\frac{\partial}{\partial t}\tilde{u} - \frac{\overline{\zeta + bf}^y}{\overline{\overline{P}^x}^y}\,\overline{\overline{\overline{P}^y\,\tilde{\tilde{v}}}^x}^y + \delta_x\Phi = 0$$

$$\frac{\partial}{\partial t}\tilde{v} + \frac{\overline{\zeta + bf}^{x}}{\overline{\overline{P}^{x}}^{y}} \overline{\overline{\overline{P}^{x}\tilde{\tilde{u}}}^{y}}^{x} + \delta_y \Phi = 0$$

This scheme conserves potential entrophy.

In the forcing terms wind stress and friction are included. The wind stress is applied as a body force in the mixed surface layer and the lateral friction is expressed in function of a Laplacian operator (note that the Laplacian operator is defined only for the scalar quantities and the correct expression for the frictional operator has to be calculated using the generalized form of the vectorial operators and the relationship :

$$\nabla^2 \mathbf{V} = \nabla(\nabla . \mathbf{V}) - \nabla \wedge (\nabla \wedge \mathbf{V})$$

The coefficient of lateral friction A_H is estimated to be $10^2\ m^2 s^{-1}$.

The extent of the domain is 11° in longitude and 15° in latitude. It is centered at 15° N and the eastern boundary is the coastline. The cape is 2° in latitude and 3° in longitude in the middle of the eastern boundary. A free slip condition is applied along all the lateral boundaries except the southern and the northern walls where open boundary conditions are applied (Camerlengo and O'Brien, 1980). Along the western wall a sponge boundary condition is used in order to damp the energy coming from the eastern part with Rossby waves.

The resolution of the grid-spacing is very high along the eastern boundary (6 km) and decreases toward the west. In latitude, the resolution is about 27 km around the cape.

The time difference scheme is a leapfrog scheme for the internal dynamics and a forward scheme for the external forces. Solutions at odd and even time-steps are regularly mixed together to avoid time splitting and the time step used is 20 mn .

MODEL OF THE GULF OF GUINEA UPWELLING. INFLUENCE OF THE COAST'S IRREGULARITIES.

Jacques C.J. NIHOUL and A. BAH

1. INTRODUCTION

Every summer, for a period from early July through September, upwelling is an important feature along the Gulf of Guinea coast (fig. 1). According to Houghton (1976), the cold water found along the northern boundary of the Gulf is local in origin and not advected into the area by the Benguela current. The coldest water is always found east of Cape Palmas and Cape Three Points. There is no correlation between local winds and nearshore temperatures or changes in the local ocean circulation.

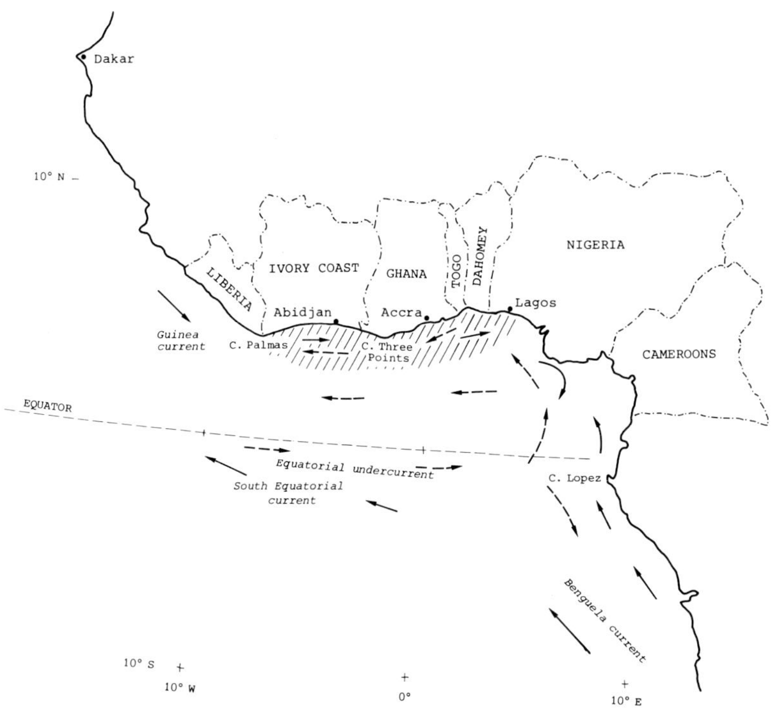

Fig. 1. Schematic map of the Gulf of Guinea, showing the region of upwelling (shaded area) and the presumed surface (——) and subsurface (— —) currents (after Houghton, 1976).

Observations, during Gate, by the Soviet Ship *RV. Parsat* seemed to suggest a forcing of oceanic origin and O'Brien (e.g. O'Brien et al., 1978; Adamec and O'Brien, 1978) hypothesized that a baroclinic Kelvin wave is excited in the western Atlantic by the onset of the southeast trades in May-June. This long wave, strong upwelling impulse, propagates eastward across the Atlantic. It then propagates poleward as a coastal Kelvin wave and produces the upwelling event along the Gulf of Guinea coast.

With a non-linear numerical model, using the β-plane approximation in an ocean initially at rest, Adamec and O'Brien (1978) calculated the first baroclinic modal response of the ocean to a sudden increase of the wind stress and succeeded in reproducing the main features of the observations with a very good agreement of both amplitude and time scales.

In all the applications of their model, however, the coast was assumed rectilinear and the possible effect of capes on the upwelling's intensity could not be studied.

The effect of capes is often discussed in terms of local conditions such as the orientation of the coast with respect to the local wind or the deflection of the coastal circulation.

An approximate calculation made by Bah (1980)*, using current results of a numerical model similar to that of Adamec and O'Brien (1978), showed a possible amplification of the upwelling east of Cape Three Points, without being strongly conclusive.

In the following, the influence of Cape Palmas and Cape Three Points is investigated, using an improved version of Bah's model, with a numerical grid which takes into account the coastal irregularities. It is shown that a Kelvin wave generated upwelling, of the type hypothesized by O'Brien et al. (1978), can explain, along a non-linear coast such as the northern coast of the Gulf of Guinea, the kind of amplification of upwelling's intensity which is observed east of the capes.

2. DESCRIPTION OF THE MATHEMATICAL MODEL

Fig. 2 shows the model geometry with the irregular coastline.

The governing equations are derived from the Boussinesq equations (e.g. Nihoul, 1982). Assuming two layers of uniform densities ρ_1 (upper layer) and ρ_2 (lower layer) and zero pressure gradient in the bottom layer, one can write, in the β-plane approximation, the equations for the transport $\mathbf{U}$ in the upper layer, in the form

$$\frac{\partial H}{\partial t} + \nabla . \mathbf{U} = 0 \qquad (1)$$

* Some misprints in the equations used for the calculation make this part of the paper difficult to understand. The reader is advised to enquire into the corrigenda.

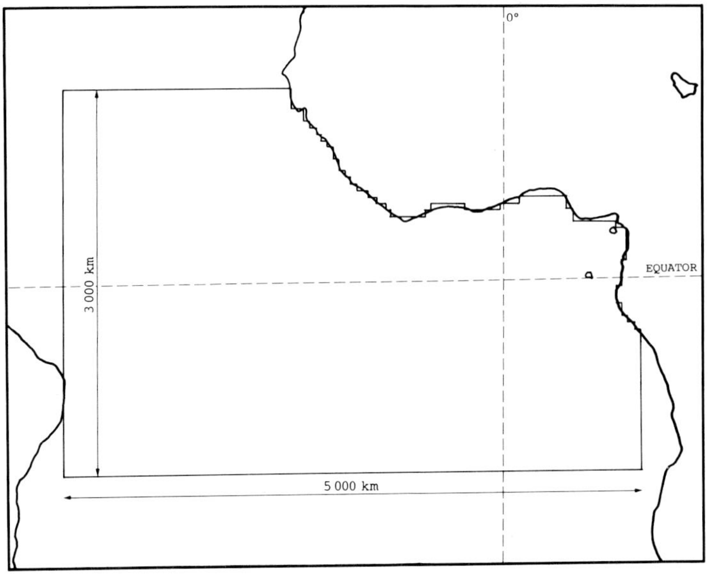

Fig. 2. The model geometry with the irregular coastline.

$$\frac{\partial \mathbf{U}}{\partial t} + \nabla .(H^{-1}\mathbf{UU}) + \beta x_2 (\mathbf{e}_3 \wedge \mathbf{U}) = - g'H \nabla h + \frac{\boldsymbol{\tau}}{\rho_1} + A \nabla^2 \mathbf{U} \tag{2}$$

where

$$H = H_0 + h \tag{3}$$

H_0 is the undisturbed thickness of the upper layer ($H_0 \sim 50$ m), h the perturbation of the upper layer's thickness and H the total (disturbed) thickness of the upper layer. A is a horizontal diffusion coefficient taking into account turbulence and shear effect, $\boldsymbol{\tau}$ is the wind stress,

$$g' = \frac{g(\rho_2 - \rho_1)}{\rho_2} \tag{4}$$

The details of the numerical model (discretization scheme, stability criteria, boundary conditions, ...) are given in (Bah, 1980).

3. APPLICATION

O'Brien et al. (1978), Adamec and O'Brien (1978) and Bah (1980) have investigated several cases of oceanic responses to impulsive changes of the wind stress, over part of the area.

One of these cases assumes a sudden increase of 0.0125 Nm^{-2} in the westward wind stress over the western 1500 km of the basin.

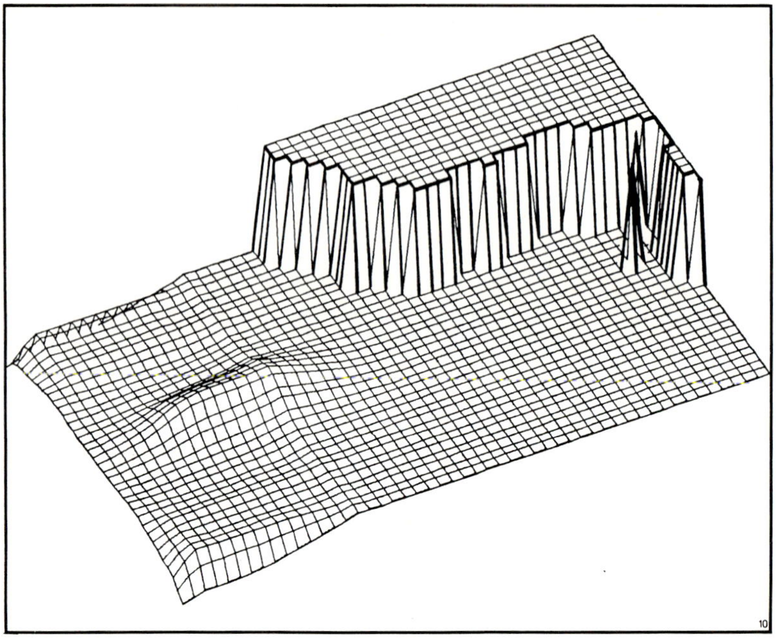

Fig. 3. Elevation of the interface 10 days after the onset of the wind.

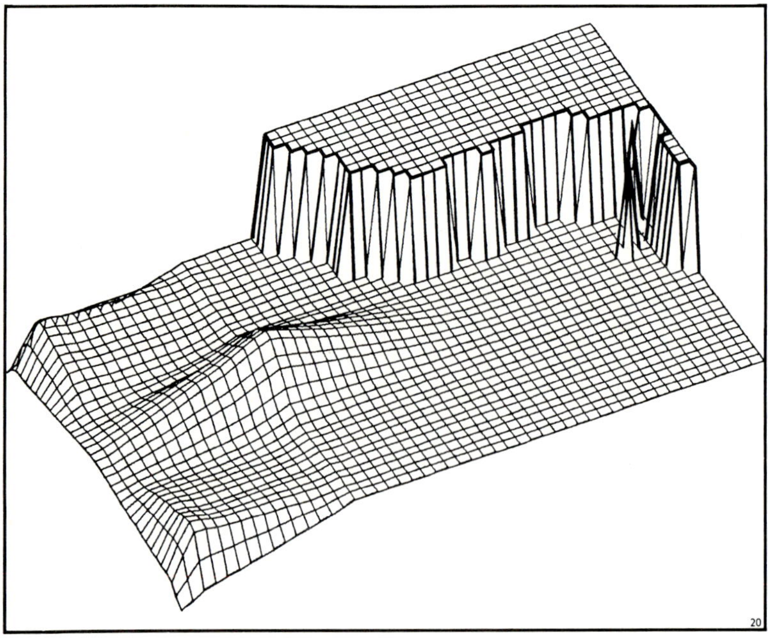

Fig. 4. Elevation of the interface 20 days after the onset of the wind.

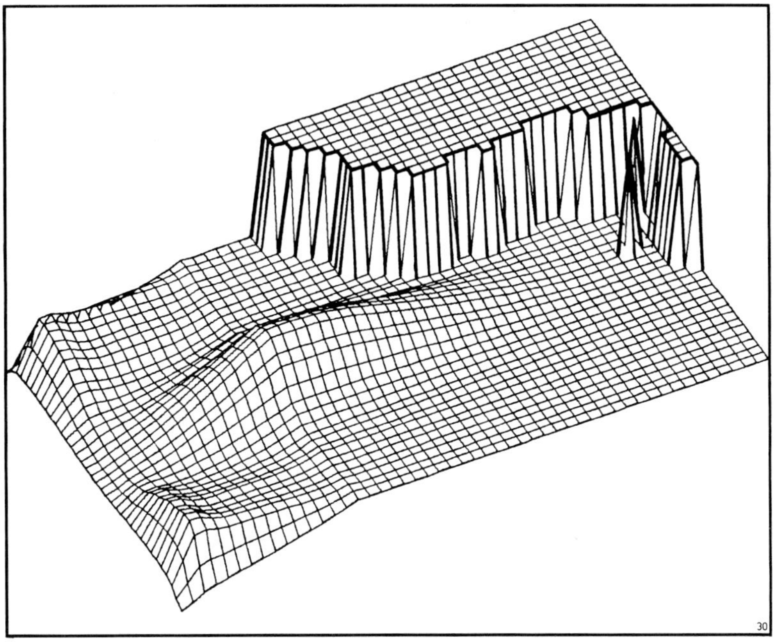

Fig. 5. Elevation of the interface 30 days after the onset of the wind.

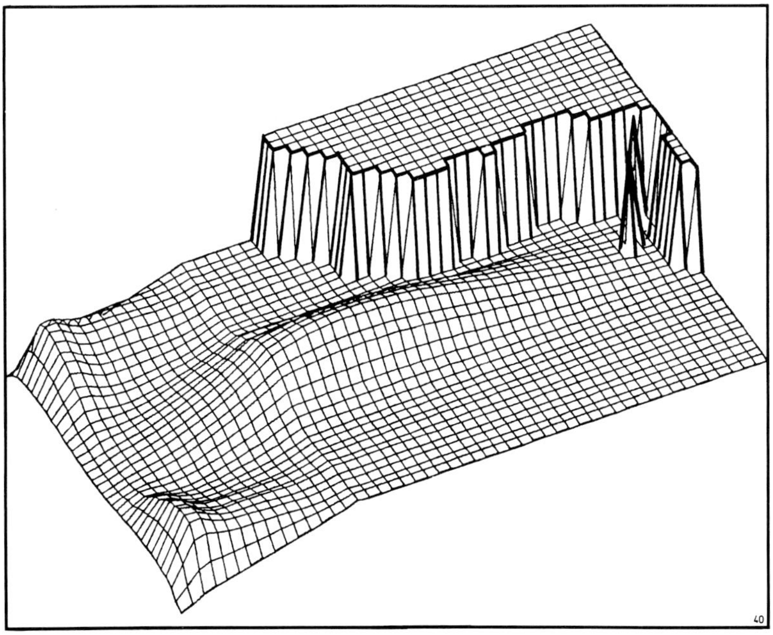

Fig. 6. Elevation of the interface 40 days after the onset of the wind.

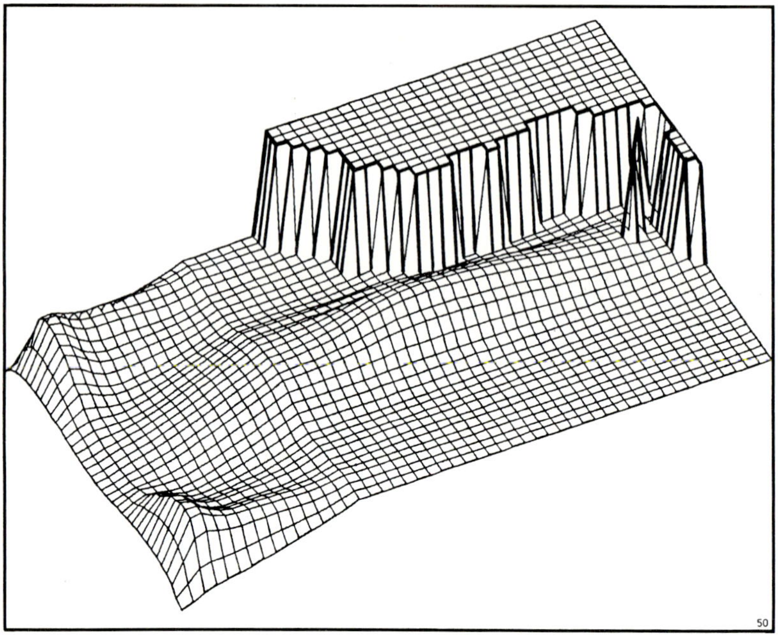

Fig. 7. Elevation of the interface 50 days after the onset of the wind.

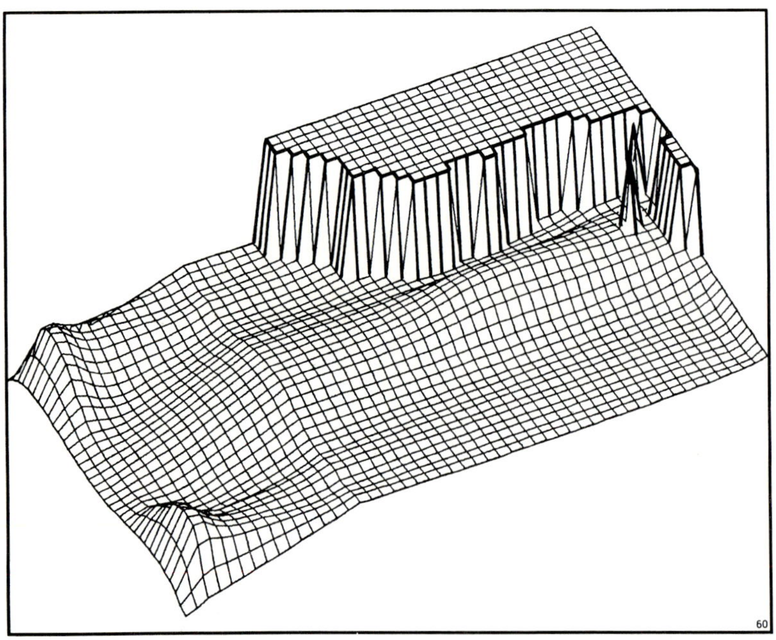

Fig. 8. Elevation of the interface 60 days after the onset of the wind.

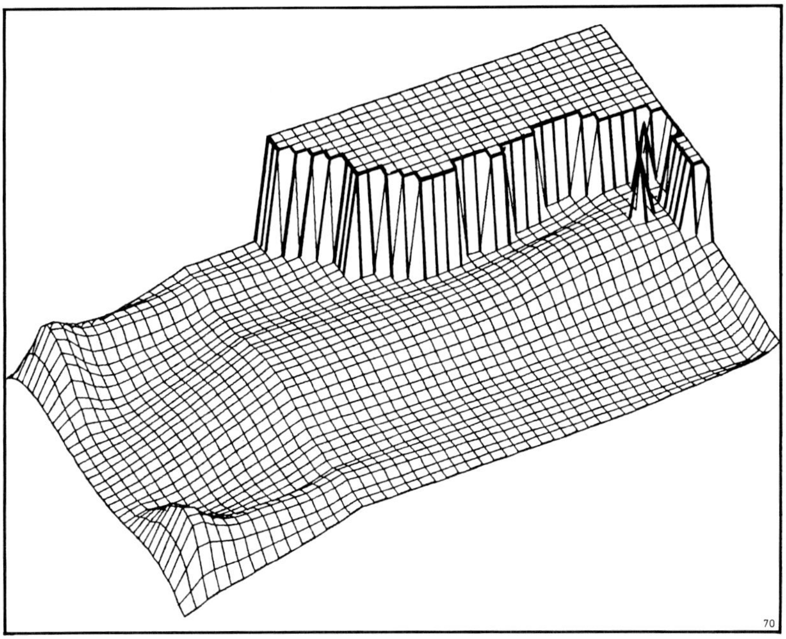

Fig. 9. Elevation of the interface 70 days after the onset of the wind.

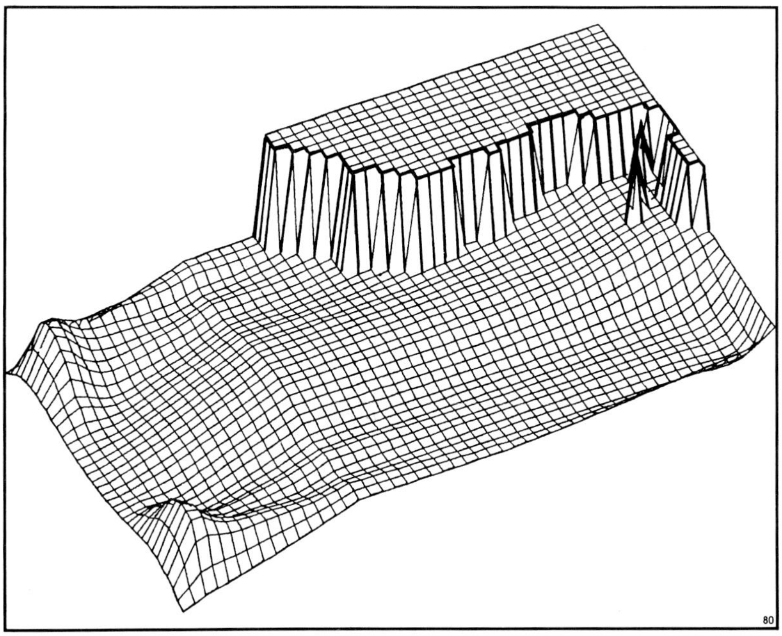

Fig. 10. Elevation of the interface 80 days after the onset of the wind.

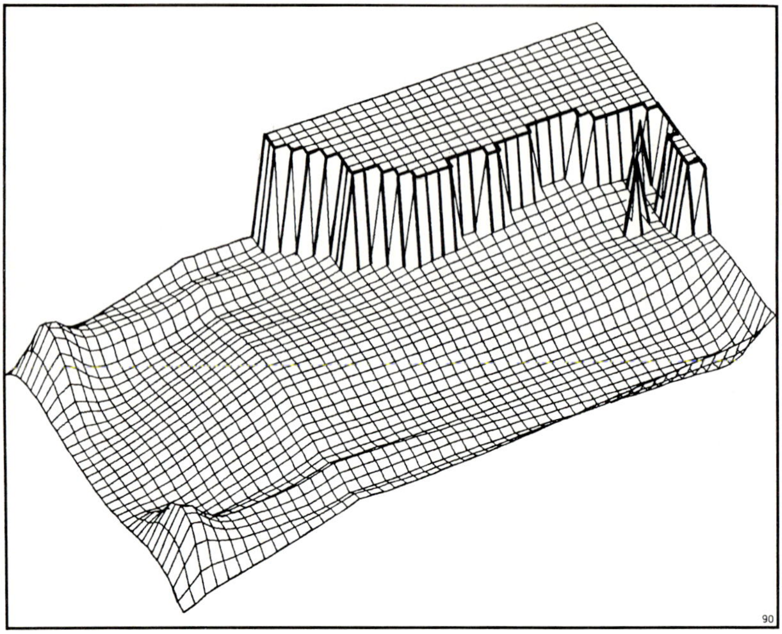

Fig. 11. Elevation of the interface 90 days after the onset of the wind.

This case has been simulated using the model described in section 2.

Figs. 3 to 11 show the propagation of the Kelvin wave generated upwelling 10, 20, ..., 90 days after the onset of the wind. An amplification of the upwelling's intensity east of Cape Three Points is apparent.

On fig. 12, the elevation of the interface in the Gulf of Guinea after 90 days is compared with the interface elevation which is found when one assumes a linear coast line. The effect of the shape of the coast on the distribution of the upwelling's intensities is indicated by the position of the 10 m elevation-curve.

The Kelvin wave model appears thus to be able to reproduce the main features of the upwelling in the Gulf of Guinea including the amplification of its intensity east of the capes on the northern boundary.

4. REFERENCES

Adamec, D. and O'Brien, J.J., 1978. The seasonal upwelling in the Gulf of Guinea due to remote forcing. J. Phys. Oceanogr., 8: 1050.

Bah, A., 1980. Upwelling in the Gulf of Guinea. In: J.C.J. Nihoul (Editor), Ecohydrodynamics, Elsevier Publ., Amsterdam, pp. 99-140.

Houghton, R.W., 1976. Circulation and hydrographic structure of the Ghana continental shelf during the 1974 upwelling. J. Phys. Oceanogr., 6: 909-924.

Nihoul, J.C.J., 1982. Hydrodynamic models of shallow continental seas, Riga Publ., Liège, 198 pp.

O'Brien, J.J., Adamec, D. and Moore, D., 1978. A simple model of upwelling in the Gulf of Guinea. Geophys. Res., 5: 641-644.

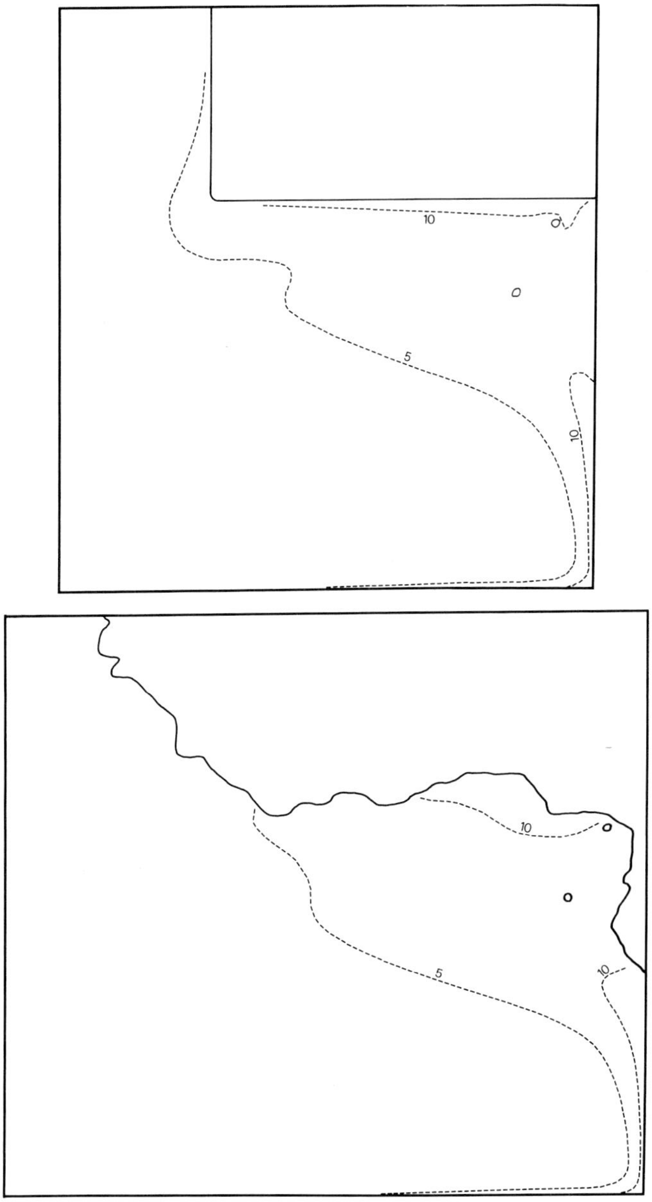

Fig. 12. Elevation of the interface calculated by the model for a rectilinear coastline (above) and a realistic coastline taking the capes into account (below). The results for a linear coastline are similar to those of O'Brien et al. (1978), Adamee and O'Brien (1978) and Bah (1980).

THE EFFECT OF ZONAL CURRENTS ON EQUATORIAL WAVES

P. Ripa and S.G. Marinone

Centro de Investigación Científica y de Educación Superior de Ensenada, Ensenada, B.C., México

Abstract

The interaction between a zonal current and equatorial waves is studied in the framework of an one-layer β-plane model. The eigenvalue problem is solved by expanding the perturbation over the solutions of the linear problem (which form an orthogonal and complete basis). Analytic expressions for the first order correction to the waves' frequency and numerical solutions of the suitable truncated eigenvalue problem are presented. Both stable and unstable solutions are described. Growing perturbations are found for flows such that only one of the two stability conditions (Miyata, 1981; Ripa, 1982b) is violated; *viz.*, a subcritical flow with potential vorticity extrema, or a jet with a monotonic potential vorticity profile which is supercritical in a neighborhood of the equator. The latter case is that of a westward jet with the meridional structure of a Kelvin wave.

1. INTRODUCTION

The equatorial oceans are characterized by a system of zonal currents with strong vertical and meridional shears and, at the same time, the equator acts as a waveguide (Matsuno, 1966). These currents and equatorial waves interact, resulting in a form of Doppler shift of the waves' frequency (Philander, 1979; McPhaden and Knox, 1979) and the currents may also become unstable (Philander, 1976, 1978; Kuo, 1978; Hughes, 1979, 1981).

Miyata (1981) and Ripa (1982b) showed that for an equatorial flow (or any current system for which the quasi-geostrophic approximation cannot be used) to be stable *two* different conditions must be met. The first one is a sort of criticaIly of the flow and the second one involves the gradient of potential vorticity. The possibility of unstable currents for which only one of the two stability conditions is violated is investigated here.

The close applicability of this model to a realistic oceanographic situation is questionable, because it has only one layer and thus it requires that the perturbations have the same vertical structure than the fundamental flow. The discussion of model predictions based on experimental data is therefore postponed until its generalization to the fully three-dimensional case is completed.

In Section 2 the model equations are reviewed and the method of solution of the eigenvalue problem is presented. The particular case of a jet with a Gaussian meridional profile is discussed in Section 3, while Section 4 is devoted to the Summary and Conclusions. An Appendix is added with the details of the evaluation of the matrix elements.

2 MODEL EQUATIONS

We work in an one-layer reduced gravity model in the unbounded equatorial β-plane. The evolution equations of the model are (see, for instance, Ripa 1982a)

$$\partial_t \underline{v} + \underline{v}\cdot\nabla\underline{v} + f\hat{z}\times\underline{v} + c^2\nabla\eta = 0, \tag{1}$$

$$\partial_t \eta + \nabla\cdot\{(1+\eta)\underline{v}\} = 0, \tag{2}$$

where $\underline{v} = (u,v)$ and $\nabla = (\partial_x, \partial_y)$ are the horizontal velocity and gradient operator, η is the vertical dilatation field, and c is the separation constant. Eastward and northward coordinates are denoted by (x,y).

The field η is defined in such a way thet the layer thickness, at any position and time, is equal to its mean value times $(1 + \eta)$; thus the mean value of η is, by definition, identically zero. (The dynamic pressure is $c^2\eta$.) The actual values of the reduced gravity and the average layer-thickness are not important: only their product, which *defines* the value c^2, is relevant. {The definition of the separation constant in terms of the *mean depth* is most appropiate, because it is time invariant. Philander (1976, 1978), though, uses the depth at the equator to define c. McPhaden and Knox (1979) start with a certain parameter, say c_0, which they equate to c while dealing with inertia-gravity waves, but they make $c_0 \neq c$ in their treatment of Kelvin waves. These facts should be born in mind when comparing the results of this work and those papers.} The values of c and β are used to evaluate the equatorial deformation radius,

$$R = (2c/\beta)^{1/2}, \tag{3}$$

which is the e-folding half-width of linear Kelvin waves. (For, say, $c = 2.4\ \mathrm{ms}^{-1}$, $R = 460$ km, $2\pi R = 2.9$ Mm, and $2\pi R/c = 14$ days.)

Perturbation expansion. The dynamic variables can be grouped in a vector field as

$$\Phi(\underline{x},t) = \begin{pmatrix} u(\underline{x},t) \\ v(\underline{x},t) \\ c\eta(\underline{x},t) \end{pmatrix}, \tag{4}$$

which is expanded in series as

$$\Phi = \Phi_0(y) + \varepsilon\Phi_1(\underline{x},t) + \varepsilon^2\Phi_2(\underline{x},t) + \ldots, \tag{5}$$

with $\varepsilon \ll 1$. The first term, Φ_0, is an exact solution of the model equations and the rest represents a small perturbation of $O(\varepsilon)$. The equations of the model at the lowest order are

$$v_0 = 0, \tag{6}$$
$$fu_0 + c^2\partial_y\eta_0 = 0, \tag{7}$$

i.e., the fundamental flow is zonal and it is in geostrophic balance. To the next order, equations (1) and (2) yield

$$D_0u_1 - (f-\partial_y u_0)v_1 + c^2\partial_x\eta_1 = 0, \tag{8}$$
$$D_0v_1 + fu_1 + c^2\partial_y\eta_1 = 0, \tag{9}$$
$$D_0\eta_1 + \nabla\cdot\{(1+\eta_0)\underline{v}_1\} = 0, \tag{10}$$

where $D_0 = \partial_t + u_0\partial_x$. The law of potential-vorticity conservation is, to first order,

$$D_0(\partial_x v_1 - \partial_y u_1 - q_0\eta_1)(1+\eta_0)^{-1} + v_1\partial_y q_0 = 0, \tag{11}$$

where

$$q_0 = (f-\partial_y u_0)/(1+\eta_0) \tag{12}$$

is the potential vorticity of the fundamental flow. Equations (8) through (10) can be represented in the form

$$(\partial_t + iL + iC)\Phi_1 = 0, \tag{13}$$

where iL contains the linear operator, and iC those proportional to u_0 and η_0. The equations at second order are

$$(\partial_t + iL + iC)\Phi_2 + \begin{pmatrix} \underline{v}_1\cdot\nabla\underline{v}_1 \\ \\ \nabla\cdot(\eta_1\underline{v}_1) \end{pmatrix} = 0. \tag{14}$$

Stability conditions. Ripa (1982b) has shown that, if there exists any number α such that

$$\left.\begin{aligned}(\alpha-u_0)^2 &\le c^2(1+\eta_0) \\ (\alpha-u_0)\,\partial_y q_0 &\ge 0\end{aligned}\right\}\ \forall\ y, \qquad \begin{aligned}(15)\\(16)\end{aligned}$$

then the first order perturbation Φ_1 is bounded in time, and thus Φ_0 is (linearly) stable. Weaker stability conditions are

$$\partial_y q_0 \ge 0\ \forall y, \tag{17}$$

and

$$\{\max(u_0)-\min(u_0)\}^2 \le c^2\{1+\min(\eta_0)\}. \tag{18}$$

Eigenvalue problem. Since the coefficients in (13) are time independent, solutions can be sought for Φ_1, of the form

$$\Phi_1 = \mathrm{Re}\{\theta(\underline{x})\exp(-i\sigma t)\}. \tag{19}$$

When (19) is substituted in (13), the following equation is obtained

$$(L + C - \sigma)\theta(\underline{x}) = 0, \tag{20}$$

i.e., the σ are the eigenvalues of the operator $L + C$. Some eigenvalues may be complex, with $\mathrm{Im}(\sigma) > 0$, indicating that the field Φ_0 is unstable.

The eigensolutions of (20) in the absence of currents $\{\Phi_0 = 0\}$, *viz.*,

$$(L - \omega_a)\tilde{\theta}_a(\underline{x}) = 0, \tag{21}$$

are the equatorial waves (Matsuno, 1966). The label 'a' denotes the zonal wave number k ($-\infty < k < \infty$), the meridional quantum number n (= -1, 0, 1, 2, ...), and a discrete index r that numbers the different solutions for each value of k & n. These eigenfunctions form a *complete* basis (Ripa, 1982a) and thus it is possible to make the expansion

$$\theta(\underline{x}) = \sum_a X_a \tilde{\theta}_a(\underline{x}), \tag{22}$$

where

$$\sum_a = (2\pi)^{-1}(\beta/c)^{1/2} \int dk \sum_{nr}, \tag{23}$$

and the expansion functions are normalized so that

$$\int d^2\underline{x}\ \tilde{\theta}_a^{\ +}\tilde{\theta}_b = (c/\beta)^{1/2}\ 2\pi\delta(k_a - k_b) \tag{24}$$

if $n_a = n_b$ and $r_a = r_b$, and vanishes otherwise (the symbol $^+$ represents transpose complex conjugate). This technique is similar to that used by Kasahara (1980), who expands in Hough functions, because he works on the sphere instead of on the β-plane.

In order to avoid confusion, hereafter we will refer to the eigensolutions of the problem with the basic flow, (20), as 'waves' and to those of (21) as 'components'. Thus, σ denotes the frequency of a wave, and ω that of a component.

Substituting of (22) in (20), premultiplying by $\tilde{\theta}_b^{\ +}$, integrating in $\underline{x}$, and using (21) and (24) we obtain

$$(\omega_a - \sigma)X_a + \sum_b C_{ab}X_b = 0, \tag{25}$$

where C_{ab} are the matrix elements of C, i.e.,

$$C_{ab} = \int d^2\underline{x}\ \tilde{\theta}_a^{\ +}C\ \tilde{\theta}_b. \tag{26}$$

It can be shown that

$$C_{ab} = 2\pi\delta(k_a - k_b)\ \ (c/\beta)^{1/2}\ \Omega_{ab}, \tag{27}$$

where

$$\Omega_{ab} = \int dy'\ \tilde{\theta}_a^{\ +}C\ \tilde{\theta}_b \tag{28}$$

is a real and (generally) non-symmetric matrix and $y' = (\beta/c)^{1/2}y$. In (27) we see that only states with the same value of k are coupled, because Φ_0 is independent of x. Moreover, if $\Phi_0(y) = \Phi_0(-y)$, then modes with different meridional parity are decoupled, i.e., $C_{ab} = 0$ for $n_a + n_b$ = odd.

With this formalism, we can represent (25) in the form

$$(\omega_a - \sigma)X_a + \sum_{n_b r_b} \Omega_{ab}X_b = 0, \tag{29}$$

and thus σ is the eigenvalue and X the eigenvector of an algebraic problem. In the Appendix it is shown that

$$\Omega_{ab} = \int dy'\{ku_0(2\tilde{\underline{v}}_a{}^*\cdot\tilde{\underline{v}}_b + c^2\tilde{\eta}_a{}^*\tilde{\eta}_b) + (\beta/\omega_a)u_0\tilde{v}_a{}^*\tilde{v}_b$$
$$+ c^2\omega_b\eta_0\tilde{\eta}_a{}^*\tilde{\eta}_b - \omega_b u_0\tilde{u}_a{}^*\tilde{\eta}_b\}. \qquad (30)$$

From the phases of the components of $\tilde{\theta}_a$, Eq. (A1), it follows on that Ω_{ab} is real. The last three terms in the integrand are not symmetric and thus may result in complex eigenvalues σ in (29). Equations (A12) through (A14) give practical formula for the evaluation of Ω_{ab}.

If σ is real, there may be *critical latitudes*, where $D_n = 0$. It follows from (11) that, at a critical latitude, either $v_1 = 0$ (regular solution) or u_1 has a logarithmic singularity. Both kinds of solutions are needed for the completeness of the eigenfunctions of (20) (Drazin and Howard, 1966). The discrete eigenmodes reported here, obtained *via* a suitable truncation of Ω_{ab} in (29), are of the first, regular, type. (Due to the truncation, the node in v_1 and η_1 is found near-but not exactly at- the critical latitude.)

The dynamic fields u_1, v_1 and η_1 are given by

$$\Phi_1 = \mathrm{Re}\{\theta(y)\,\exp(ikx-i\sigma t)\}, \qquad (31)$$

where

$$\theta = \begin{pmatrix} U(y) \\ V(y) \\ Z(y) \end{pmatrix}, \qquad (32)$$

and the normalization is such that

$$\int dy'(|U|^2 + |V|^2 + |Z|^2) = 1. \qquad (33)$$

First order solutions. If the basic state is 'weak' enough, the solutions of (29) are equatorial waves slightly modified by the fundamental current and layer depth structure. In the limit u_0, $\eta_0 \to 0$, the eigensolutions are given by {see for instance Courant and Hilbert (1953), pp. 343-346}

$$\sigma \sim \omega_a + \Omega_{aa} + O(\Omega^2) \qquad (34a)$$

$$\theta \sim \tilde{\theta}_a + \sum \tilde{\theta}_b\Omega_{ba}/(\omega_a-\omega_b) + O(\Omega^2) \qquad (34b)$$

for each state 'a', where the summation is carried over $(n_b,r_b) \neq (n_a,r_a)$. Notice that σ is real in (34a). This approximation is valid as long as the correction to the free fields in (34b) is small; i.e., for $|\Omega_{ba}| << |(\omega_a - \omega_b)|$.

With (34a), it follows that if the first order solutions are valid, the correction to the free-frequency is small compared to the separation between frequencies.

In order to illustrate the meaning of (34a) let us consider a few examples:

a) *Short inertia-gravity waves and Kelvin-waves.* Taking the limit $sc \to \pm 1$ in (A5) we get $N(\pm) \sim 1$, $N(\mp) \sim 0$, $N(0) \sim 0$, where the upper (lower) signs correspond to eastward (westward) propagating inertia-gravity waves. Using this in (A12) and (A13) we get for (34a)

$$\omega \sim \pm kc\{1 + 1/2\langle n\pm 1|\eta_0|n\pm 1\rangle\} + k\langle n\pm 1|u_0|n\pm 1\rangle + O(\Omega^2). \tag{35}$$

We recall that $\langle m|\eta_0|m\rangle$ and $\langle m|u_0|m\rangle$ are average values of η_0 and u_0, weighted with the square of the parabolic cylinder function of order m [*cf.* (A6b) and (A14)]. For a Kelvin wave ($sc \equiv 1$) we obtain (35) with the + sign, $n+1=0$; thus, at the first order, Kelvin waves remain non-dispersive, σ/k = constant.

The second term in the right hand side of (35) is clearly an *effective* Doppler shift, whereas $c(1 + 1/2\langle\eta_0\rangle)$ is an effective value of the separation constant (because the mean depth 'seen' by the wave is $\langle 1+\eta_0\rangle$ times the average over the entire β-plane of the layer-thickness, and $\langle 1+\eta_0\rangle^{1/2} \sim 1 + 1/2\langle\eta_0\rangle$).

McPhaden and Knox (1979) made, for the Kelvin wave, an expansion of Φ_1, σ & c_0^2 in powers of the strength of u_0 & η_0 {see their Eq. (14)}, obtaining

$$\begin{aligned}\sigma &\sim k\{c_0 + \langle 0|u_0|0\rangle\},\\ c^2 &\sim c^2\{1 + \langle 0|\eta_0|0\rangle\},\end{aligned} \tag{36}$$

which is equivalent to (32) up to $O(\Omega^2)$ {Eq. (36) corresponds to their Eqs. (24d) and (24e)}. In this expression, the effective Doppler shift and change of the value of c are presented separately.

b) *Short Rossby waves.* These correspond to $sc \ll -2n-1$, which yields $N(\pm) \sim 0$, $|N(0)| \sim 1$ in (A5). Using this result and $\omega \sim -\beta/k$ in (A12), (A13) and (34a) we get

$$\sigma \sim \omega + k\langle n|u_0|n\rangle + O(\Omega^2). \tag{37}$$

c) *Mixed Rossby-gravity wave.* Using (A4) in (A12) and (A13) we obtain, for $n = 0$,

$$\begin{aligned}\sigma \sim \omega + \omega\,(4-2sc)^{-1}\{&(3sc-1)\langle 1|u_0|1\rangle/c + (1-sc)(2sc+1)\langle 0|u_0|0\rangle/c\\ &+ \langle 1|\eta_0|1\rangle\},\end{aligned} \tag{38}$$

where

$$\omega^2 = \beta c/(1-sc), \quad k = s\omega \quad (-\infty < sc < 1). \tag{39}$$

Notice that (38) reduces to (35), with the + sign, for $sc \sim 1$, and to (37) for $sc << -1$.

Equations (37) and (38) predict $\omega \sim k\langle n|u_0|n\rangle$ for $sc \to \infty$ (ultra short Rossby waves). If this were valid, these waves could have an eastward phase-propagation under the action of a westerly. However, $\Omega_{ab} \sim k\langle u_0\rangle = O(k)$ and $\omega_a - \omega_b \sim 2\beta^2 c^{-1}(n_a - n_b)k^{-3} = O(k^{-3})$, the approximation (34) breaks down $k^{-4} < O(\beta^{-2}c|u_0|)$.

3 SOLUTIONS FOR A GAUSSIAN PROFILE

Solutions to (29) were found for the symmetric zonal current field

$$u_0(y) = Ac\exp(-y^2/L^2) \tag{40}$$

and, in virtue of Eq. (7),

$$\eta_0(y) = AB\exp(-y^2/L^2), \tag{41}$$

where

$$B = \beta L^2/(2c) \equiv (L/R)^2. \tag{42}$$

Any Gaussian jet is then parametrized by two non-dimensional numbers: $A = u_0(0)/c$ and $B = (L/R)^2$. The parameters of the stable jets, as guaranteed by (17) and (18), are shown in Fig. 1; the details of the calculation are given by Ripa (1982b) (the parameters λ and μ used in that paper are equal to $\lambda = AB$, $\mu = B^{-1}$).

We approximate the solutions of (29) by those of a finite submatrix of Ω, choosing the basic states so that any added component has less than 0.5% of the total energy. Typically, we use 40 components. If each of the three equations (8), (9) and (10) is written as $R_1 + R_2 + R_3 = 0$, then the accuracy of the solution was further checked by assuring that

$$\frac{\int|R_1 + R_2 + R_3|^2 dy}{\int\{|R_1|^2+|R_2|^2+|R_3|^2\}dy} < 10^{-2},$$

with the integration carried over a region of appreciable amplitude.

Stable solutions

For a *Kelvin wave*, $\omega = kc$ and $n = -1$. The first order solution (34a) yields

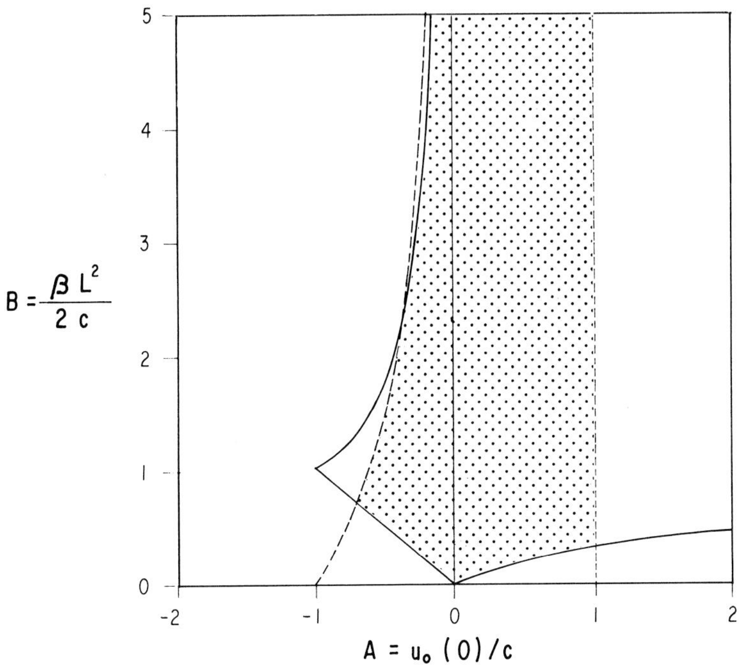

Fig. 1.- Stability zone of a Gaussian jet centered at the equator. $u_0(0)$ is the equatorial zonal velocity and L is the e-folding half-width of the jet. For values of A and B in the shaded area the flow is stable, as guaranteed by equations (17) and (18); these two conditions are violated beyond the solid and dashed lines, respectively.

$$\sigma = kc\{1 + AC(1+B/2)\} + O(A^2) \tag{43}$$

where

$$C = \{2B/(1+2B)\}^{1/2}. \tag{44}$$

The expansion (36), made by McPhaden and Knox (1979), takes the form

$$\sigma = kc_0(1 + AC) + O(A^2), \tag{45a}$$
$$c_0^2 = c^2(1+ABC) + O(A^2), \tag{45b}$$

which coincides with (43) up to $O(A^2)$. Unfortunately, McPhaden and Knox used c_0, instead of c, to nondimensionalize σ and k and thus one of the nonlinear effects is hidden in the plot-scale of their dispersion diagram.

Figure 2 shows the dispersion relation of Kelvin waves, for different values of A & B. The solid curves are the solution of (29) (calculated taking a sufficiently high number of states), whereas the dashed lines give the first order approximation (43). The latter is a reasonable approximation for moderate values of k & A, say, $|kRA| < 0.8$, but for larger slopes it is not, and in fact the Kelvin wave becomes slightly dispersive.

For a *mixed Rossby-Gravity wave*, n = 0, the first order correction (38), yields

$$\sigma \sim \omega + \omega\, AC\{(3sc-1)c^2 + (1-sc)(2sc+1) + B\}/\{4-2sc\}. \tag{46}$$

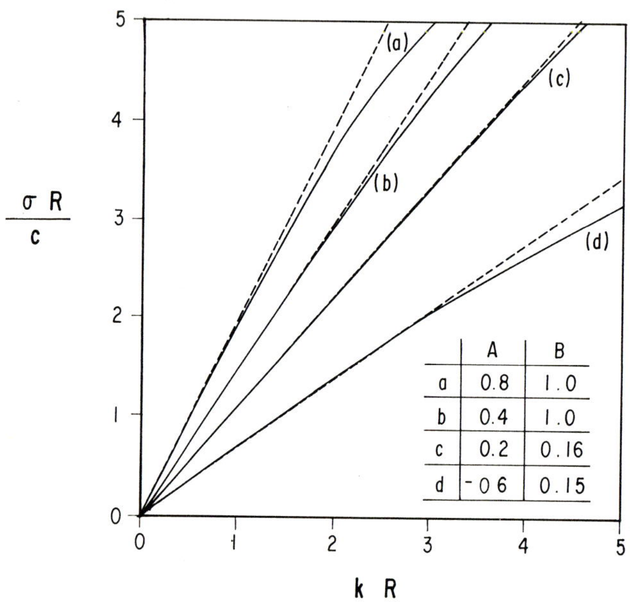

	A	B
a	0.8	1.0
b	0.4	1.0
c	0.2	0.16
d	-0 6	0.15

Fig. 2.- Dispersion relation of a Kelvin wave, for different values of A and B. The dashed and solid lines correspond to the exact, (29), and approximate, (43), solutions respectively. Note the dispersive character for $|kRA| > 0.8$.

For large $|k|$, short waves, we have

$$\sigma = \omega + kc\, AC \text{ for } sc << 1, \tag{47}$$

which corresponds to the Rossby mode limit, Eq. (37), or

$$\sigma = kc\{1 + AC(1 + B/2)\} \quad \text{for } sc \to 1, \tag{48}$$

which corresponds to the gravity mode limit, Eq. (35). Figure 3 shows the dispersion relation, comparing the results of (29) and (46), for two values of A with B fixed. In the Rossby wave limit the phase propagates westward, as found by Philander (1979), even for $A > 0$, unlike what the first order approximation (47) indicates. At low frequencies, the currents have a strong effect due the comparable values of the phase speed and the particles velocity in the basic state (Philander 1979, Philander and Pacanowski 1981), but the effect is not strong enough to reverse the direction of phase-propagation.

Figure 4 shows the dispersion relation of the gravest waves, obtained by solving (29) for the Φ_0 field given by $A = B = 0.15$ {for these values, the flow is stable according to (17) and (18)}. We see that the effect of the current is stronger for the Rossby than for the inertia-gravity waves, because the former have phase velocities that are comparable to $u_0(y)$, and for small B the change of

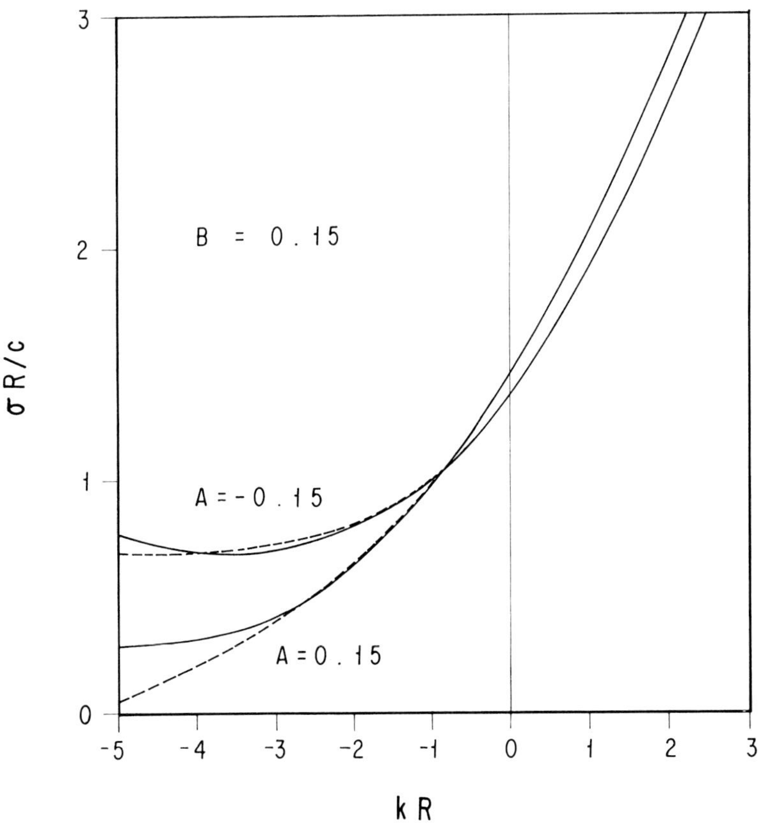

Fig. 3.- Dispersion relation of the mixed Rossby-gravity waves for B = 0.15. The solid and dashed lines correspond to the exact, (29), and approximate, (46), solutions respectively.

depth is not important. The Rossby waves present a reduction of their maximum frequency.

Figure 5 shows the dispersion relation for A = -0.15 and B = 0.15. The effect on the Rossby waves is stronger than in the case of an eastward current (Fig. 4).

Dynamic fields. Examples of the amplitudes $\theta = \{U(y), V(y), Z(y)\}$ are shown in Figs. 6, 7 and 8 for the Rossby (n = 2), mixed Rossby-gravity (n = 0) and Kelvin n = -1) waves for A = ±0.15 and B = 0.15. Also shown, for comparison, are the corresponding structures for A = 0. The amplitudes U and Z are imaginary and V is real. The largest departure from the A = 0 case is found on the maximum shear zone of the basic state {the normalization is fixed by Eq. (33)}. Note too the appearence of a meridional velocity component for the Kelvin wave (Fig. 8).

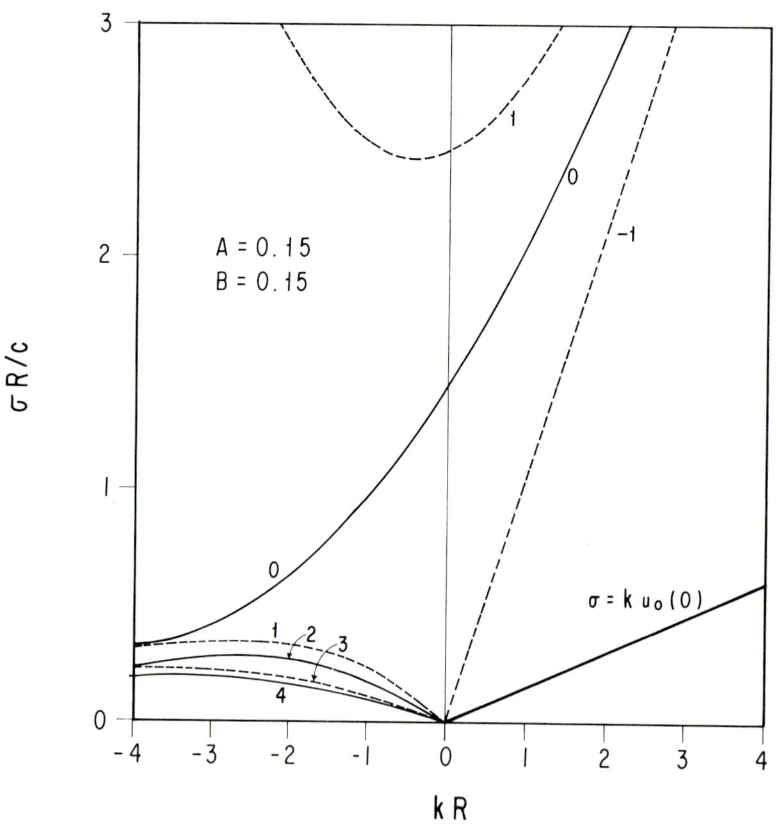

Fig. 4.- Dispersion relation of the equatorial waves modified by the effect of an eastward narrow current (A = B = 0.15). The parameter n corresponds to the meridional quantum number of each mode in the limit $A \to 0$. Solutions with (k,σ): real between the heavy solid line and the $\sigma = 0$ axis have critical latitudes.

Unstable solutions

Growth rate. The maximum of Im(σ) of the unstable waves is shown in Fig. 9 for two values of B (0.15 and 0.5) for A between -2 and 0. Unstable solutions with $A < 0$ & $B < 2^{-1/2}$ must satisfy $A < -B$; at least for narrow easterlies, (17) seems to be also a *necessary* stability condition, in the sense that $A = -B$ is the threshold for instability. Eastward currents are more stable than the westward ones in the sense that: a) the minimum value of $|u_0(0)|$ for instability is greater for the eastward currents, and b) the maximum growth rate is greater for $A < 0$ than for $A > 0$ (not shown), for the same value of $|A|$. These results coincide with those presented by Philander (1976).

The curvature, or β-effect, has a stabilizing influence in the sense that, for fixed A, Im(σ) is greater for smaller values of B (Fig. 9). This effect has been noted before by Charney (1947) and Pedlosky (1964) in analitical models for mid-

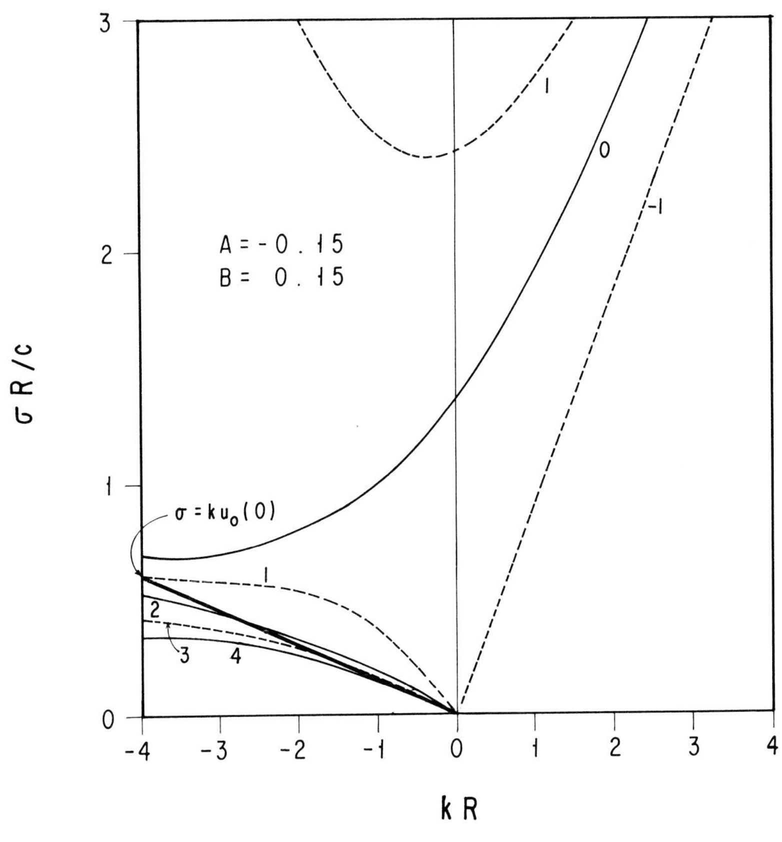

Fig. 5.- As in Fig. 4, for a westward current ($-A = B = 0.15$).

latitudes, and Kuo (1978) in a numerical model.

Dispersion diagram. Here we present dispersion relations which include unstable solutions, for A = -0.6 & B= 0.15 (this represent a jet with a width equal to 0.4 times that of a Kelvin mode), Fig. 10, and for A = -0.9 & B = 1 (i.e., the same width as that of a Kelvin wave), Fig. 11.

If the flow is unstable, there is no value of α such that conditions (15) and (16) are simultaneously satisfied. For B = 0.15, condition (15) (with, say, α = 0) is satisfied, the flow is subcritical, but condition (16) is violated because the gradient of potential vorticity is negative in a neighborhood of the equator (more precisely, in $|y| < .5606L$ for A = -0.6). For B = 1, on the other hand, $\partial_y q_0 \equiv \beta$ for any value of A, and therefore condition (16) is satisfied for $\alpha \geq \max(u_0)$; unstable flows are related to the violation of (15): *e.g.*, at the equator, $u = Ac = -0.9c$ but $c(1+\eta)^{1/2} = c(1+AB)^{1/2} = 0.32c$.

In the first case (Fig. 10), the complex eigenvalues are found where two curves (which in the limit $A \to 0$ correspond to the mixed Rossby-gravity and second Rossby waves) coalesce. However, the instability is not solely due to the interaction of these two modes *via* the fundamental flow, as Fig. 10 might suggest. Namely, if the matrix (29) is truncated down to just these two states, the solution is stable (Fig. 12): In order to obtain unstable solutions it is necessary to have several components interacting with the basic state.

Ripa (1982b) showed that if energy and pseudomomentum *were* exactly quadratic (as they are in the quasi-geostrophic approximation), then the flow with B = 1 would be stable for *any* value of A. He further argued that if the flow is unstable (which can only happen for A < -0.62 or A > 1) then: 1) The amplitude of the flow must be large enough for the terms higher than the quadratic to be important, in the energy and pseudomomentum integrals. 2) The Kelvin component in the expansion (22) of the perturbation must be relatively energetic. We did not find complex eigenvalues for A > 1; the distortion of the dispersion diagram of A = -0.9, Fig. 11, as compared with the A = 0 (no currents) case is an indication that the amplitude required for instability is indeed quite large. {Notice that in Fig. 11 there is little indication of the Kelvin wave. In fact, considering only the interaction among Kelvin modes, (Boyd 1980, Ripa 1982a, or Eq. (43) with B = 1) the phase speed of the Kelvin wave would be $\{1+A(3/2)^{1/2}\}c$, which is equal to -0.1c for A = -0.9.}

Dynamic fields. The amplitudes U, V and Z corresponding to the fastest growing solution shown in Fig. 10 (B = 0.15) are presented in Fig. 13. The relative phase is fixed by making V real at the equator. Similary, in Fig. 14 we show the amplitudes of the most unstable solution from Fig. 11 (B = 1). Since the solution is made up of odd states, the phase is fixed by making U imaginary at the equator.

$$B = \frac{L^2}{R^2} = 0.15 \quad , \quad k = -2.56\ cR^{-1}$$

$$\textit{Curve a} : \ A = \frac{u_0(0)}{c} = 0.15 \quad , \quad \sigma = 0.28\ cR^{-1}$$

$$\textit{Curve b} : \ A = 0 \quad , \quad \sigma = 0.31\ cR^{-1} \quad , \quad n = 2$$

$$\textit{Curve c} : \ A = -0.15 \quad , \quad \sigma = 0.39\ cR^{-1}$$

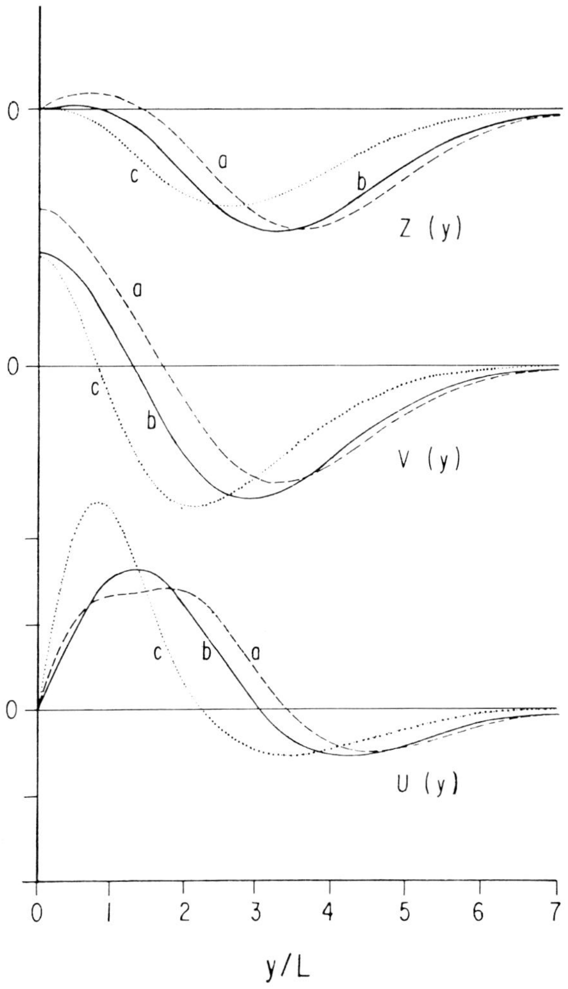

Fig. 6.- Amplitudes U, V and Z as a function of latitude, for the second Rossby mode modified by Gaussian flows and in the absence of currents. The normalization of the fields is given by Eq. (33).

Note that for small values of B (i.e., currents narrower than the Kelvin mode) the strongest instability happens through even perturbations in v (meanders), while for B = 1 the fastest growing disturbance has even u (varicose mode); the former was also found by Philander (1976), whereas the latter is con-

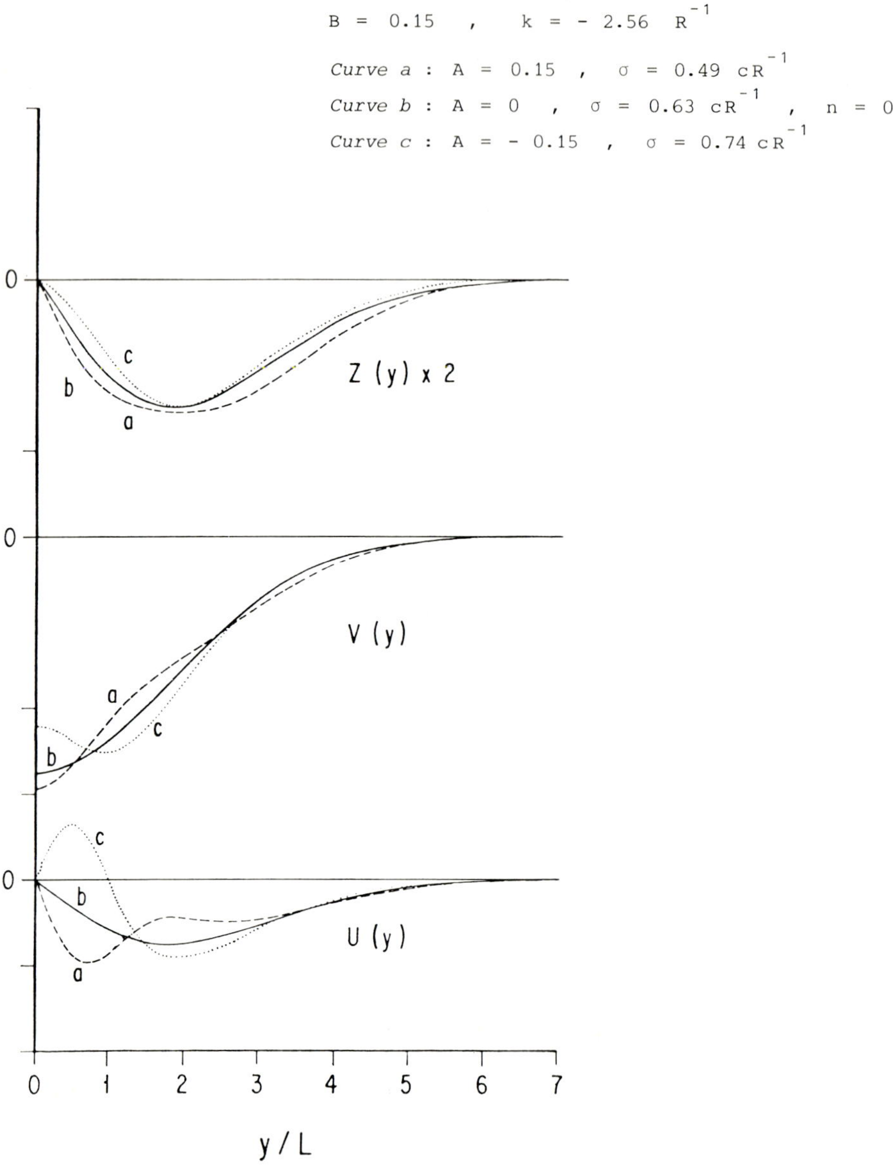

Fig. 7.- As in Fig. 6 for the mixed Rossby-gravity wave.

sistent with the need for a Kelvin component in the growing perturbation, as mentioned above.

In the dispersion relation of Fig. 11, the complex eigenvalues appear in the modes that, in the limit $A \to 0$, are the $n = 7$ & 9 Rossby waves, but the corresponding dynamic fields look more like those of a Kelvin wave, and the major part

$B = 0.15$, $k = - 2.56$ R

Curve a : $A = 0.15$, $\sigma = - 2.76$ cR

Curve b : $A = 0$, $\sigma = - 2.56$ cR , $n = - 1$

Curve c : $A = 0.15$, $\sigma = - 2.36$ cR

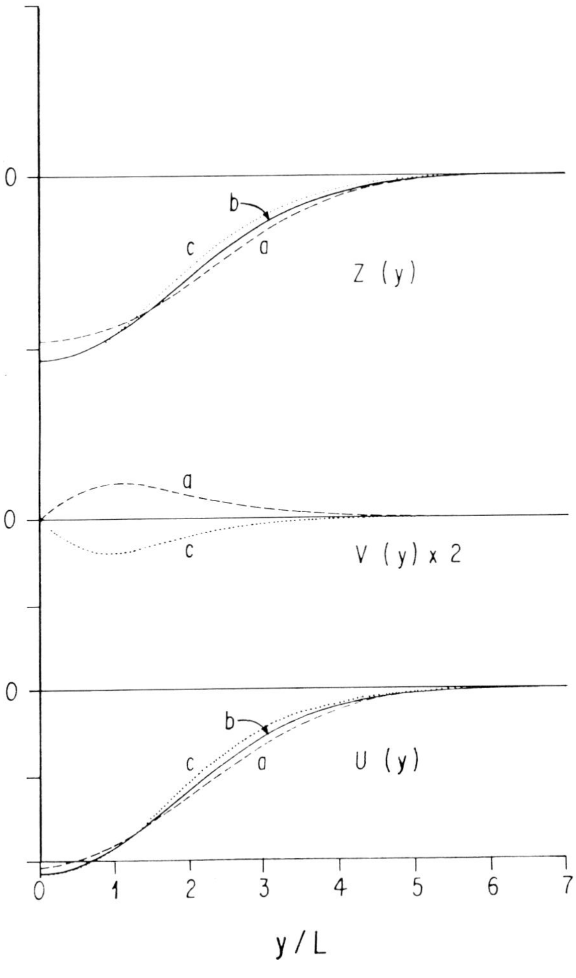

Fig. 8.- As in Fig. 6 for the Kelvin wave.

of the energy is also concentrated in the Kelvin component. (Here, though, we have a finite v_1 field, absent in a pure Kelvin wave.) This is similar to the wave discussed by Hughes (1979, 1981).

4 SUMMARY AND CONCLUSIONS

The effect of a zonal current on the equatorial waves is studied in a one-lay-

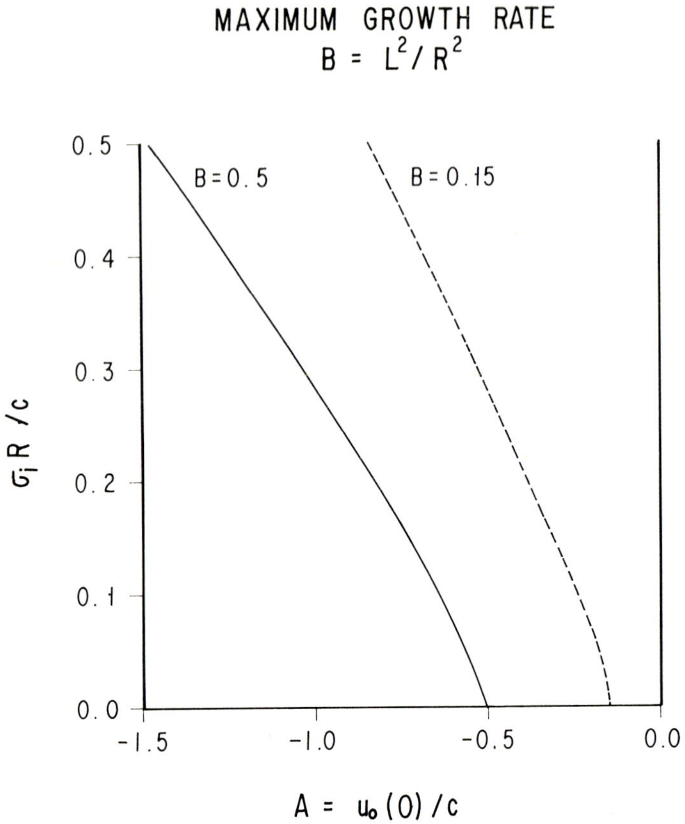

Fig. 9.- Maximum growth rate of the equatorial instability as a function of the current amplitude, for two values of its width.

er reduced gravity model in the equatorial β-plane. The normal-modes problem is solved expanding the perturbation in the solutions of the linear system. Both stable and unstable solutions are found for Gaussian jets with amplitude and width in agreement with the stability criterion derived by Ripa (1982b).

An analitical expression is derived for the frequency shift to first order in the amplitude of the jet. This correction is real and (in the cases of Kelvin, short Rossby and short gravity waves) proportional to the zonal wavenumber. However, the solution of the complete equation shows that the Kelvin waves becomes dispersive in the presence of strong currents. It is also found that, for all the waves, westward currents produce, in general, a stronger Doppler shift than eastward ones do.

Unstable solutions are found for jets such that only one of the two stability conditions is violated; *viz.*, a subcritical and narrow jet with potential vorticity extrema, or a supercritical jet with the shape of a Kelvin mode.

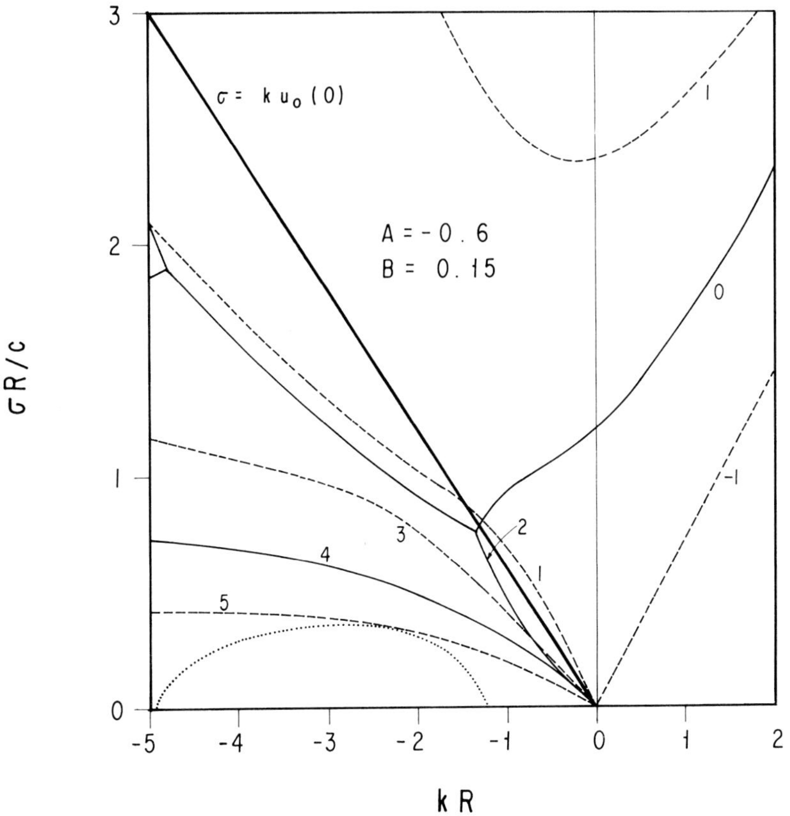

Fig. 10.- As in Fig. 4 for B = 0.15 and A = - 0.6 . The dotted line corresponds to the imaginary part of σ .

Comparing westward and eastward jets with the same strength and width, it is found that the former are more unstable (larger maximum growth rate for the perturbation). The energetics of both types of instability are discussed in a companion publication (Marinone and Ripa, 1983).

Acknowledgements

We are grateful to Sergio Ramos for drafting the figures, to Antoine Badan and Cuauhtémoc Nava for comments, and to Dolores Peralta for typing the manuscript.

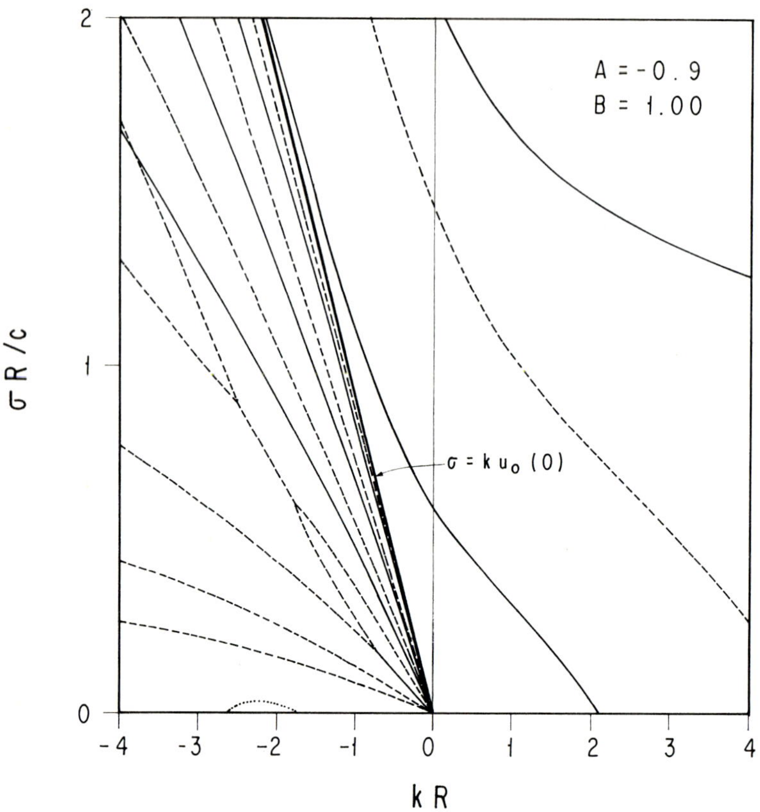

Fig. 11.- As in Fig. 10 for A = - 0.9 and B = 1 (*i.e.*, a current with the same width of a Kelvin mode).

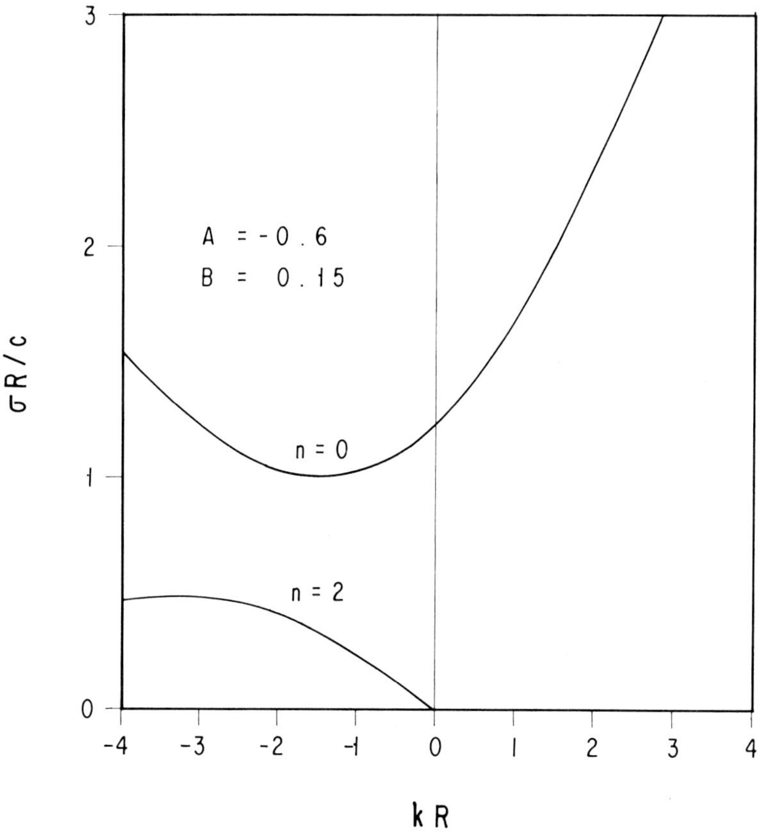

Fig. 12.- Eigenvalues of (29) including only the mixed Rossby-gravity and the second Rossby components. The parameters A and B are the same that those of Fig. 10; however, the frequencies are real, indicating that more than two components are needed in order to get the instability.

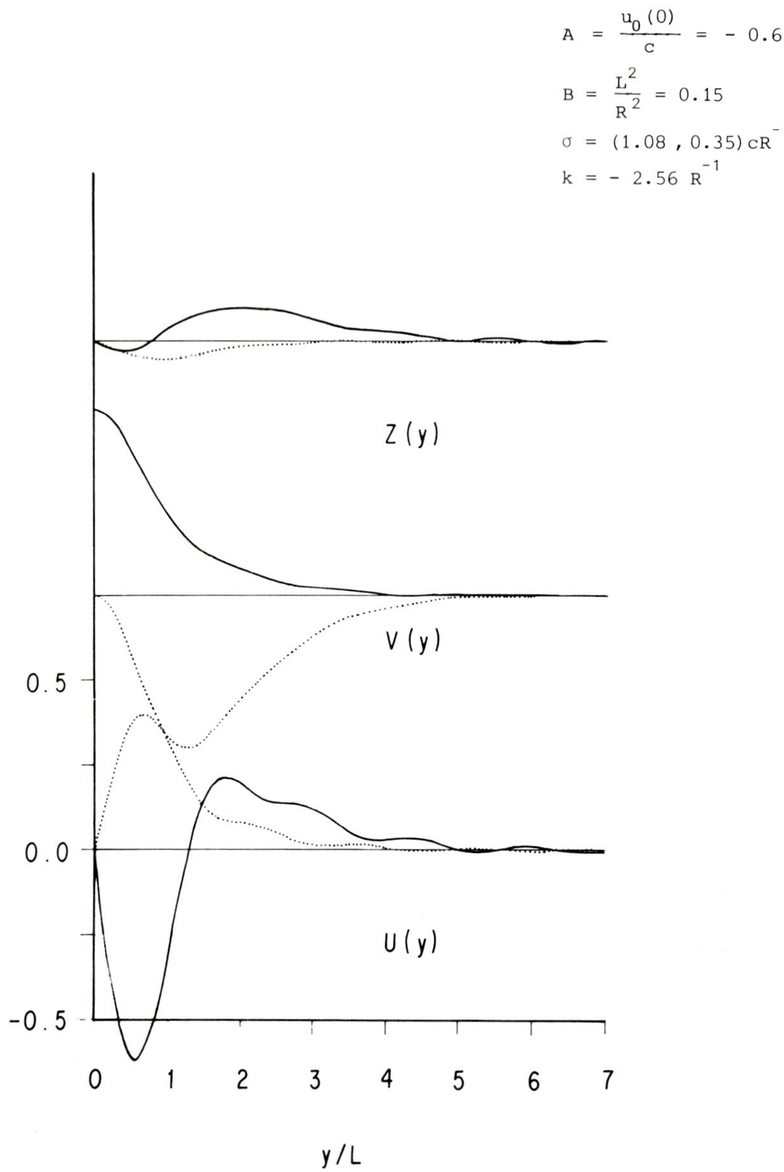

Fig. 13.- Complex amplitude of the dynamic fields u_1, v_1 and η_1 corresponding to the fastest growing solution of Fig. 10. The solid and dashed lines correspond to the real and imaginary parts of the amplitude.

$$A = \frac{u_0(0)}{c} = -0.9$$

$$B = \frac{L^2}{R^2} = 1$$

$$\sigma = (0.77\ ,\ 0.034)\,cR^{-1}$$

$$k = -2.22\ R^{-1}$$

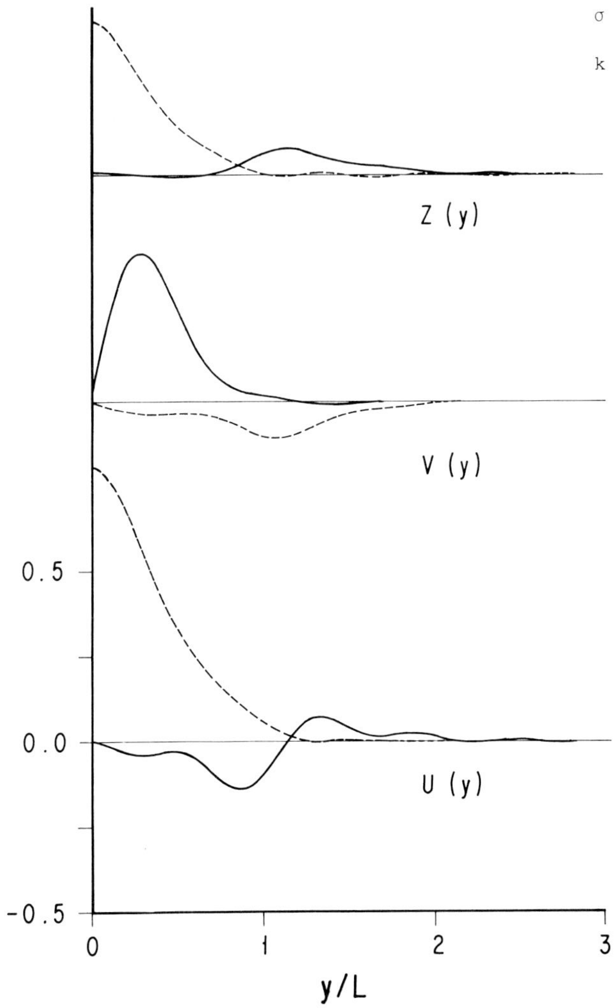

Fig. 14.- As in Fig. 13, for the fasted growing solution of Fig. 11.

APPENDIX. MATRIX ELEMENTS OF C.

The expansion functions take the form

$$\{\tilde{u}_a(\underline{x}) \pm c\tilde{\eta}_a(\underline{x})\}/2^{1/2} = iN_a(\pm)\Psi(n\pm 1,y')\exp(ikx),$$
$$\tilde{v}_a(\underline{x}) = N_a(0)\Psi(n,y')\exp(ikx), \tag{A1}$$

where $y' = (\beta/c)^{1/2}y$, and Ψ are the normalized parabolic cylinder functions (Miller, 1966). The eigenvalues ω and eigencoefficients N are more easily written using the zonal slowness,

$$s = k/\omega \tag{A2}$$

as a parameter. The expression are (Ripa, 1982a)

$$\omega = kc$$
$$(n = -1;\ sc = 1), \tag{A3}$$
$$N(+) = 1,\ N(-) = N(0) = 0$$

for the Kelvin component,

$$\omega^2 = \beta c/(1-sc),\ k = s\omega,$$
$$N(+) = (2-sc)^{-1/2},\ N(-) = 0 \quad (n = 0;\ sc < 1), \tag{A4}$$
$$N(0) = \mathrm{sgn}(\omega)(1-sc)^{1/2}N(+)$$

for the mixed Rossby-gravity components $(sc < 1)$, and

$$\omega^2 = \beta c(2n+1+sc)/(1-s^2c^2),\ k = s\omega,$$
$$N(\pm) = \mathrm{sgn}(1+sc)(1\pm sc)(n+ 1/2\pm 1/2)^{1/2}/Q \quad (n \geq 1), \tag{A5}$$
$$N(0) = \mathrm{sgn}(\omega)\{(2n+1+sc)(1-s^2c^2)\}^{1/2}/Q$$

for the inertia-gravity $(-1 < sc < 1)$ and Rossby $(sc \leq -2n-1)$ components, where $Q = \{4n+2 + sc\{3-(sc)^2\}\}^{1/2}$. The normalization is

$$N_a(+)^2 + N_a(-)^2 + N_a(0)^2 = 1,\quad \int_{-\infty}^{\infty}\Psi(m,y')^2dy' = 1, \tag{A6a,b}$$

which implies

$$\int_{-\infty}^{\infty}(|\tilde{u}_a|^2 + |\tilde{v}_a|^2 + c^2|\tilde{\eta}_a|^2)\,dy' = 1. \tag{A7}$$

The matrix elements Ω_{ab} are defined by (28), where iC is the operator involving u_0 and η_0 in (8) through (10). Expanding the integrand in (28) we obtain

$$i\Omega_{ab} = \int dy' \{\tilde{\theta}_a^{+} u_0 \partial_x \tilde{\theta}_b + \tilde{u}_a^{*} \partial_y u_0 \tilde{v}_b + c^2 \tilde{\eta}_a^{*} \nabla\cdot(\eta_0 \tilde{\underline{v}}_b)\}. \tag{A8}$$

The first term in the expression between { } comes from the advection by the flow u_0; the second one is due to the contribution of the shear $\partial_y u_0$ to the Coriolis term in (8); finally, the third one corresponds to the effect of the topography of the basic flow, η_0, in the mass conservation law (10). The integral in (A8) may be simplified using the following properties of the eigenfunctions of (21):

$$i\omega_a(-ik\tilde{v}_a^{*} - \partial_y \tilde{u}_a^{*} - f\tilde{\eta}_a^{*}) + \beta \tilde{v}_a^{*} = 0, \tag{A9}$$

$$-i\omega_b \tilde{\eta}_b + ik\tilde{u}_b + \partial_y \tilde{v}_b = 0, \tag{A10}$$

to eliminate explicit derivatives. (These two equations are the linearized conservation laws of potential vorticity and mass.)

In the first term in (A8), ∂_x may be replaced by ik. The second term may be written as $\partial_y(\tilde{u}_a^{*} u_0 \tilde{v}_b)$, whose meridional integral obviously vanishes, minus $(\partial_y \tilde{u}_a^{*}) u_0 \tilde{v}_b + \tilde{u}_a^{*} u_0 \partial_y \tilde{v}_b$; the y-derivatives of $\tilde{u}_a^{*}$ and $\tilde{v}_b$ are then evaluated from (A9) and (A10), respectively. In the third and last term, $\nabla.(\eta_0 \tilde{\underline{v}}_b)$ is equal to $\partial_y \eta_0 \tilde{v}_b + \eta_0 \nabla\cdot\tilde{\underline{v}}_b$, which is simplified using (7) and (A10). The three terms in (A8) are thus replaced by

$$\begin{aligned} &1:\ iku_0(\tilde{\underline{v}}_a^{*}\cdot\tilde{\underline{v}}_b + c^2\tilde{\eta}_a^{*}\tilde{\eta}_b), \\ &2:\ u_0\{i(k+\beta/\omega_a)\tilde{v}_a^{*} + f\tilde{\eta}_a^{*}\}\tilde{v}_b + u_0\tilde{u}_a^{*}(-i\omega_b\tilde{\eta}_b + ik\tilde{u}_b), \\ &3:\ -fu_0\tilde{\eta}_a^{*}\tilde{v}_b + c^2\eta_0 i\omega_b\tilde{\eta}_a^{*}\tilde{\eta}_b. \end{aligned} \tag{A11}$$

Adding these three expressions we finally get Eq. (30).

The expressions (A1) are replaced in (30) for the practical evaluation of the matrix elements, which yields

$$\begin{aligned} 2\Omega_{ab} = &\ (3k-\omega_b/c)U(+,+) + (3k+\omega_b/c)U(-,-) \\ &+ (k+\omega_b/c)U(+,-) + (k-\omega_b/c)U(-,+) \\ &+ (4k+2\beta/\omega_a)U(0,0) \\ &+ \omega_b\{D(+,+) - D(+,-) - D(-,+) + D(-,-)\}, \end{aligned} \tag{A12}$$

where

$$U(i,j) = N_a(i)N_b(j) \langle n_a+i|u_0|n_b+j\rangle,$$
$$D(i,j) = N_a(i)N_b(j) \langle n_a+i|\eta_0|n_b+j\rangle \tag{A13}$$

and

$$\langle m'|u_0|m\rangle = \int_{-\infty}^{\infty} dy' u_0(y)\Psi(m',y')\Psi(m,y'),$$
$$\langle m'|\eta_0|m\rangle = \int_{-\infty}^{\infty} dy' \eta_0(y)\Psi(m',y')\Psi(m,y'). \tag{A14}$$

For the Gaussian jet (40) these integrals are evaluated using the analytic expressions derived by Busbridge (1948), in the form

$$\langle m'|u_0|m\rangle = Ac\,\{\pi m'!m!(1+1/2B)\}^{-1/2}\,(-\mu)^{-\lambda} \tag{A15}$$
$$\Gamma(\lambda + 1/2)\; F(-m',-m;\; -\lambda + 1/2;\; \mu),$$

where $\mu = 1/2+B$ and $\lambda = (m'+m)/2$ (λ must be a non-negative integer), and Γ & F are the gamma & hypergeometric functions. Finally, it follows from (41) that $D(i,j) = (B/c)U(i,j)$.

REFERENCES

Boyd, J.P., 1980. The nonlinear equatorial Kelvin wave. J. Phys. Ocean., 10: 1-11.

Busbridge, I.W., 1948. Some integrals involving Hermite polynomials. J. London Math. Soc., 23: 135-141.

Charney, J.G., 1947. The dynamics of long waves in a baroclinic westerly current. J. Meteor., 4: 135-163.

Courant, R. and D. Hilbert, 1953. Methods of mathematical physics. Interscience, New York, 561 pp.

Drazin, P.G., and L.N. Howard, 1966. Hydrodynamic Stability of Parallel Flow of Inviscid Fluid. In 'Advances in applied Mechanics', Academic Press, New York, 9: 1-89.

Hughes, R.L., 1979. On the dynamics of the equatorial undercurrent. Tellus, 31: 447-455.

______, 1981. On inertial instability of the equatorial undercurrent. Tellus, 33: 291-300.

Kasahara, A., 1980. Effect of zonal flows on the free oscillations of a barotropic atmosphere. J. Atmos. Sci., 37: 917-929.

Kuo, H.L., 1978. A two-layer model study of the combined barotropic and baroclinic instability in the tropics. J. Atmos. Sci. 35: 1840-1860.

Marinone, S.G., and P. Ripa., 1983. Energetics of the instability of a depth independent equatorial jet. In preparation.

Matsuno, T., 1966. Quasi-geostrophic motions in the equatorial area. J. Meteor. Soc. Japan, 44: 25-43.

McPhaden, M.J., and R.A. Knox, 1979. Equatorial Kelvin and Inertio-Gravity waves in zonal shear flow. J. Phys. Ocean., 9: 263-277.

Miller, J.C.P., 1966. Parabolic cylinder functions. In 'Handbook of Mathematical Functions', M. Abramowitz and I.A. Stegun, editors. Dover Publication, Inc. Chap. 19, pp. 685-720.

Miyata, M., 1981. A criterion for barotropic instability at the equator. Tropical Ocean-Atmosphere Newsletter, 5, January. (Unpublished manuscript).

Pedlosky, J., 1964. The stability of currents in the atmosphere and the oceans. Part. I. J. Atmos. Sci., 27: 201-219.

Philander, S.G.H., 1976. Instabilities of zonal equatorial currents. J. Geophys. Res., 81: 3725-3735.

________, 1978. Instabilities of zonal equatorial currents, 2. J. Geophys. Res., 83: 3679-3682.

________, 1979. Equatorial waves in the presence of equatorial undercurrent. J. Phys. Ocean., 9: 254-262.

________, and R.C. Pacanowski, 1981. Response of equatorial oceans to periodic forcing. J. Geophys. Res., 86: 1903-1916.

Ripa, P., 1982a. Nonlinear wave-wave interactions in a one-layer reduced gravity model on the equatorial β-plane. J. Phys. Ocean., 12: 97-111.

________, 1982b. General stability conditions for zonal flows in a one-layer model on the β-plane or the sphere. J. Fluid. Mech., in press.

THEORY OF FREQUENCY SPECTRA OF EQUATORIALLY TRAPPED WAVES.

W. FENNEL and H.U. LASS

1. INTRODUCTION

In the late 1960's theoretical investigations of baroclinic motions in a linear inviscid equatorial ocean on a β-plane revealed the possibility of equatorial trapped waves. For a historical review of the theories see e.g. Wunsch (1978).

By separation of variables in case of free motion, the structure of meridional modes and the dispersion relation, classifying Rossby-, inertial-gravity and the mixed Rossby-gravity waves, can be discussed.

Current and water level measurements at various points in the equatorial oceans, performed during the 1970's, gave experimental evidence of equatorial trapped motion coherent over large zonal distances. Since most of the measured time series are expressed in terms of frequency spectra at a fixed position, it would be desirable to calculate the frequency spectra of the dynamic variables from the equation of motion for an adequate comparison of theory and measurements. This has been tried by Wunsch and Gill (1976). However, their results depend on a pseudo-dissipation parameter which does not appear in the basic equation they used and, moreover, they did not derive closed expressions for the frequency spectra in the whole range of possible frequencies.

The aim of the present paper is to determine the power spectra of Eulerian measurements of velocity and pressure in the equatorial wave guide, on the basis of a Greenfunction technique.

We consider vertical standing mode or equivalently a two layered linear, hydrostatic and inviscid ocean on an equatorial β-plane. Although including dissipation processes should be possible, we consider an inviscid theory just for simplicity. Moreover, the main dissipation processes of the ETW's have not been identified at present. This is a further reason to consider a theory which avoids more or less artifical dissipation terms.

In the second section we discuss some aspects of the free motions. In section three, some general properties of the Green functions and their spectral representations will be investigated. Clearly, in an inviscid theory, the causality is not accounted for explicitly. By means of the spectral representations, which are compact representations of the initial conditions the causality will be insured.

In the fourth section, the results will be applied to the calculation of power spectra. This implies the assumption of forcing functions. The ocean fluctuations can be excited in principle by deterministic or stochastic atmospheric fluctuations. We confine ourselves to two deterministic examples of forcing, a broad band and a small band forcing in the zonal wavenumber domain. This is motivated by the following facts :

(i) The description of stochastic fluctuations in terms of their moments requires the stochastic process to be stationary and homogeneous. However, Weisberg and Hortgan (1981) found evidence for a seasonal modulation of the ETW's. Moreover, the long wave length of the ETW's makes any assumption of homogeneity questionable.

(ii) Very little information is available for the wavenumber structure of the atmospheric forcing field above the equatorial ocean. Therefore, it seems to be reasonable to try some examples of forcing functions.

The calculation of the spectrum of the $n = 0$ mode on an equatorial β - plane will be discussed in section five. In the sixth section, we summarize the results of our calculations and compare them with measurements.

2. FREE MOTION

Firstly we consider some special aspects of the free motion problem. For the meridional current we have the well-known boundary value problem (see e.g. Moore and Philander, 1977).

$$D(x,y,t)\ v(x,y,t) = 0 \tag{1}$$

with $v \to 0$ for $y \to \pm\infty$. $D(x,y,t)$ is the differential operator

$$D(x,y,t) = \frac{\partial^3}{\partial t^3} - \frac{\partial}{\partial t}\Delta_h - \frac{\partial}{\partial x} + y^2\frac{\partial}{\partial t} \tag{2}$$

and Δ_h is the horizontal Laplacian

$$\Delta_h = \frac{\partial^2}{\partial x^2} + \frac{\partial^2}{\partial y^2}$$

The zonal component u and the pressure p are determined by v according to

$$u_{tt} - u_{xx} = y\ v_t + v_{xy} \tag{3}$$

$$p_{tt} - p_{xx} = -\ y\ v_x - v_{ty} \tag{4}$$

or, equivalently

$$-\ y\ u_t + u_{xy} = v_{tt} - v_{yy} \tag{5}$$

$$y\ p_x - p_{ty} = v_{tt} - y^2\ v \tag{6}$$

For v we make the usual separation

$$v(x,y,t) = \sum_{n=0}^{\infty} \int \frac{d\omega}{2\pi} \frac{dk}{2\pi} e^{-i\omega t + ikx} S_n(\omega,k) \psi_n(y) \tag{7}$$

where $\psi_n(y)$ are the normalized Hermite functions

$$\psi_n(y) = \frac{e^{-y^2/2} H_n(y)}{\pi^{1/4} \sqrt{2^n n!}}$$

From (1) and (7) follows

$$-i D_n(\omega,k) S_n(\omega,k) = 0 \tag{8}$$

where $D_n(\omega,k)$ is the dispersion polynomial

$$D_n(\omega,k) = \omega^3 - (2n+1)\omega - k^2\omega - k \tag{9}$$

It can be shown that $D_n(\omega,k)$ has three distinct real roots $\omega_{nj}(k)$ $(j = 1,2,3)$ with the properties

$$\omega_{n1} + \omega_{n2} + \omega_{n3} = 0$$

$$\omega_{n1} \cdot \omega_{n2} \cdot \omega_{n3} = k$$

$$\frac{1}{\omega_{n1}} + \frac{1}{\omega_{n2}} + \frac{1}{\omega_{n3}} = -\frac{(2n+1) + k^2}{k}$$

Therefore we may rewrite (9) in the form

$$D_n(\omega,k) = (\omega - \omega_{n1})(\omega - \omega_{n2})(\omega - \omega_{n3}) \tag{10}$$

The graphical representation of the roots ω_{nj} may be found using the roots of (9) with respect to k

$$k_{n1,2} = -\frac{1}{2\omega} \left[1 \pm 2\sqrt{(\omega^2 - \omega_{nG}^2)(\omega^2 - \omega_{nR}^2)}\,\right] \tag{11}$$

The values

$$\omega_{nR,G} = \sqrt{\frac{n+1}{2}} \mp \sqrt{\frac{n}{2}} \tag{12}$$

are the points of zero group velocity (fig. 1).

Generalizing the relation $x\,\delta(x) = 0$ of Dirac's δ-function we find from (8) and (10) the spectral function $S_n(\omega,k)$ to be

$$S_n(\omega,k) = A\,\delta(\omega - \omega_{n1}) + B\,\delta(\omega - \omega_{n2}) + C\,\delta(\omega - \omega_{n3}) \tag{13}$$

Since we deal with a third order differential equation with respect to t, three constants A, B and C appear in the solution (13). Choosing the

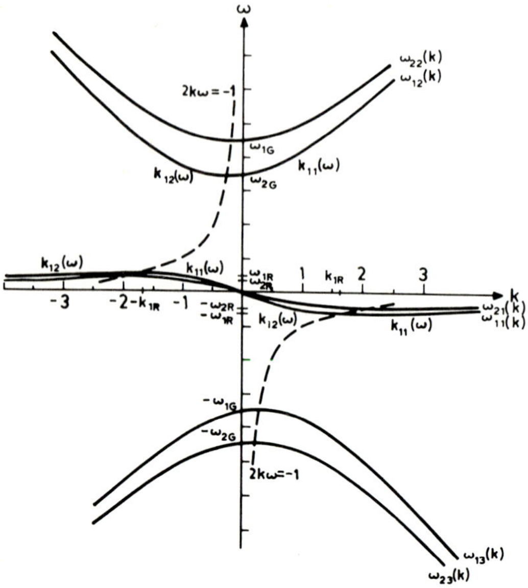

Fig. 1. The dispersion diagram of equatorially trapped waves for $n \geq 1$.

pseudo-initial conditions

$$v_n(x,0) = v_{nt}(x,0) = 0 \tag{14}$$

$$v_{ntt}(x,0) = \delta(x) \tag{15}$$

or, equivalently

$$\int \frac{d\omega}{2\pi} S_n(\omega,k) = \int \frac{d\omega}{2\pi}(-i\omega) S_n(\omega,k) = 0 \tag{16}$$

$$\int \frac{d\omega}{2\pi}(-\omega^2) S_n(\omega,k) = 1 \tag{17}$$

we find, inserting (13) in (16) and (17), a linear system

$$\begin{pmatrix} 1 & 1 & 1 \\ \omega_{n1} & \omega_{n2} & \omega_{n3} \\ \omega_{n1}^2 & \omega_{n2}^2 & \omega_{n3}^2 \end{pmatrix} \begin{pmatrix} A \\ B \\ C \end{pmatrix} = \begin{pmatrix} 0 \\ 0 \\ -2\pi \end{pmatrix}$$

and it follows that

$$S_n(\omega,k) = -2\pi \left\{ \frac{\delta(\omega-\omega_{n1})}{(\omega_{n2}-\omega_{n1})(\omega_{n3}-\omega_{n1})} - \frac{\delta(\omega-\omega_{n2})}{(\omega_{n2}-\omega_{n1})(\omega_{n3}-\omega_{n2})} + \frac{\delta(\omega-\omega_{n3})}{(\omega_{n3}-\omega_{n1})(\omega_{n3}-\omega_{n2})} \right\}$$

or

$$S_n(\omega,k) = -2\pi \sum_{j=1}^{3} \frac{\delta(\omega - \omega_{nj})}{\frac{\partial}{\partial\omega} D_n(\omega,k)\Big|_{\omega=\omega_{nj}}} \tag{18a}$$

In the usual discussion of the free motion, v_n is given in the form

$$v_n(x,t) \sim e^{-i\omega t + ikx}$$

where ω and k are connected by a dispersion relation $\omega = \omega_{nj}(k)$. It is easy to show that such a representation implies initial conditions of the type

$$v_n(x,0) = v_{nt}(x,0) = 0 \quad , \quad v_{ntt}(x,0) = e^{ik_0x}$$

in the ω-k-domain we have

$$\int v_n(\omega,k) \frac{d\omega}{2\pi} = \int - i\omega\, v_n(\omega,k) \frac{d\omega}{2\pi} = 0$$

$$\int - \omega^2\, v_n(\omega,k) \frac{d\omega}{2\pi} = 2\pi\, \delta(k - k_0)$$

This corresponds to an initial small band forcing in the k-domain. The result is

$$v_n(\omega,k) = 2\pi\, \delta(k - k_0)\, S_n(\omega,k) \tag{18b}$$

After Fourier transformation, one obtains

$$v_n(x,t)\, \psi_n(y) = \psi_n(y) \sum_{j=1}^{3} \frac{e^{-i\,\omega_{nj}(k_0)\,t + ik_0x}}{\frac{\partial}{\partial\omega} D_n(\omega,k_0)\Big|_{\omega=\omega_{nj}(k_0)}}$$

Thus, we see that the usual solutions of the free motion problem describe the response of the ocean to a pulse like forcing with a spatial pattern like e^{ik_0x}.

The physical meaning of the spectral function $S_n(\omega,k)$ according to eq. (18a) is the following : $S_n(\omega,k)$ is the Fourier transform of a wave process $v_n(x,t)$ which is analytically continued in the negative time domain. This function contains no information about the disturbances of the "switch on" process at $t = 0$ and, therefore, $S_n(\omega,k)$ mirrors the pure properties of the wave process. In other words we may consider $S_n(\omega,k)$ to be the spectral function of a process that was forced at $t = -\infty$. This becomes evident in the discussion of the Green function in the following section. In the following we consider only the modes $n \geq 1$. The mode $n = 0$ will be discussed in section 5. For $n \geq 1$ we may express the u- and p-field in terms of $S_n(\omega,k)$ in the usual manner

$$u_n(\omega,k,y) = \frac{i}{2}\left\{\frac{\sqrt{2n+2}}{\omega - k}\, \psi_{n+1}(y) + \frac{\sqrt{2n}}{\omega + k}\, \psi_{n-1}(y)\right\} S_n(\omega,k) \tag{19}$$

$$p_n(\omega,k,y) = \frac{i}{2}\left\{\frac{\sqrt{2\,n+2}}{\omega - k}\,\psi_{n+1}(y) - \frac{\sqrt{2n}}{\omega + k}\,\psi_{n-1}(y)\right\} S_n(\omega,k) \tag{20}$$

3. GREEN FUNCTIONS AND SPECTRAL REPRESENTATIONS

Including a forcing function $F(x,y,t)$, we have, instead of eq. (1) :

$$D(x,y,t)\; v(x,y,t) = F(x,y,t) \tag{21}$$

We confine ourselves to two examples of forcing functions, first a broad band forcing given by Dirac's δ-functions

$$F(x,y,t) = \delta(x-x')\;\delta(y-y')\;\delta(t-t') \tag{22}$$

and, second, to a small band forcing in the wave number domain

$$F(x,y,t) = e^{ik_0(x-x')}\;\delta(y-y')\;\delta(t-t') \tag{23}$$

Clearly, with eqs. (21) and (22) we have to determine the Green function of our boundary value problem

$$D(x,y,t)\; G(x,y,t;x',y',t') = \delta(x-x')\;\delta(y-y')\;\delta(t-t') \tag{24a}$$

This function is of special importance since it leads in principle to a formal solution of the general problem (21) by means of

$$v(x,y,t) = \int dx'\;dy'\;dt'\;G(x,y,t;x',y',t')\;F(x',y',t')$$

Let $K(x,y,t;x',y',t')$ be the solution of the problem (21) and (23). This function is also a Green function

$$D(x,y,t)\; K(x,y,t;x',y',t') = \delta(y-y')\;\delta(t-t')\;e^{ik_0(x-x')} \tag{24b}$$

To calculate this function, we expand G, K and $\delta(y-y')$ in terms of Hermite-functions $\psi_n(y)$

$$G(x,y,t;x',y',t') = \sum_n G_n(x,t;x',t')\;\psi_n(y)\;\psi_n(y')$$

$$K(x,y,t;x',y',t') = \sum_n K_n(x,t;x',t')\;\psi_n(y)\;\psi_n(y')$$

$$\delta(y-y') = \sum_n \psi_n(y)\;\psi_n(y')$$

From eq. (2), it follows that G and K depend only on the differences $x-x'$ and $t-t'$. After Fourier transformation with respect to $x-x'$ and $t-t'$, we have

$$G_n(\omega,k) = \frac{i}{D_n(\omega,k)} \tag{25a}$$

$$K_n(\omega,k) = \frac{2\pi i\ \delta(k-k_0)}{D_n(\omega,k_0)} \tag{25b}$$

where $D_n(\omega,k)$ is the dispersion polynomial (9).

The transformation of eqs. (25a) and (25b) in the x,t-domain constitutes a problem since these functions possess poles on the real ω-axis at $\omega = \omega_{nj}$ $(j = 1,2,3)$. Due to the different kinds of Green functions which resolve eqs. (24a) and (24b) there are various possibilities to go around these poles.

The prescription how to go around the poles follows from the principle of causality. To show this we consider the retarded Green functions

$$G^r(\tau) = \Theta(\tau)\ G_{hom}(\tau) \tag{27a}$$

$$(\tau = t - t')$$

$$K^r(\tau) = \Theta(\tau)\ K_{hom}(\tau) \tag{27b}$$

$\Theta(\tau)$ is the step function $[\Theta(\tau) = 1$ for $\tau > 0$ and $\Theta(\tau) = 0$ for $\tau < 0]$. Inserting eqs. (27a,b) in eqs. (24a,b), respectively, we see that G_{hom} and K_{hom} are the solutions of the homogeneous problems $DG_{hom} = 0$ and $DK_{hom} = 0$ with the initial conditions

$$G_{hom}(0) = \frac{\partial}{\partial\tau} G_{hom}(\tau)\Big|_{\tau=0} = 0 \quad , \quad \frac{\partial^2}{\partial\tau^2} G_{hom}(\tau)\Big|_{\tau=0} = -\ \delta(x-x')\ \delta(y-y') \tag{28a}$$

$$K_{hom}(0) = \frac{\partial}{\partial\tau} K_{hom}(\tau)\Big|_{\tau=0} \quad , \quad \frac{\partial^2}{\partial\tau^2} K_{hom}(\tau)\Big|_{\tau=0} = -\ e^{ik_0(x-x')}\ \delta(y-y') \tag{28b}$$

Fourier transformation of eq. (27a) yields

$$G_n^r(\omega,k) = \int \frac{d\omega'}{2\pi}\ S_n(\omega',k) \int d\tau\ e^{i(\omega-\omega')\tau}\ \Theta(\tau) \tag{29a}$$

We see that the Fourier transform of G_{hom} is just the spectral function (18a) discussed in the previous section. This is clear since the initial value problem for G_{hom} is identical with that given by eqs. (1) and (14), (15).

Similarly, we find, using eq. (18b),

$$K_n^r(\omega,k) = \int \frac{d\omega'}{2\pi}\ S_n(\omega',k)\ 2\pi\ \delta(k-k_0) \int d\tau\ \Theta(\tau)\ e^{i(\omega-\omega')\tau} \tag{29b}$$

Now we make use of the contour integral representation of the step function

$$\Theta(\tau) = \frac{i}{2\pi} \int d\omega\ \frac{e^{-i\omega\tau}}{\omega + i\varepsilon} \qquad (\varepsilon \to +0) \tag{30}$$

ε is a positive infinitesimal quantity. Consider ω to be a complex variable. The function $\frac{1}{\omega + i\varepsilon}$ has a simple pole in the lower half plane at $\omega = -\ i\varepsilon$ (see fig. 2). For $\tau > 0$ according to Jordan's Lemma, we can complete the integration path by a semicircular contour of radius $R \to \infty$ in the lower

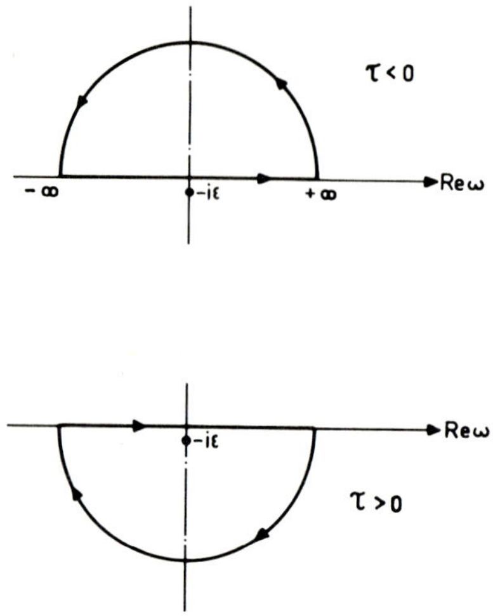

Fig. 2. Integration path of the contour integral representation of the step function for $\tau < 0$ and $\tau > 0$.

half plane. Applying the residue theorem, the integral is found equal to 1. For negative τ, we must use a semicircular contour in the upper half plane and the integral vanishes. Therefore eq. (30) is in fact a representation of the step function. With eq. (30), it follows from eqs. (29a) and (29b)

$$G_n^r(\omega,k) = -\int \frac{d\omega'}{2\pi i} \frac{S_n(\omega',k)}{\omega - \omega' + i\varepsilon} \tag{31a}$$

$$K_n^r(\omega,k) = -\int \frac{d\omega'}{2\pi i} \frac{2\pi\ \delta(k - k_0)\ S_n(\omega',k)}{\omega - \omega' + i\varepsilon} \tag{31b}$$

These are the spectral representations of the retarded Green functions G^r and K^r.

Using Dirac's formula

$$\frac{1}{\omega \pm i\varepsilon} = \frac{P}{\omega} \mp i\pi\ \delta(\omega) \tag{32}$$

where the letter P denotes a Cauchy principle value, we have

$$\frac{1}{i}\ G_n^r(\omega,k) = P\int \frac{d\omega'}{2\pi} \frac{S_n(\omega',k)}{\omega - \omega'} - \frac{i}{2}\ S_n(\omega,k) \tag{33a}$$

$$\frac{1}{i}\ K_n^r(\omega,k) = P\int \frac{d\omega'}{2\pi} \frac{S_n(\omega',k)\ 2\pi\ \delta(k - k_0)}{\omega - \omega'} - i\pi\ S_n(\omega,k)\ \delta(k - k_0) \tag{33b}$$

These are the general solutions of eqs. (24a) and (24b), respectively, which consist of the general solutions of the homogeneous problems $S_n(\omega,k)$ and $2\pi\ \delta(k - k_0)\ S_n(\omega,k)$ and the special solutions of the non-homogeneous problems in terms of the principle value integrals. From eqs. (33a) and (33b), we see that

$$2\ \mathrm{Im}\ \frac{1}{i}\ G_n^r(\omega + i\varepsilon\ ,\ k) \ = \ -\ S_n(\omega,k) \qquad (34a)$$

$$(\varepsilon \rightarrow 0)$$

$$2\ \mathrm{Im}\ \frac{1}{i}\ K_n^r(\omega + i\varepsilon\ ,\ k) \ = \ -\ 2\pi\ \delta(k - k_0)\ S_n(\omega,k) \qquad (34b)$$

Now we can see how the formal solutions (25a,b) must be modified in order to be consistent with the initial conditions (28a,b) and the principle of causality

$$G_n^r(\omega,k) \ = \ \frac{i}{D_n(\omega + i\varepsilon\ ,\ k)}$$

Partial fraction expansion yields

$$G_n^r(\omega,k) \ = \ i\ \sum_{j=1}^{3}\ \frac{1}{(\omega + i\varepsilon - \omega_{nj})\ \frac{\partial}{\partial\omega}\ D_n(\omega,k)\ \Big|_{\omega=\omega_{nj}}} \qquad (35)$$

Taking from eq. (35), $\mathrm{Im}\ \frac{1}{i}\ G_n^r$, one finds using eq. (32) the relation (34a). The result for K_n^r is simply found by multiplying eq. (35) by $2\pi\ \delta(k - k_0)$. We note that the spectral functions are already given by eqs. (18a,b). Clearly the prescription how to go around the poles is given by means of the positive infinitesimal ε .

The discussion along those lines is well known in other branches of theoretical physics [e.g. quantum field theory or many particle theory, see e.g. Wu and Omura (1962) or Martin and Schwinger (1959)]. Of central importance is the fact that causality and spectral representations are intimatly connected. Therefore we can conclude that such representations also exist for the zonal velocity and for the pressure field. Using the spectral functions $u_n(\omega,k,y)$ and $p_n(\omega,k,y)$ given by eqs. (19) and (20) and denoting the corresponding Green functions U_n and P_n , we have

$$U_n(\omega,k,y) \ = \ -\int \frac{d\omega'}{2\pi i}\ \frac{u_n(\omega',k,y)}{\omega\ -\ \omega'\ +\ i\varepsilon} \qquad (36)$$

$$P_n(\omega,k,y) \ = \ -\int \frac{d\omega'}{2\pi i}\ \frac{p_n(\omega',k,y)}{\omega\ -\ \omega'\ +\ i\varepsilon} \qquad (37)$$

for the broad band forcing case (22). In the small band forcing case, eqs. (36) and (37) must be multiplied by $2\pi\ \delta(k - k_0)$.

4. SPECTRA OF EULERIAN TIME SERIES

Now we consider a time series of a meridional current component, say $G_n^r(x,t)$ at fixed a . We choose $x = x'$ without loss of generality. In the ω-k-domain this is equivalently to a k-integration of $G_n^r(\omega,k)$

$$\hat{G}_n^r(\omega) = \int \frac{dk}{2\pi} G_n^r(\omega,k)$$

$\hat{G}_n^r(\omega)$ is just the frequency spectrum of the current amplitude. Introducing the correlation function

$$R_n^v(\tau) = \frac{1}{T}\int_{-T/2}^{+T/2} \hat{G}_n^r(t'+\tau)\, \hat{G}_n^r(t')\, dt'$$

where the length of the time series is assumed to be large $(T \to \infty)$, we have after Fourier transformation

$$R_n^v(\omega) = \frac{1}{T}\,|\hat{G}_n^r(\omega)|^2 = \frac{1}{T}\left|\int \frac{dk}{2\pi} G_n^r(\omega,k)\right|^2$$

Now we may insert G_n^r according to eq. (33a) and find

$$R_n^v(\omega) = \frac{1}{T}\left[P\int\frac{d\omega'}{2\pi}\int\frac{dk}{2\pi}\,\frac{S_n(\omega',k)}{\omega-\omega'}\right]^2 + \frac{1}{4T}\left[\int\frac{dk}{2\pi}\, S_n(\omega,k)\right]^2 \tag{38}$$

For our further discussion, it is necessary to estimate

$$\hat{S}_n(\omega) = \int \frac{dk}{2\pi} S_n(\omega,k)$$

To this end, we use the spectral function (18a). It is readily seen that nothing is changed by the introduction of step functions

$$\begin{aligned} S_n(\omega,k) = \Theta(\omega_{nR}-|\omega|) & \frac{2\pi\,\delta(\omega-\omega_{n1})}{|(\omega_{n2}-\omega_{n1})(\omega_{n3}-\omega_{n1})|} \\ & -\Theta(\omega-\omega_{nG})\frac{2\pi\,\delta(\omega-\omega_{n2})}{|(\omega_{n2}-\omega_{n1})(\omega_{n3}-\omega_{n2})|} \\ & -\Theta(-\omega-\omega_{nG})\frac{2\pi\,\delta(\omega-\omega_{n3})}{|(\omega_{n3}-\omega_{n1})(\omega_{n3}-\omega_{n2})|} \end{aligned}$$

Here the relations $\omega_{n2}-\omega_{n1} > 0$, $\omega_{n3}-\omega_{n1} < 0$ and $\omega_{n3}-\omega_{n2} < 0$ which are valid for all k (fig. 1) were used. Now, we may write

$$S_n(\omega,k) = 2\pi\,[\Theta(\omega_{nR}-|\omega|) - \Theta(|\omega|-\omega_{nG})]\,\delta[D_n(\omega,k)] \tag{39}$$

where the δ-function relation

$$\delta[f(x)] = \sum_n \frac{\delta(x-x_n)}{f'(x_n)}, \quad f(x_n) = 0$$

was used.

This gives, respecting eq. (11),

$$\hat{S}_n(\omega) = [\Theta(\omega_{nR}-|\omega|) - \Theta(|\omega|-\omega_{nG})]\int dk\,\delta\{[k-k_{n1}(\omega)][k-k_{n2}(\omega)]\}$$

$$\hat{S}_n(\omega) = \frac{\Theta(\omega_{nR} - |\omega|) - \Theta(|\omega| - \omega_{nG})}{|\omega(k_{n1} - k_{n2})|} \left[\int dk\ \delta(k - k_{n1}) + \int dk\ \delta(k - k_{n2}) \right]$$

With

$$|\omega(k_{n1} - k_{n2})| = 2\sqrt{(\omega^2 - \omega_{nR}^2)(\omega^2 - \omega_{nG}^2)}$$

and

$$\int dk\ [\delta(k - k_{n1}) + \delta(k - k_{n2})] = 2\ [\Theta(\omega_{nR} - |\omega|) + \Theta(|\omega| - \omega_{nG})]$$

we have

$$\hat{S}_n(\omega) = \frac{\Theta(\omega_{nR} - |\omega|) - \Theta(|\omega| - \omega_{nG})}{\sqrt{(\omega^2 - \omega_{nR}^2)(\omega^2 - \omega_{nG}^2)}} \qquad (40a)$$

The spectrum of eq. (40a) is the frequency spectrum of the meridional current components of the ETW which are excited by a broad band forcing, long before our "time series" started. In this case we may assume that the ETW forgot the disturbances of the "switch on" processes. Then the corresponding power spectrum is

$$R_n^v(\omega) = \frac{1}{4T} \frac{\Theta(\omega_{nR} - |\omega|) + \Theta(|\omega| - \omega_{nG})}{|(\omega^2 - \omega_{nR}^2)(\omega^2 - \omega_{nG}^2)|} \qquad (41)$$

If the forcing occured within our time series, the power spectrum will be smeared out. This process is accounted for by the principle value integral in eq. (33a) and, therefore, in eq. (38).

Thus

$$R_n^v(\omega) = \frac{1}{T} \left[P \int \frac{d\omega'}{2\pi} \frac{\Theta(\omega_{nR} - |\omega'|) - \Theta(|\omega'| - \omega_{nG})}{\sqrt{(\omega'^2 - \omega_{nR}^2)(\omega'^2 - \omega_{nG}^2)}\ (\omega - \omega')} \right]^2 + \frac{1}{4T} \frac{\Theta(\omega_{nR} - |\omega|) + \Theta(|\omega| - \omega_{nG})}{|(\omega^2 - \omega_{nR}^2)(\omega^2 - \omega_{nG}^2)|} \qquad (42)$$

Now we determine the principle value integral in eq. (42). With respect to

$$P \int \frac{d\omega'}{2\pi} \frac{\hat{S}_n(\omega')}{\omega - \omega'} = P \int \frac{d\omega'}{2\pi} \int \frac{dk}{2\pi} \frac{S_n(\omega',k)}{\omega - \omega'} = P \int \frac{dk}{2\pi} \frac{1}{D_n(\omega,k)}$$

one can write

$$P \int \frac{d\omega'}{2\pi} \frac{\hat{S}_n(\omega')}{\omega - \omega'} = \begin{cases} -\frac{1}{2} \frac{1}{\sqrt{(\omega^2 - \omega_{nR}^2)(\omega_{nG}^2 - \omega^2)}} & \text{for} \quad \omega_{nR} < |\omega| < \omega_{nG} \\ 0 & \text{otherwise} \end{cases}$$

Thus the power spectrum is explicitly

$$R_n^v(\omega) = \frac{1}{4T} \frac{1}{|(\omega^2 - \omega_{nR}^2)(\omega^2 - \omega_{nG}^2)|} \qquad (43a)$$

We see that the effect of the "switch on" process is to fill the gap between the two peaks which occur in the spectrum of eq. (41) of the quasi steady wave process which was forced at $t = -\infty$. We note some properties of eq. (43). For large ω the spectrum decreases like ω^{-4}

$$R_n^v(\omega) \approx \frac{1}{4T} \frac{1}{\omega^4} \qquad \text{large } \omega \tag{44}$$

and for small ω it follows the law

$$R_n^v(\omega) \approx \frac{1}{T} \qquad \text{small } \omega \tag{45}$$

Eqs. (44) and (45) are valid for all mode numbers n . Obviously, the spectrum of the meridional component is peaked at the frequencies of zero group velocity for all mode numbers $n \geq 1$. The y-dependence of the power spectra (43) or (41) may be obtained by multiplying these formulas by $\psi_n^2(y)$

$$R_n^v(\omega,y) = R_n^v(\omega)\ \psi_n^2(y) \tag{46}$$

Now we estimate the power spectra of the zonal current and of the pressure field.

$$R_n^u(\omega,y) = \frac{1}{T} \left| \int \frac{dk}{2\pi} U_n(\omega,k,y) \right|^2 \tag{47}$$

$$R_n^p(\omega,y) = \frac{1}{T} \left| \int \frac{dk}{2\pi} P_n(\omega,k,y) \right|^2 \tag{48}$$

From eqs. (36), (37) together with eqs. (19) and (20), we see that the following integrals must be calculated

$$I_1(\omega) = \int \frac{dk}{2\pi} \sqrt{2n+2}\ \frac{S_n(\omega,k)}{\omega - k} = \frac{2\omega^2 + 1}{2\omega\sqrt{2n+2}}\ \hat{S}_n(\omega) \tag{49}$$

$$I_2(\omega) = \int \frac{dk}{2\pi} \sqrt{2n}\ \frac{S_n(\omega,k)}{\omega + k} = \frac{2\omega^2 - 1}{2\omega\sqrt{2n}}\ \hat{S}_n(\omega) \tag{50}$$

We have

$$I_3(\omega) = P \int \frac{I_1(\omega')}{\omega - \omega'} \frac{d\omega'}{2\pi} = \begin{cases} \dfrac{2\omega^2 + 1}{4\omega\sqrt{2n+2}\sqrt{(\omega_{nR}^2 - \omega^2)(\omega^2 - \omega_{nG}^2)}} & \text{for } \omega_{nR} \leq |\omega| \leq \omega_{nG} \\ 0 & \text{otherwise} \end{cases} \tag{51}$$

and

$$I_4(\omega) = P \int \frac{I_2(\omega')}{\omega - \omega'} \frac{d\omega'}{2\pi} \quad \begin{cases} \dfrac{2\omega^2 - 1}{4\omega\sqrt{2n}\sqrt{(\omega_{nR}^2 - \omega^2)(\omega^2 - \omega_{nG}^2)}} & \text{for } \omega_{nR} \leq |\omega| \leq \omega_{nG} \\ 0 & \text{otherwise} \end{cases} \tag{52}$$

where $\hat{S}_n(\omega)$ is given by eq. (40a).

Inserting eqs. (49) to (52) in eqs. (47) and (48), one finds

$$R_n^{u,p}(\omega,y) = \frac{1}{16\,T}\,\frac{1}{\omega^2\,|(\omega^2-\omega_{nR}^2)(\omega^2-\omega_{nG}^2)|}\left\{\frac{(2\,\omega^2+1)^2}{2n+2}\,\psi_{n+1}^2(y) \pm \frac{4\,\omega^4-1}{\sqrt{n(n+1)}}\,\psi_{n+1}(y)\,\psi_{n-1}(y) + \frac{(2\,\omega^2-1)^2}{2n+2}\,\psi_{n-1}^2(y)\right\} \tag{53a}$$

Both for small and for large ω, eq. (53a) varies like ω^{-2}. Clearly, eqs. (43a) and (53a) are due to the broad band forcing case.

In the case of the small band forcing [eq. (23)], the calculations are considerably simplified. The k-integration of eq. (18b) yields

$$\hat{S}_n(\omega) = -\,2\pi\sum_{j=1}^{3}\frac{\delta[\omega-\omega_{nj}(k_0)]}{\frac{\partial}{\partial\omega}\,D_n^2(\omega,k_0)} \tag{40b}$$

Thus, for a time series of a process excited at $t\to-\infty$, we have a line spectrum. The peaks occur at $\omega=\omega_{nj}(k_0)$. If the forcing occurs within the considered time series, we find the power spectrum by inserting eq. (18b) in eq. (38)

$$R_n^{v}(\omega) = \frac{1}{4T}\,\frac{1}{D_n^2(\omega,k_0)} \tag{43b}$$

The power spectrum is peaked at the frequencies $\omega=\omega_{nj}(k_0)$ where the dispersion curves intersect the line $k=k_0$. Clearly, these are generally not the points of zero group velocity. The power spectrum (43b) decreases like ω^{-6} in the frequency limit and is constant for small ω. Thus the peak frequencies as well as the spectral slope at high frequencies differ significantly in the cases of small and broad band forcing.

The corresponding power spectra of the zonal current and the pressure in the small band forcing case follow simply from eqs. (47) and (48) using eqs. (36) and (37) with the spectral functions (19) and (20) multiplied by $2\pi\,\delta(k-k_0)$. The result is

$$\begin{aligned} R_n^{u,p}(\omega,y) = \frac{1}{2T}\Big\{&(n+1)\,\psi_{n+1}^2(y)\,\Big[\frac{1}{D_n(\omega,k_0)}-\frac{1}{D_n(k_0,k_0)}\Big] \\ &+ n\,\psi_{n-1}^2(y)\,\Big[\frac{1}{D_n(\omega,k_0)}-\frac{1}{D_n(-k_0,k_0)}\Big] \\ &\pm 2n\sqrt{n+1}\,\psi_{n+1}(y)\,\psi_{n-1}(y)\,\Big[\frac{1}{D_n(\omega,k_0)}-\frac{1}{D_n(k_0,k_0)}\Big]\times \\ &\Big[\frac{1}{D_n(\omega,k_0)}-\frac{1}{D_n(-k_0,k_0)}\Big]\Big\} \end{aligned} \tag{53b}$$

For large ω as well as for small ω, $R_n^{u,p}$ becomes independent of ω.

5. THE MODE $n = 0$

Now we discuss some properties of the mode $n = 0$. Here we consider only the broad forcing case. With $n = 0$, from eq. (40a), we get

$$\hat{S}_0(\omega) = \frac{2}{1 - 2\omega^2} \tag{54}$$

and from eq. (43a), we have

$$R_0^v(\omega) = \frac{1}{T} \frac{1}{(1 - 2\omega^2)^2} \tag{55}$$

Although this result seems to be very reasonable, we must be aware that the three roots of the dispersion polynomal

$$D_0(\omega,k) = \omega^3 - \omega - k^2\omega - k$$

$$\omega_{01} = -k \tag{56}$$

$$\omega_{02,3} = \frac{k}{2} \pm \sqrt{\frac{k^2}{4} + 1} \tag{57}$$

must be taken into account in order to find eqs. (54) and (55). The correspon-ding spectral functions follows from eqs. (18a), (56) and (57).

$$S_0(\omega,k) = \frac{-2\pi\ \delta(\omega - \omega_{01})}{(\omega_{02} - \omega_{01})(\omega_{03} - \omega_{01})} + \frac{2\pi\ \delta(\omega - \omega_{02})}{(\omega_{02} - \omega_{01})(\omega_{03} - \omega_{02})} - \frac{2\pi\ \delta(\omega - \omega_{03})}{(\omega_{03} - \omega_{01})(\omega_{03} - \omega_{02})}$$

or, equivalently,

$$S_0(\omega,k) = -\frac{2\pi\ \delta(\omega + k)}{2\omega^2 - 1} - \frac{2\pi\ \delta(\omega - \frac{1}{\omega} - k)}{2\omega^2 - 1} \tag{58}$$

Defining

$$S_0^{(1)} = \frac{2\pi\ \delta(\omega + k)}{1 - 2\omega^2}$$

the corresponding velocity and pressure fields are

$$u_0^{(1)}(\omega,k,y) = \frac{\sqrt{2}\ \pi\ i\ \delta(\omega + k)}{\omega(1 - 2\omega^2)} [\psi_1(y) + (\omega^2 - \tfrac{1}{2})\ \phi(y)] \tag{59}$$

$$p_0^{(1)}(\omega,k,y) = \frac{\sqrt{2}\ \pi\ i\ \delta(\omega + k)}{\omega(1 - 2\omega^2)} [\psi_1(y) - (\omega^2 - \tfrac{1}{2})\ \phi(y)] \tag{60}$$

where

$$\phi(y) = \frac{\pi^{1/4}}{\sqrt{2}} e^{y^2/2} (1 - \frac{2}{\sqrt{\pi}} \int_0^y e^{-s^2} ds)$$

vanishes only for $y \to +\infty$. For $y \to -\infty$, $\phi(y)$ varies like $e^{y^2/2}$ and, therefore, the equatorial β-plane approximation fails. The only exceptions are the points

$$\omega = \pm\sqrt{\frac{1}{2}}$$

which define the intersections of the dispersion curves (55) and (57) and, moreover, the curve

$$\omega = -\frac{1}{2k}$$

which is the locus of zero group velocity points.

On the other hand it is known that the introduction of channel boundary conditions resolve this problem (see e.g. Philander, 1977 and Dickey, 1978)

$$v = 0 \quad \text{for} \quad y = \pm L$$

Then the eigenvalues ν_n are no longer integers. The eigenfunctions may be expressed in terms of Kummer-functions (Dickey, 1978). For $L \to \infty$ the ν_n tends to n and the usual problem follows. For ν_n with $n \geq 1$ the dispersion curve are only slightly modified. But for $n = 0$ a dispersion diagram follows quite similary to the higher modes $(n \geq 1)$ which approaches the dispersion relations (56) and (57) for $\nu_0 \to 0$. Then the zero group velocity points $\omega_{\nu_0 G}$ and $\omega_{\nu_0 R}$ will nearly come together and a spectrum like eq. (55) follows.

Therefore we may conclude, that the difficulties arising from eqs. (59) and (60) are due to the limit $n \to 0$ or, equivalently, $\nu_0 \to 0$. In the meridional velocity the limit $n \to 0$ produces no problems and, therefore, we may postulate the power spectrum of eq. (55).

6. SUMMARY AND COMPARISON WITH MEASUREMENTS

Employing the linear inviscid equations of motion on an equatorial β-plane, closed expressions for the Eulerian power spectra of the dynamical variables at a fixed point have been calculated using a Greenfunction technique under consideration of the principle of causality. A vertical modal structure or at least a two layered ocean has been assumed.

The shape of the spectra depends on the type of forcing (small or broad band forcing in the wave number domain) and whether or not the forcing occurs during the time of observation. Forcing events which occured before the time of measurements cause the spectra to have narrow peaks and spectral gaps governed by the dispersion relation. Forcing events during the time of observation fill the spectral gaps and spread the peaks in the frequency domain.

In the case of small band forcing in the wave number domain, the peaks of the frequency spectra are located at the frequencies of the free waves for the given wave number. During broad band forcing the peaks of the spectra occur at frequencies of vanishing horizontal group velocity and, as in the case $n = 0$ at other higher order zeroth of the dispersion relation. Spectra of broad band forcing have been discussed in detail. The power spectra for different meridional modes of the meridional current component, shown in fig. 3, are characterized by a constant level at low frequencies, resonant peaks at frequencies of zero group velocity for Rossby- and inertial gravity waves, a resonant peak at

$$\omega = \sqrt{\frac{1}{2}}$$

for the Yanai-wave and a steep decrease of spectral energy at high frequencies like ω^{-4} .

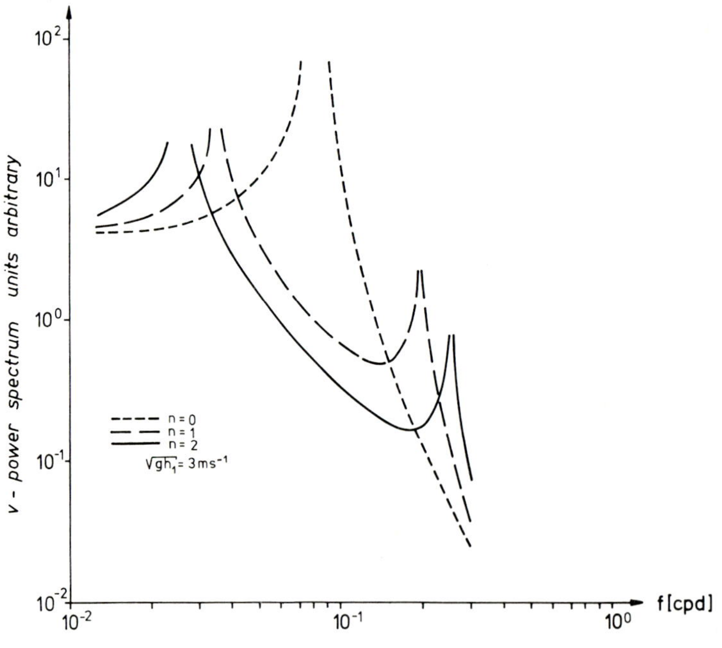

Fig. 3. Theoretical power spectra of the first three meridional modes independent of the location.

Power spectra of different modes of the zonal velocity component and the pressure field are red for low and high frequencies like ω^{-2} and have, at least for modes $n \geq 1$, the same peaks as the meridional component spectra (fig. 4).

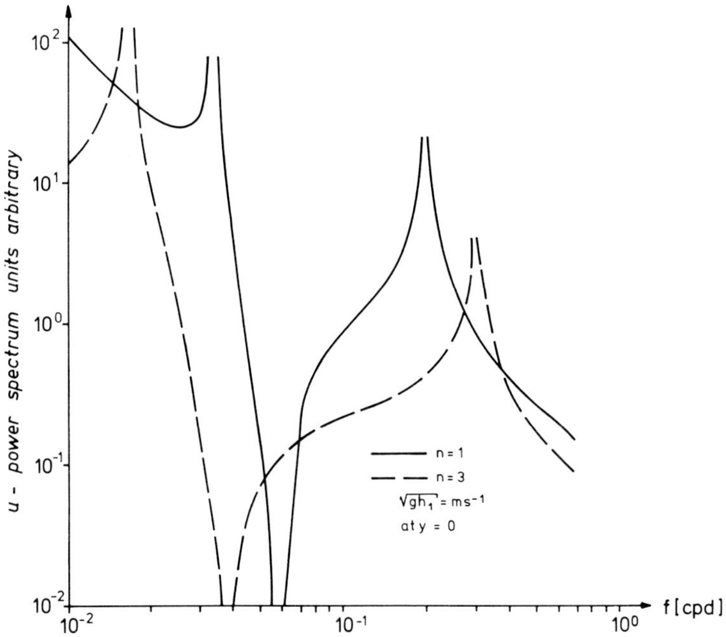

Fig. 4. Theoretical power spectra of the first and third modes of the zonal current component at the equator.

Comparing measured current spectra with our results we may expect the best agreement with current spectra in the surface layer. In order to dimensionalize the frequency of the calculated spectra we must assume values for the first vertical eigenvalue. For the Atlantic we follow Moore and Philander (1977) and take $c_1 = 2{,}4$ m/s , for the Pacific we adopt the value of Ripa and Hayes (1981) with $c_1 = 3$ m/s .

The general shape of the pressure spectra given by Wunsch and Gill (1976) and Ripa and Hayes (1981) agrees quite well with that of our calculated spectra assuming a summation of the spectra of different meridional modes. The shape of the v - spectrum given by Weisberg (1979) also agrees with that given in fig. 3. The u - spectra of Weisberg (1979) and Lass and Fennel (1981) behave in the high frequency range like ω^{-3} which is a somewhat steeper slope than that predicted by our calculated u - spectrum. Eriksen (1980) reports a ω^{-2} to ω^{-3} high frequency slope of energy spectra. The shape of kinetic energy spectra is dominated by the most energetic component which is, in our case, the u - spectrum. However, caution must be given comparing our spectra with that of Eriksen (1980) because his spectrum is based on observations in the deep sea where vertically propagating waves are observed.

Concerning the peak frequencies at zero group velocity for meridional modes $n \geq 1$ and at

$$\omega = \sqrt{\frac{1}{2}}$$

(which corresponds to a period of 14 days in the Atlantic and 12 days in the Pacific respectively) for the Yanai-mode there is a lot of experimental evidence. See Wunsch and Gill (1976), Weisberg (1979), Fennel and Lass (1979), Horigan and Weisberg (1981), Ripa and Hayes (1981), Düing et al. (1975), Fahrbach and Meincke (1981), Jarrige and Rual (1981) and Bubnov (1981) for details of the observations. Qualitativly we have a quite good agreement between our theory and the spectra from a 770 day record in the Pacific given by Halpern (1982) [figs. 5 and 6; D. Halpern, private communication].

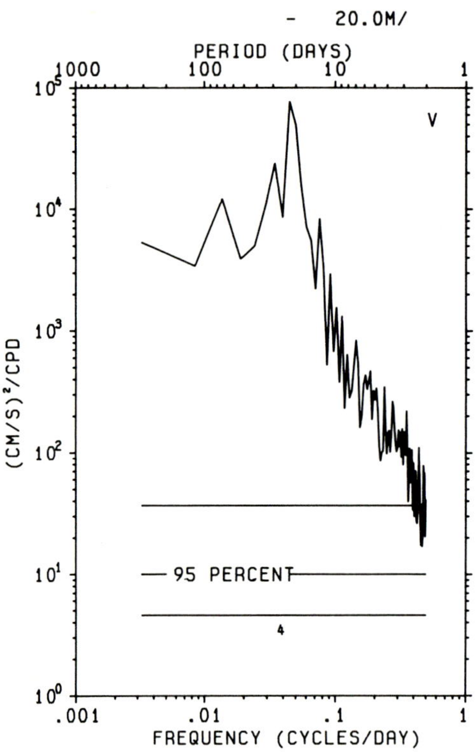

Fig. 5. Spectrum of the zonal current component at 110° W on the equator at a depth 20 m. (Courtesy of D. Halpern.)

From the comparison of the observed spectra with our calculated power spectra for small and broad band forcing in the wave number domain we conclude that the real equatorial ocean seems to be excited by forces which have a broad band character in the frequency-wave number domain.

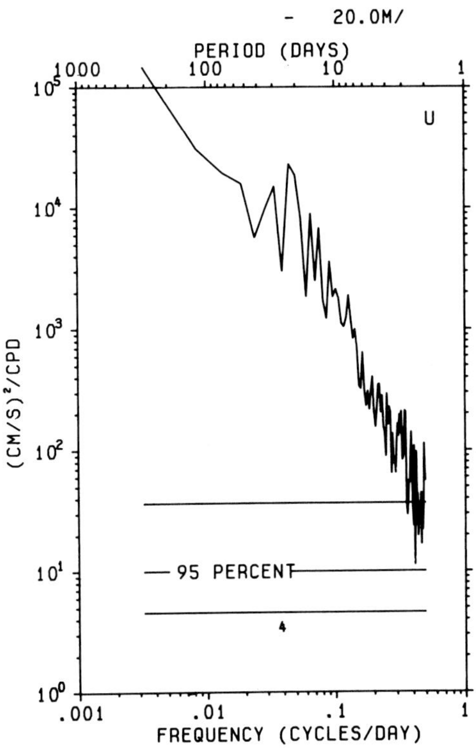

Fig. 6. Spectrum of the meridional current component at 110° W on the equator at a depth of 20 m. (Courtesy of D. Halpern.)

Acknowledgement

The authors wish to express their thank to Dr. D. Halpern for the delivery of the spectra of the two year time series in the tropical Pacific.

REFERENCES

Düing, W., Hisard, P., Katz, E., Meincke, J., Miller, L., Moroshikin, K.V., Philander, G., Rybnikov, A., Voigt, K. and Weisberg, R., 1975. Meanders and long waves in the equatorial Atlantic. Nature, 257: (5524), 280-284.

Eriksen, C., 1980. Evidence for a continuous spectrum of equatorial waves in the Indian Ocean. Journal of Geophysical Research, 85: 3285-3303.

Eriksen, C., 1981. Equatorial waves with periods from a few days to a few weeks in the Pacific and Indian Oceans. Paper presented at the SCOR WG 47 Meeting in Venice/Italy from 27-30 April 1981.

Eriksen, C., 1981. Deep currents and their interpretation as Equatorial trapped waves in the Western Pacific Ocean. J. Phys. Oceanogr., 11: 1, 48-70.

Fahrbach, E. and Meincke, J., 1981. Moored currentmeter measurements in the Equatorial Atlantic during FGGE. Paper presented at the SCOR WG 47 Meeting in Venice/Italy from 27-30 April 1981.

Fennel, W. and Lass, H.U., 1979. On the vertical eigenvalue problem of equatorial trapped waves. Gerlands Beiträge zur Geophysik (Leipzig), 88: 279-293.

Garrett, C. and Munk, W., 1972. Space time scales of internal waves. Geophysical Fluid Dynamics, 2: 225-264.
Jarrige, F. and Raul, P., 1981. Measurements in the equatorial currents of the Atlantic and Pacific Oceans. Paper presented at the SCOR WG 47 Meeting in Venice/Italy from 27-30 April 1981.
Lass, H.U. and Fennel, W., 1981. Current observations in the equatorial wave guide at 28°40' W during FGGE (unpublished manuscript).
Lighthill, M.J., 1969. Dynamic response of the Indian Ocean to onset of the southwest monsoon. Phil. Trans. Roy. Soc. London, A 265: 45-92.
Martin, P. and Schwinger, J., 1959. Theory of many particle systems. Physical Review, 115: 1342.
Moore, D. and Philander, S.G., 1977. Modelling of the tropical oceanic circulation. In: The Sea, Wiley, New York, pp. 319-361.
Pedlosky, J., 1979. Geophysical Fluid Dynamics. Springer, New York, 624 pp.
Ripa, P. and Hayes S.P., 1981. Evidence for equatorial trapped waves the Galapagos Islands. Journ. Geophys. Res., 86: 6509-6516.
Weisberg, R.H., 1979. Equatorial waves during GATE and their relation to the mean zonal circulation. Deep Sea Research, 26 (Supplement II): 179-198.
Weisberg, R.H. and Horigan, A.M., 1981. Low-frequency variability in the equatorial Atlantic. Journal of Physical Oceanography, 11: 913-920.
Wunsch, C. and Gill, A.E., 1976. Observations of equatorial trapped waves in Pacific sea level variations. Deep Sea Research, 23: 371-390.
Wyld, H.W., 1976. Mathematical methods for physics. Benjamin Inc., New York, 628 pp.
Wu, T.Y. and Ohmura, T., 1962. Quantum Theory of Scattering. Prentice-Hall Inc., New York.

INSTABILITY ON THE EQUATORIAL BETA-PLANE

JOHN P. BOYD AND ZAPHIRIS D. CHRISTIDIS

1. INTRODUCTION

Since both the tropical atmosphere and the equatorial ocean are regions of strong latitudinal shear, it is important to study the hydrodynamic instabilities which are possible at these low latitudes. Philander (1978) has studied the traditional barotropic, Rossby wave-like, Rayleigh-Kuo instabilities. Dunkerton (1981) and Stevens (1982) have independently analyzed the inertial instabilities and found the critical shear for zero wavenumber. Ripa (1982) has generalized the Rayleigh-Kuo criterion for instability. In spite of these interesting beginnings, however, the present authors are embarked upon the first systematic and comprehensive study of shear instabilities in equatorial flows. In time, we hope to apply our results to both the ocean and atmosphere in articles planned for both the Journal of Physical Oceanography and the Journal of the Atmospheric Sciences. This paper is only a modest progress report. Nonetheless, enough is already known to show that equatorial instabilities are a likely prospect and that their dynamics is strikingly different from those of the middle latitudes.

Our results to date are a mixture of the analytical and numerical. For small zonal wavenumber k, it is possible to employ the "long-wave" and "gamma-plane" approximations to clearly identify the different stable and unstable equatorial modes and for some to calculate the growth rates in approximate closed form. This analysis is given in our joint letter (Boyd and Christidis, 1982) and will not be repeated here. Instead, we will merely summarize its conclusions and concentrate instead upon explaining and displaying some of our most recent numerical results.

2. MODEL AND METHODS

The equations we solve are those obtained by separating variables in a continuously stratified flow under the assumption that the mean wind U is a function of latitude only. Nondimensionalizing as in Boyd (1980) so that the equivalent depth is equal to 1, we obtain

$$i\,k\,[U(y) - c]\,u - [y - dU/dy]\,v + ik\phi = 0 \tag{1}$$

$$i\,k\,[U(y) - c]\,v + yu + \phi_y = 0 \tag{2}$$

$$i\,k\,[U(y) - c]\,\phi + iku + v_y = 0 \tag{3}$$

where k is the zonal wavenumber, c the phase speed, subscript y denotes differentiation with respect to y, ϕ is the height and U(y) the mean wind or current. These are almost the nonlinear shallow water wave equations (linearized about a zonal current U(y)) except for the absence of some terms in (3) which are proportional to the variable mean depth. The set (1)-(3) is the natural choice for modelling the atmosphere, and while it lacks the pseudomomentum conservation law of the shallow water wave equations (linearized about a zonal current U(y)) except for the absence of some terms in (3) which are proportional to the variable mean depth. The set (1)-(3) is the natural choice for modelling the atmosphere, and while it lacks pseudomomentum conservation law of the shallow water set, these "stratified flow" equations can be analyzed via the "gamma-plane" approximation while the shallow water wave equations cannot. Our numerical results indicate instabilities of roughly the same qualitative nature for either set (although there are the expected quantitative differences and our comparisons are still in progress), so in the rest of this paper, as in Boyd and Christidis(1982), we shall discuss the "stratified flow" equations (1)-(3) only.

We employed two different algorithms to solve (1)-(3): (i) a Chebyshev pseudospectral/QZ method and (ii) a shooting method. The Chebyshev method is quite sensitive to critical latitude singularities and is therefore ineffective for near-neutral modes. It is also rather costly since the operation count is proportional to the cube of N where N is the number of polynomials in the expansion. However, the QZ eigenvalue matrix algorithm does not require a first guess, so our method is useful for verifying that we have not missed modes or overlooked strong instabilities in some region of parameter space.

The shooting method involved marching in from either computational boundary and matching the two numerical solutions at the equator. Since equatorially trapped waves are exponentially increasing as we march towards the equator, this double shooting method avoids numerical "stiffness" and a simple fourth-order Runge-Kutta algorithm could be used for the marching. Because the expense is only linearly proportional to the number of grid points, this shooting method is very cheap. Newton's method was used to vary c until the two solutions matched at the equator, and this has the disadvantage that it will not converge unless a good first quess for c is available. We met this need by employing the "continuation" method. Beginning with a known equatorial wave for either zero shear or zero wavenumber, or occasionally from c's calculated by the Chebyshev program, we systematically increased the shear or wavenumber in small steps, using previous results to linearly extrapolate a first guess for the next value of shear or wavenumber in the sequence.

To cope with critical latitude singularities (the set (1)-(3) is singular where $U(y) = c$, which occurs for real y when c is real), the differential equations were integrated using a path of integration in the complex y-plane which dipped below the real y-axis as shown in Fig. 1. This makes it possible to follow the modes right up to the neutral curve without difficulty. One can in fact compute the modes even when the imaginary part of the phase speed is negative. As shown in the figure, however, the path of integration implicitly forces us to draw the branch-cut from the critical latitude singularity to infinity in such a way that the branch-cut crosses the real y-axis when the imaginary part of the phase speed is negative. (This is necessary because the shooting procedure computes a solution which is smooth and continuous on the path of integration, which is possible if and only if the branch-cut does not cross the path of integration). The fact that the branch-cut crosses the real y-axis implies that the modes we compute are discontinuous for some real latitude y, and therefore are unphysical.

Fig.2 shows the real and imaginary parts of the phase speeds of the n=1 Rossby wave in a linear shear as computed using this complex path of integration. The imaginary part of the phase speed is everywhere negative, which shows that these <u>discrete</u> modes are unphysical. What happens if one integrates along the real y-axis instead is that, as is known from studies of Rossby waves in shear in the middle latitudes, a <u>continuous</u> spectrum of neutral Rossby waves occurs.

The Kelvin wave, however, is a different class of wave with a phase speed of the opposite sign. Our calculations show that it is unstable, i.e., has an imaginary part of the phase speed which is positive. Such modes have critical latitudes above rather than below the real y-axis and thus can be smooth and continuous on the real axis and have a physical existence as discrete modes.

3. NUMERICAL RESULTS

Three major species of instability have been identitified so far: (i) barotropic instability (ii) inertial instability and (iii) Kelvin wave instability, which has no counterpart for non-equatorial flows. Barotropic instability has already been studied (Philander, 1978), so our discussion will be confined to the remaining two types. Because the most interesting effects are observed for linear shear

$$U(y) = S\,y \qquad (4)$$

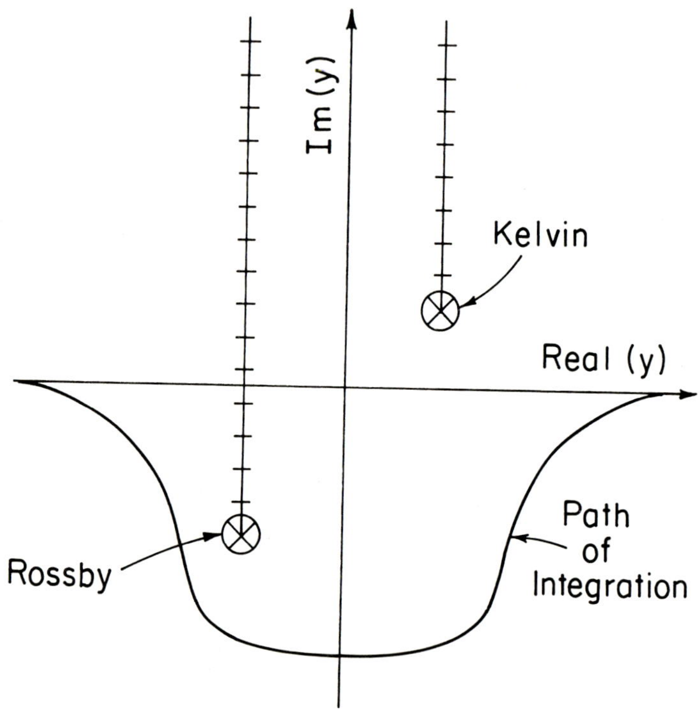

Fig. 1 The complex y-plane with the path of numerical integration for linear shear shown looping below the real axis. The branch cuts from the critical latitude to infinity are also shown schematically as cross-hatched lines for a Rossby and a Kelvin wave.

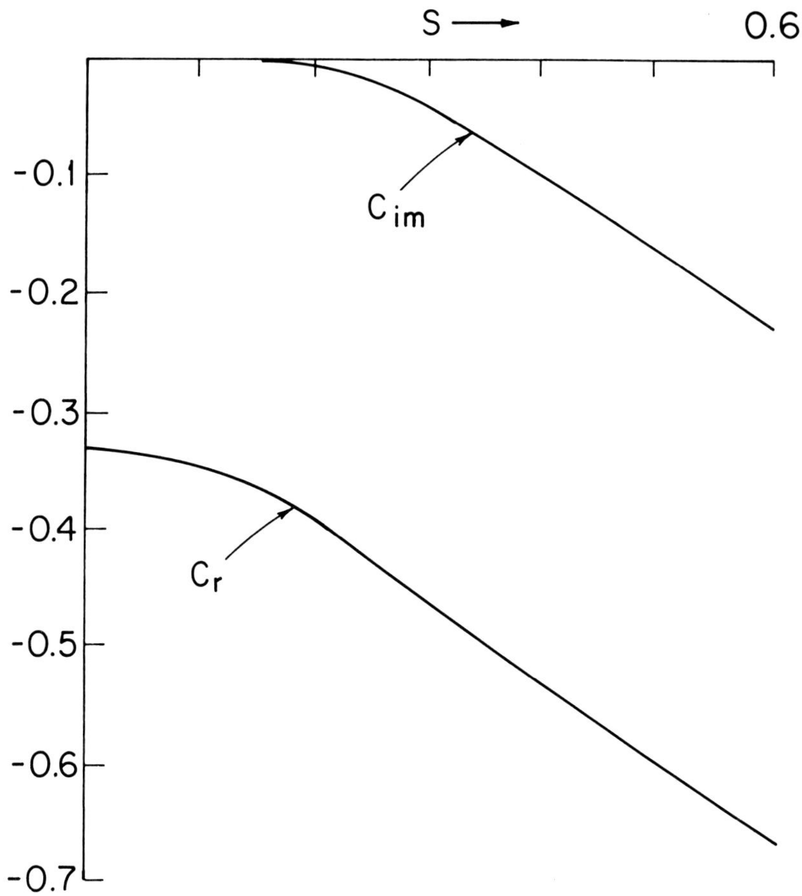

Fig. 2 The variation of the real and imaginary parts of the phase speed c for the n=1 equatorial Rossby wave with linear shear U(y)=Sy as computed using the complex path of integration shown in Fig. 1. Since the imaginary part of c is negative, this computed mode is discontinuous at the latitude where the branch-cut crosses the real y-axis and is not physical. The computation shows that no discrete Rossby modes exist which are continuous at all latitudes.

where S is a constant, we shall concentrate upon this simple case, but instability is observed for a wide range of shear profiles and we shall display results for a parabolic wind shear at the end.

Inertial instabilities have been known since the early part of the century and normally take the form of axisymmetric disturbances, i.e., zero wavenumber. Independently, Dunkerton (1981) and Stevens (1982) showed that one could use the equatorial "gamma-plane" approximation (Boyd, 1978) to exactly solve (1)-(3) for the special case of (a) linear shear and (b) zero wavenumber. Stevens (1982) gives a particularly clear and careful description of the physical causes of the instability. Both found that instability occurs only for $|S| > 2$ and that the unstable/damped pair of modes both have the structure of n=0 equatorial gravity waves.

Boyd and Christidis (1982) show that, in contrast to all other known cases of inertial instability, on the equatorial beta-plane the growth rates are largest for non-zero values of the zonal wavenumber k. In addition, instability is possible for finite wavenumber below the critical shear S=2 calculated by Dunkerton and Stevens for k=0. These surprises show that equatorial instability has to be considered as a separate sub-area of hydrodynamic instability theory; blind application of misconceptions based on other stability problems can only lead to embarrassment. However, the wavenumber of maximum growth rate is relatively small (k = 0.27 for S=2 and smaller values for larger S) and growth rates for non-zero k are only 30% or so larger than for k=0 except for shears close to 2. Further, graphs of the eigenfunctions show shapes very similar to those given in Boyd (1978) for the "gamma-plane" approximation, so there is no doubt that these unstable modes are indeed inertial instabilities.

Equatorial instability, however, is far more than just this single mode. There are two unstable modes, each of which has at least two distinct regimes with the modes exchanging identities in the transition regime, and there are at least two physically distinct mechanisms of instability.

In strong shear, it is therefore necessary to be careful about precisely what one means by "Kelvin" wave and "gravity" wave. Boyd and Christidis (1982) apply these labels in accordance with the so-called "long-wave" and "gamma-plane" approximations which apply for small wavenumber to different classes of waves. The "long-wave" equations, for example, filter out gravity waves to permit only Kelvin and Rossby waves, which have finite phase speeds in the limit $k \to 0$. The Rossby modes have negative real and imaginary parts of the phase speed and therefore are always stable as discussed in Sec. 2, so there is no danger of confusing them with the Kelvin wave, for which the imaginary part and (usually) the real part of c are both positive. Thus, it is sensible to define "Kelvin" wave to mean that solution of the "long-wave" equations which is unstable.

Similarly, the "gamma-plane" approximation is applicable only to gravity waves (including inertially unstable ones), so it is reasonable to define "gravity wave" to mean a wave that can be approximated, however crudely, via the "gamma-plane", or as the analytic continuation of such a wave as the wavenumber k is increased numerically in small steps. The "continuation" method makes it possible to define a mode through the simple process of computing its phase speed through small steps in S or k from some known solution for either zero shear or wavenumber. Thus, there is no real ambiguity in either labelling a wave regime as "Kelvin" or "gravity" or in identifying a single continuous mode as having distinct regimes where different labels must apply.

The dominant mode of instability is what Boyd and Christidis (1982) name the "mixed Kelvin/inertial mode". For small shears ($S < 2$), this mode is an unstable Kelvin wave width the real part of the phase speed roughly equal to one as for a neutral Kelvin wave in the absence of shear. One remarkable discovery is that the imaginary part of the phase speed for the Kelvin wave can be approximated to within an error of 8% on the interval $0.4 < S < 1.15$ by the curve-fit formula

$$\mathrm{Im}(c) = 13.56 \exp(-5.34/S) \tag{5}$$

For $S < 0.4$, the imaginary part of c is too small to be accurately computed, but the form of (5) suggests that there is instability for all values of the shear, however small, although growth rates are very small unless $S > 1$. This absence of a "critical shear" has not been seen in any other similar instability problem that the authors are aware of, but it is suggestive that the Kelvin wave instability is a critical latitude effect (like ordinary Rayleigh-Kuo barotropic instability) since the location of the critical latitude is inversely proportional to S so that the amplitude of the Kelvin wave at the critical latitude is an exponential function of 1/S. The inertial instability, in contrast, occurs (for small wavenumber) only for waves with very large c so the critical latitude is very distant from the equator at a location where the wave has no appreciable amplitude.

Thus, there appear to be two distinct, physically different mechanisms for instability on the equatorial beta-plane for linear shear, and three for those other current profiles for which barotropic instability is also possible.

Boyd and Christidis(1982) show using the "uniform gamma-plane" approximation that in the vicinity of $S = 2$, which is the critical shear for the zero wavenumber inertial instability found by Dunkerton and Stevens, the wave types exchange identities. In particular, for very small k, the Kelvin wave suddenly becomes neutral with a negative real phase speed for $S > 1.9$ and for larger shear merges with the negative frequency, n=0 neutral gravity wave to become the inertially unstable mode and its complex conjugate. For larger wavenumber, the two modes do not actually merge, but instead the imaginary part of the phase speed of the Kelvin wave dips to a non-zero value and then begins to rise rapidly as the shear is increased. It is still true, however, that for $S > 2$, this mode is well-approximated via the "gamma-plane" and equally well by the "long-wave" equations for $S < 2$. In the transition zone, $S \approx 2$, this mode is truly a hybrid of Kelvin and inertial wave, and neither label is appropriate alone.

The other unstable mode is the "mixed gravity/Kelvin" mode. For small shear, this is the positive frequency n=0 gravity wave. Like the Kelvin wave (and unlike the negative frequency gravity wave which is always neutral), this mode is weakly unstable even for weak shear. Growth rates are always very small, however, and this mode would likely be suppressed by viscous damping in most physical situations. For $S > 2$, this mode becomes an unstable Kelvin wave and growth rates grow rapidly for $S > 2.4$. However, for all values of S, growth rates are small in comparison to those for the Kelvin/inertial mode, so the latter is always the dominant mode of instability.

Great care is thus needed in interperting numerical results correctly, The "long-wave" equations reveal two branches of Kelvin wave instability with a region of neutrality around $S = 2$ in between them as shown in Fig. 1 of Boyd and Christidis(1982). What can only be seen by careful analysis and numerical continuation is that these two branches belong to physically distinct modes: the lower branch Kelvin wave cannot be smoothly continued to the upper branch Kelvin wave (by repeated numerical solution with small steps in S, for example) because the "long-wave" approximation inevitably fails in a neighborhood of $S = 2$.

In summary, we have a weakly unstable "mixed gravity/Kelvin wave" whose real part of c is always large and positive and whose imaginary part is always small plus a strongly unstable "mixed Kelvin/inertial" wave whose real part of phase speed is small but may be of either sign. Fig. 3 displays contour plots of the nondimensional growth rate for the dominant "mixed Kelvin/inertial" in linear shear. The mode does not exist on the equatorial beta-plane for large values of the wavenumber because when $S > 1/k$, one can show by analyzing the asymptotic behavior as y goes to infinity of the set (1)-(3) that this counterbalances the equatorial trapping created by the variations of the Coriolis parameter, and the mode is no longer latitudinally confined. Consequently, a portion of the figure is left blank.

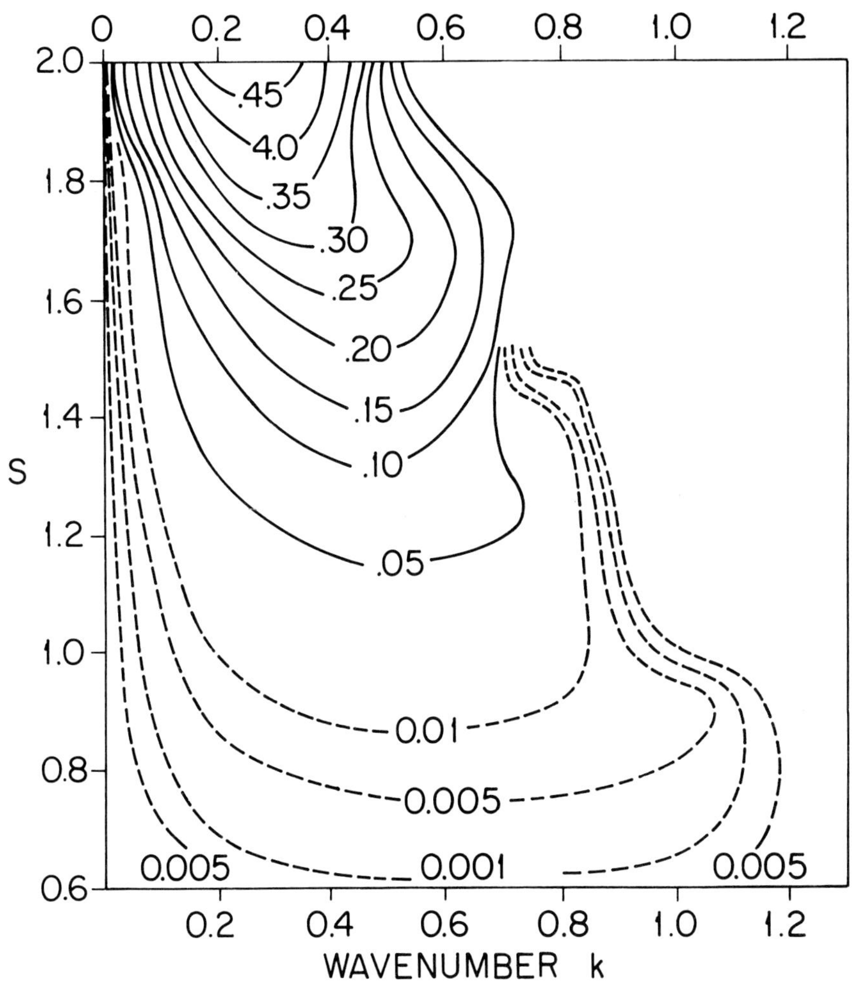

Fig. 3 Contours of constant growth rate (imaginary part of nondimensional frequency) for the mixed Kelvin-inertial mode in linear shear, plotted as a function of shear strength S and zonal wavenumber k. This is the Kelvin-like regime; the inertial-like regime is S>2 and is not shown.

The shear profile

$$U(y) = S\ (\ 1 - e^{-y^2/10}\ ) \tag{6}$$

is approximately equal to the simple parabola

$$U(y) = (S/10)\ y^2 \tag{7}$$

for small y, but is bounded for large y so that the modes are always equatorially trapped. The profile is barotropically unstable when $S > 5$. Fig. 4 shows the growth rates for the unstable Kelvin wave for positive S. (No instability is found for negative S.) The Kelvin wave appears to be unstable for all values of the shear strength (so long as the sign is right), but growth rates are rather small until one has crossed the barotropic stability barrier. We have not yet made detailed comparisons of the Kelvin and barotropic instability in the parametric region where they co-exist, but it is striking that Kelvin waves can again grow with time where classical stability predicts only a continuous spectrum of barotropically stable Rossby waves.

Neither of the two profiles we have discussed is unduly realistic, but they suggest that more complicated currents which are neither symmetric nor antisymmetric about the equator will have unstable Kelvin waves, too, and if the shear is strong enough, barotropic and inertial instabilities as well. In this model, we have ignored vertical shear, but when it is included, baroclinic instability is possible also. It is clear that much further research is needed.

6. VISCOSITY AND INSTABILITY IN THE OCEAN

Dunkerton(1981) points out that the nondimensional shear S is related to the dimensional shear S* via

$$S = E^{1/4}\ S^*\ (2\Omega) \tag{7}$$

where $E = 4\ \Omega^2 a^2\ /(g\ H)$ is "Lamb's parameter" with H the "equivalent depth", Ω the angular frequency of the earth's rotation, and a the radius of the planet. Since E is proportional to the square of the vertical mode number m, it follows that the nondimensional shear is a monotonically increasing function of m. This in turn leads to the absurd prediction that the (inviscid) instability will have largest dimensional growth rates for <u>infinite</u> m and therefore infinitely small vertical and horizontal length scales.

Dunkerton noted that the resolution of this contradiction is that viscous effects increase not as the quarter power of E but rather as E itself. Therefore, as the vertical mode number m increases, S increases as the square root of m but viscous effects as m squared. For sufficiently large m -- how large

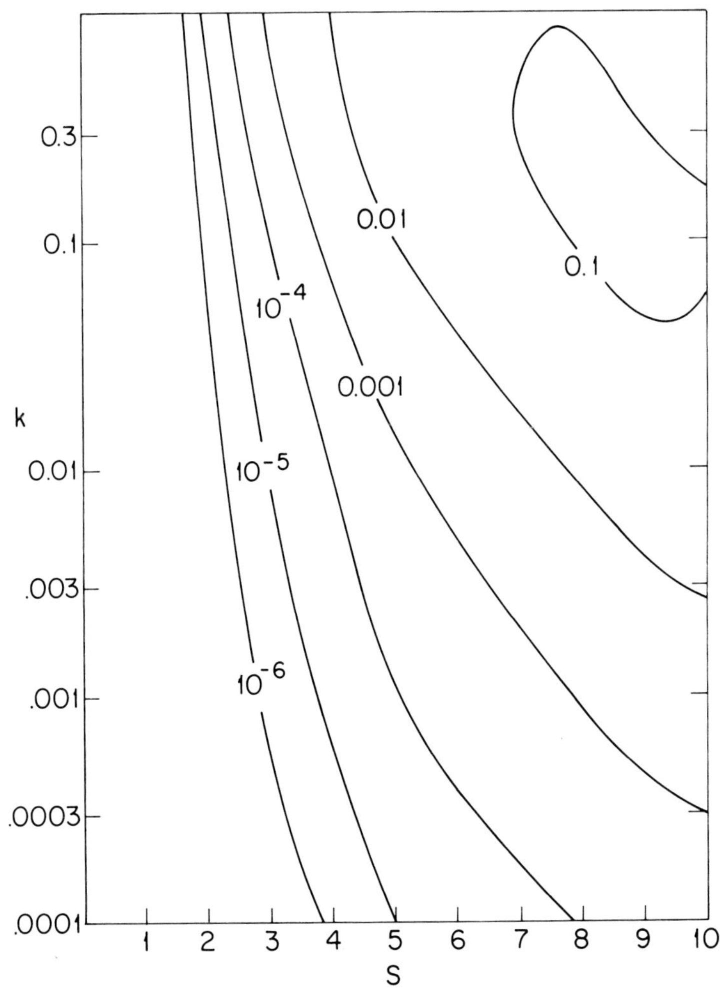

Fig. 4 Contours of constant growth rate for the Kelvin mode in the quasi-parabolic shear described in the text. The wavenumber is plotted on a logarithmic scale; the shear strength on a linear scale. The flow is barotropically unstable for S>5.

depends on both the dimensional shear and the viscosity -- the damping must win and dimensional growth rates must level off and then decrease as the mode number increases still further.

Thus, equatorial instabilities will occur at finite vertical and horizontal scales, but the reason is depressing in that one cannot know exactly what these scales or the e-folding time of the wave are unless one knows both the shear and the viscosity. The former is highly variable and mostly concentrated in the Equatorial Undercurrent (in the oceanic case) instead of being independent of depth all the way to the botton, as implicitly assumed in our model and those of Dunkerton, Stevens, and Ripa. The viscosity is also highly variable, even harder to measure, and raises the unpleasant issue of whether the dissipation can be approximated by a linear eddy viscosity at all, or whether some more complex nonlinear damping might be more appropriate.

The results of our inviscid model must therefore be interperted with much caution when applied to the real ocean. It is still worthwhile, however, to attempt at least an order-of-magnitude estimate.

For the first baroclinic mode, let us take H=40 cm so that the nondimensional units have dimensional equivalents of 1.7 days (time), 300km (length), and 2.0 m/s (velocity). S = 2 for linear shear, which is the dividing line between the "Kelvin" and "inertial" instability regimes, then requires a current of 1 m/s (-1 m/s) 75 km north (south) of the equator. This is a very strong shear, so it seems unlikely that inertial instability will be seen for the lowest baroclinic mode. At the depth where the latitudinal shear is a miximum, the Equatorial Undercurrent can be crudely modelled by a constant plus a parabola of the form of (6) with a nondimensional shear of S = 20! This is enough to allow barotropic instability, but it is far more than is needed to create unstable Kelvin waves as well as shown in Fig. 4. Thus, barotropic and Kelvin wave "instability" are likely for low baroclinic modes; inertial instability can occur only for high order baroclinic modes. Only further work can determine which form of instability is dominant.

5. SUMMARY

The simple, latitudinal-shear-only model has shown that unstable Kelvin waves will occur for rather arbitrary wind profiles at shear strengths below those needed for the previously studied inertial and barotropic instabilities. In linear shear, the dominant unstable mode is a "mixed Kelvin/inertial" wave which we have so named because it is an unstable Kelvin wave for weak shear but an inertially unstable n=0 gravity wave for strong shear. In parabolic shear, inertial instability does not occur, but unstable Kelvin waves form in shears too weak for barotropic instability.

The neglect of vertical shear and viscosity are serious limitations on the model. Nonetheless, there is every reason to expect that unstable Kelvin waves can be generated by the Equatorial Undercurrent. Only nulti-level, three-dimensional numerical models can hope to resolve the competion between the unstable Kelvin waves discussed here, the familiar inertial and barotropic instabilities, and baroclinic instability for the role of generating unstable waves in the equatorial ocean. It is altogether possible that two or more of these mechanisms operate simultaneously.

ACKNOWLEDGEMENTS. This work was supported by the National Science Foundation through grant OCE8108530. We thank S. Jacobs, T. Dunkerton, J. Kindle, and D. Stevens for helpful comments.

REFERENCES

Boyd, J.P., 1978. The effects of latitudinal shear on equatorial waves. Part I: Theory and methods. J. Atmos. Sci., 35: 2236-2258.

Boyd, J.P., 1980. The nonlinear equatorial Kelvin wave. J. Phys. Oceangr., 10: 1-11.

Boyd, J.P., and Z.D. Christidis, 1982. Low wavenumber instability on the equatorial beta-plane. Geophys. Res. Lett., 9: 769-772.

Dunkerton, T.J., 1981. On the inertial stability of the equatorial middle atmospher. J. Atmos. Sci., 38: 2354-2364.

Philander, S.G.H., 1978. Instabilities of zonal equatorial currents, 2. J. Geophys. Res., 83: 3679-3682.

Ripa, P., 1982. A generalization of the Rayleigh-Kuo criterion to the equatorial beta-plane. To be published.

Stevens, D.E., 1982. On symmetric stability and instability of zonal mean flows near the equator. J. Atmos. Sci., 39: in press.

ON THE GENERATION OF ROSSBY SOLITONS DURING EL NINO

JOHN C. KINDLE

Environmental Simulation Branch (Code 322), Naval Ocean Research and Development Activity, NSTL Station, MS 39529 USA

ABSTRACT

The excitation of equatorial Rossby solitons during El Nino is studied numerically using a one-layer reduced-gravity formulation. The model basin - an idealization of the tropical Pacific - extends 15,000 km zonally and 3,000 km meridionally. Forcing is applied as a patch of zonal wind stress representing the relaxation of the equatorial easterlies preceding El Nino. It is shown that a relaxation of the equatorial winds generates internal Rossby solitons subsequent to the reflection of an equatorial Kelvin wave front from the eastern boundary. The meridional structure of the soliton is that of a first latitudinal mode Rossby wave. The simple El Nino simulations result in the excitation of two or three solitons separated by approximately 3,000 km. The results of this study suggest that internal Rossby solitons are most likely to be observed in the central and western Pacific during a major El Nino year.

1. INTRODUCTION

If El Nino is initiated by a remotely forced internal Kelvin wave front (Wyrtki, 1975; Hurlburt et al., 1976; McCreary, 1976) then the reflection of this wave should result in a set of westward propagating Rossby waves emanating from the eastern boundary (Moore, 1968). Along the equator the first latitudinal mode Rossby wave would have the largest amplitude. Theoretical studies of remotely forced El Nino events reveal such a wave response (e.g., McCreary, 1977; Kindle, 1979; Busalacchi and O'Brien, 1981). Indirect observational evidence exists in the form of westward propagating sea surface temperature (SST) anomalies at a speed consistent with a first latitudinal mode Rossby wave (Barnett, 1977; Ramussen and Carpenter, 1982; Weare, 1982). It is the thesis of this paper that the first latitudinal mode Rossby wave resulting from El Nino may evolve into one or more equatorial Rossby solitons. These robust nonlinear waves may provide an important mechanism for the westward propagation of El Nino effects along the equator. Brief introductions to El Nino and Rossby solitons are provided below.

1.1 El Nino

Bjerknes (1966) was the first to associate El Nino with the large scale weakening of the southeast trades. Wyrtki (1975) suggested that the relaxation of the zonal component of the equatorial trades in the central and western Pacific initiated the event by exciting an equatorially trapped internal Kelvin wave. This hypothesis was supported by the theoretical work of Hurlburt, et al. (1976) and McCreary (1976, 1977). Busalacchi and O'Brien (1981) forced a linear numerical model with observed winds and found that El Nino events in the 1960's were most likely initiated by wind events in the western Pacific. Kindle (1979) examined El Nino from its onset to the termination phase using a nonlinear model forced by an idealized representation of the seasonal and inter-annual wind variability associated with El Nino. In that study, the first mode Rossby wave resulting from the incidence of the El Nino Kelvin wave at the eastern boundary propagated across the basin and reflected as an equatorial Kelvin wave. The relative timing of the reflection of the wave from the western boundary and the re-intensification of the trade winds determined the character of the solution at the eastern boundary during the year following a major El Nino. Due to the spatial averaging applied to the solution, the effects of the Rossby wave's nonlinearity were obscured and no solitons were observed.

Rasmussen and Carpenter (1982) and Weare (1982) recently analyzed historical data sets of tropical sea surface temperatures. The results of both these studies, which are consistent with previous descriptions (e.g., Hickey, 1975; Barnett, 1977; Wyrtki, 1977), show that the warming associated with El Nino at the eastern boundary begins near the first of the year and reaches a maximum around May. Subsequently, the warm SST anomalies propagate westward at a speed of .5 to 1 m/sec. By the end of the El Nino year, warm SST anomalies along the equator are well established from the coasts of Ecuador and Peru to the dateline. The above normal equatorial surface temperatures are believed to be related to climate anomalies throughout the tropics and higher latitude regions (e.g., Bjerknes, 1966; Namais, 1976; Julian and Chervin, 1978; Wright, 1979; Horel and Wallace, 1981).

1.2 Equatorial Rossby Solitons

Solitons exist when the tendency of a wave to disperse is balanced by its nonlinearity. The theoretical existence of equatorial Rossby solitons was demonstrated by Boyd (1980) who examined the nonlinear form of the shallow water wave equations on an equatorial beta-plane. He utilized the method of multiple scales to study only Rossby waves in an unbounded ocean.

The latitudinal dependence of the eigenfunctions of the system are the classic equatorial modes (e.g., Moore and Philander, 1977). The zonal-temporal behavior is governed by the one-dimensional Korteweg-de Vries (KdV) equation. It has been well established that the solution to the KdV equation for a virtually arbitrary initial condition yields a finite number of solitons and a dispersive wavetrain (Gardener, et al., 1967). Hence, as $t \rightarrow \infty$, solitons are the only stable solution to the KdV equation. Boyd demonstrated that internal Rossby solitons could be generated by very general initial conditions provided that the initial feature represents a depression of the thermocline. Along the equator, the largest amplitude soliton has a meridional dependence of a first latitudinal mode Rossby wave.

Kindle (1982) examined the generation of Rossby solitons by equatorial wind events. Using a one-layer reduced gravity numerical model of the tropical Pacific, he found that even a modest relaxation of the equatorial trades could generate equatorial Rossby solitons. The time scale of the wind event could vary from a few weeks to months. If the duration of the event is longer than two months, more than one soliton can evolve from the first latitudinal mode Rossby wave front emanating from the eastern boundary. The ease with which equatorial Rossby solitons develop in the analytical work of Boyd and the numerical experiments of Kindle suggest that these curious nonlinear waves may play an important role in the dynamics of the equatorial Pacific.

The approach of this paper is similar to that of Kindle (1982). In Section 2 the numerical model is described. Section 3 demonstrates the ability of equatorial wind events to generate Rossby solitons. The final experiments are simple representations of El Nino events; it will be shown that Rossby solitons are most easily excited by a prolonged weakening of the equatorial trades such as during El Nino.

2. The Model

The numerical model is that which is described by Kindle (1979, 1982). The governing system is the equatorial beta-plane, shallow water wave equations with the inclusion of eddy viscosity. The equations are solved for one active layer. The model geometry - an idealization of the tropical Pacific - is rectangular in shape and extends 15,000 km in the zonal direction. The northern boundary, which is open in order to permit the exit of coastal Kelvin waves, is located 1,500 km from the equator. Because only latitudinally symmetric modes are examined, the solution in the southern hemisphere is the reflection of that in the northern half.

The explicit numerical scheme employs leapfrog time differencing and centered space differencing; the viscous terms are lagged in time. A staggered grid which incorporates the "C" stencil (Mesinger and Arakawa, 1976) is utilized. The grid spacing in the zonal direction varies from 20 km near the boundaries to 165 km in the interior; the meridional spacing is 25 im throughout the domain.

3. Numerical Results

In this section, the results of seven experiments are described. In Cases 1 and 2, the model is initialized with the analytical solution for an equatorial Rossby soliton derived by Boyd (1980). In Cases 3 and 4, a patch of zonal wind stress (meridionally uniform) is impulsively applied for 30 days in order to demonstrate the generation of Rossby solitons by equatorial wind events. Cases 5, 6, and 7 are very simple representations of El Nino events. Except for the first two experiments, the model is started from rest. The initial thickness of the upper layer is 150 meters and the density stratification between the two layers is $3\sigma_t$ units. In Cases 1 and 2, the initial layer thickness is 100 meters and $4\sigma_t$ units.

The model is initialized with Boyd's solution for the first latitudinal mode equatorial Rossby soliton (Fig. 1). Case 1 is the linear solution. The effects of dispersion cause the initial feature to evolve into a wave-train (Fig. 2a). In Case 2 the nonlinear terms are included; Fig. 2b reveals that the initial disturbance propagates across the basin with very little change in shape or amplitude. Clearly, the nonlinear terms balance the tendency to disperse, thus demonstrating that the model reproduces the behavior of the equatorial Rossby soliton derived analytically by Boyd.

In the experiments described below, wind stress forcing is applied to the model ocean in order to demonstrate the excitation of solitons by wind events. The forcing is in the form of a patch of zonal wind stress with a uniform meridional distribution and a top-hat zonal shape. The wind is directed from west to east and represents a relaxation of the mean equatorial trades. As described by McCreary (1977) and Kindle (1979), the wind patch radiates packets of Kelvin and Rossby waves. The Kelvin waves, which have a much larger amplitude than the Rossby waves, propagate a downwelling pulse towards the eastern boundary. The reflection of this downwelling pulse initiates westward-propagating Rossby waves. The subsequent description focuses on the response of the first latitudinal mode Rossby wave created by the eastern boundary reflection.

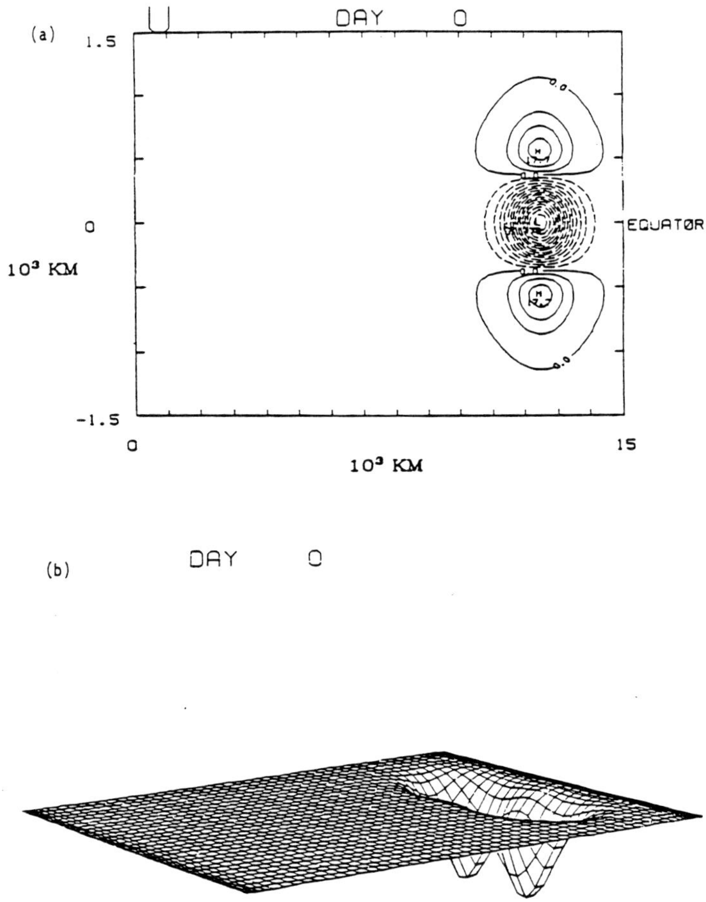

Figure 1. Initial condition for Cases 1 and 2 - the prototype Rossby soliton derived by Boyd. (a) Zonal velocity component. The contour interval is 5 cm/sec. (b) Model pycnocline surface as viewed from the southwest Pacific. The maximum amplitude of the initial depression is 25 meters. (from Kindle, 1982)

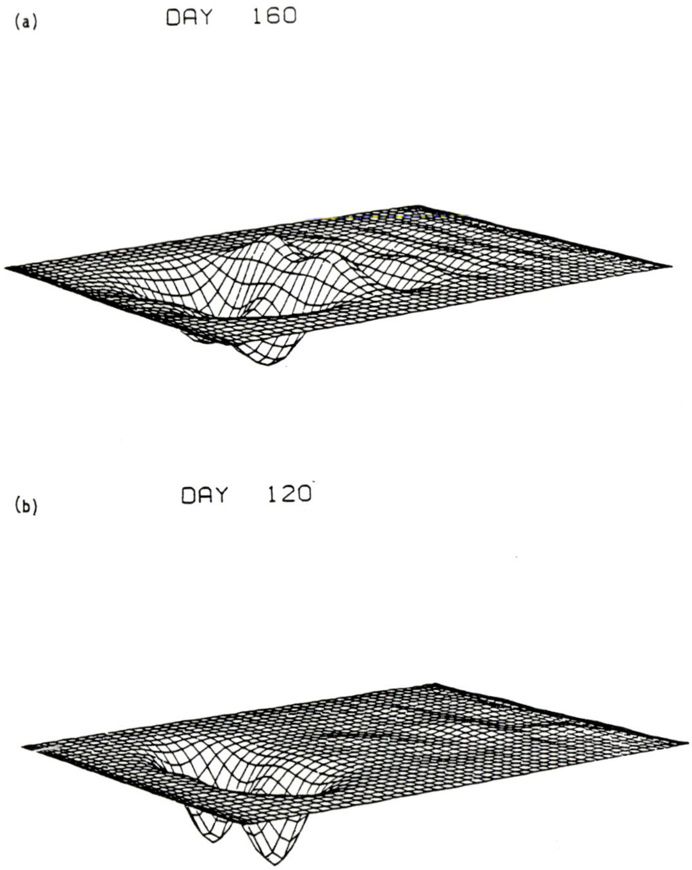

Figure 2. (a) Model pycnocline surface for Case 1 (linear solution). (b) Case 2 (nonlinear solution). (from Kindle, 1982)

In Case 3, the wind stress has a zonal width scale of 2,000 km and an amplitude of .5 dyne/cm^2. The eastern edge of the patch is located 6,500 km from the eastern boundary. The forcing is applied impulsively and switched off at Day 30. Figure 3 presents the results of the linear calculation. As the first mode Rossby wave propagation westward, it evolves into a wavetrain. Although not as pronounced as in Case 1, the effects of dispersion are very evident. The nonlinear counterpart to this experiment (Case 4) is shown in Figure 4. The first mode Rossby wave propagates across the basin and exhibits only slight changes in its shape and amplitude. Clearly, the nonlinear effects are balancing the tendency of the wave to disperse and causing the peak to be larger and sharper than in the linear experiment. It is in this sense that we describe the first mode Rossby wave as having evolved into a single Rossby soliton.

The preceding experiments demonstrated that equatorial solitons could develop from wind events in which the equatorial trades relax for periods of approximately one month. Kindle (1982) revealed that Rossby solitons could be excited by wind events whose durations vary from a few weeks to months. In the following experiments we will examine the response to a prolonged relaxation of the equatorial wind. It has been hypothesized (Wyrtki, 1975; Hurlburt, et al., 1976; McCreary, 1976) that just such events are responsible for initiating El Nino; the coastal Kelvin wave caused by the incident downwelling equatorial Kelvin wave creates an intrusion of abnormally warm surface water along the coasts of Peru and Ecuador. Hence, it is of interest to examine whether equatorial Rossby solitary waves are associated with this event. In fact, Boyd (1980) suggests that El Nino would be a likely mechanism for the generation of Rossby solitons.

The extended period of anomalously weak equatorial trades associated with El Nino are represented very simply in the experiments below. They are identical to Case 2 except that the forcing is not switched off. This permits a clearer examination of the response subsequent to the reflection of a Kelvin wave front.

In the first experiment, the nonlinear terms are neglected. The zonal velocity field is shown in Fig. 5. Although, no wavetrain has developed as in the linear prototype experiments (Fig. 2a), the effects of dispersion are evident. The wavefront associated with the first mode Rossby wave has elongated and evolved into a double peak separated by a slight saddle. The solutions are similar to the linear one-dimensional initial-value experiments conducted by Boyd (Fig. 7, 1980).

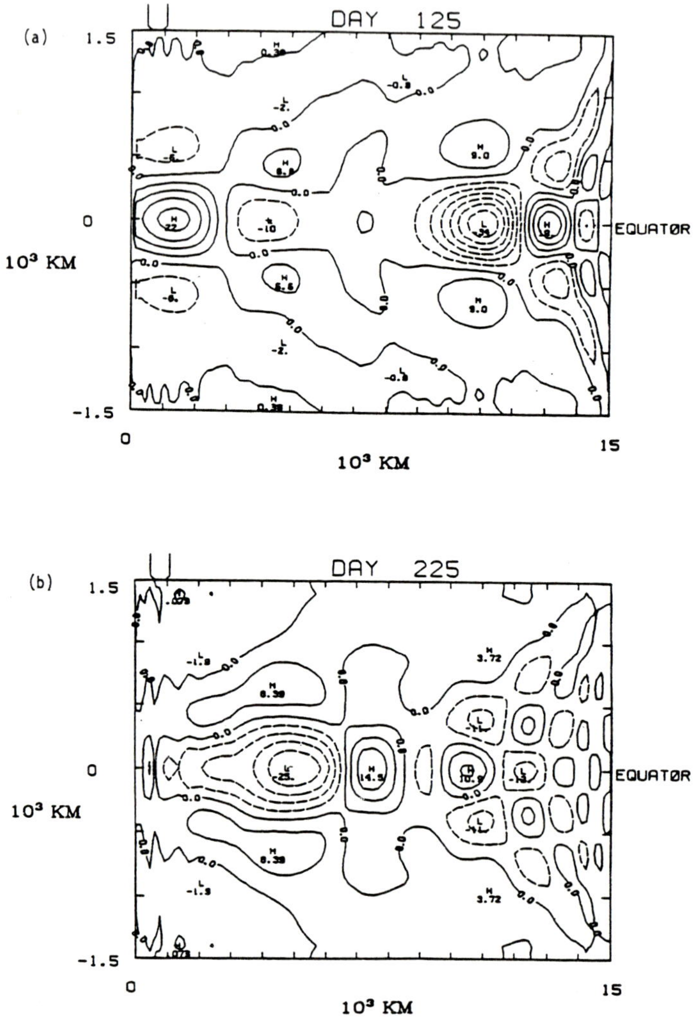

Figure 3. (a) The zonal velocity component at Day 125 for Case 3 (linear solution). The first mode Rossby wave resulting from the Kelvin wave reflection is located 4000 km from the eastern boundary. The contour interval is 5 cm/sec. (b) Same as (a) except at Day 225. (from Kindle, 1982)

(a)

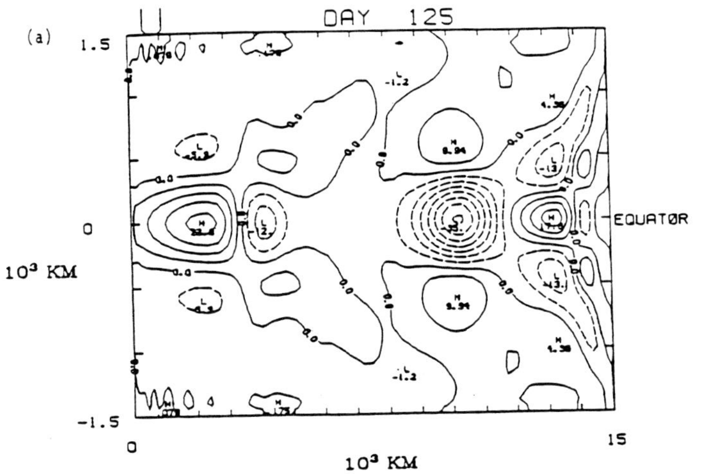

(b)

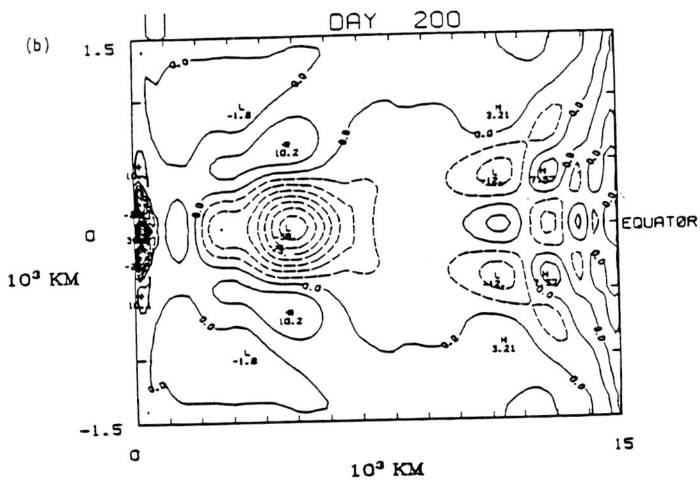

Figure 4. (a) Same as Figure 3 except for Case 4 (nonlinear solution). (b) Zonal velocity at Day 200 for Case 4. Equatorial soliton is clearly evident at x = 4000 km. (from Kindle, 1982)

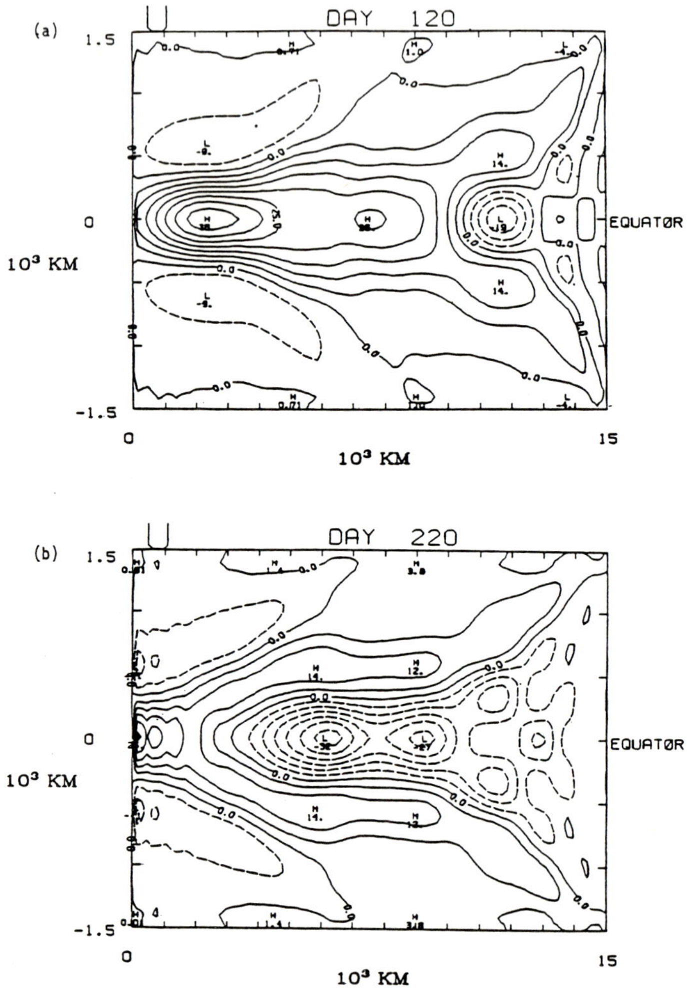

Figure 5. Zonal velocity component for linear El Nino experiment (Case 5). (a) Solution at Day 120. The first mode Rossby wave resulting from the Kelvin wave reflection is at x ~ 12,000 km. (b) Solution at Day 220. (from Kindle, 1982)

Figure 6 illustrates the results of the corresponding nonlinear case. Note that at Day 120, the solution is virtually identical to the linear case. Hence, this representation of El Nino is virtually linear through the onset of the event. If the incoming Kelvin wave and its reflection are essentially linear, this may help to explain the relatively close agreement between the linear experiments of Busalacchi and O'Brien (1981) and the observations of the onset of El Nino. But equatorial Rossby solitons require only a balance between weak nonlinearity and weak dispersion.

Figure 6b, reveals that the presence of weak nonlinearity can alter the character of the solution dramatically. The first mode Rossby wave has evolved into two distinct Rossby solitons. The largest amplitude soliton travels the most rapidly; as time progresses the two solitons will continue to separate. This is also consistent with the work of Boyd (1980) who found that for an initial condition with a "top-hat" zonal shape, multiple solitons could develop. The amplitudes of the individual peaks decrease monotonically away from the lead soliton.

In the final experiment, the amplitude of the El Nino event is increased by extending the zonal width scale of the wind patch by 50%. The patch of zonal wind stress now extends from x = 5,500 km to x = 8,500 km. The amplitude of the wind event remains at .5 dynes/cm^2. Figure 7 depicts the zonal velocity at Days 120 and 220. During this period the solution is very similar to that of Case 6 except that the amplitude is approximately 50% greater. Figure 8, however, reveals the development of a third soliton which, by Day 280, is well in front of the m = 3 Rossby wave (at x = 9,000 km). Clearly, the number and amplitude of the solitons are proportional to the amplitude of El Nino event. It should be noted that nonlinearity appears to have a dramatic effect on the solutions at the western boundary. Even though an intense eddy field develops in that region, numerical experiments (not shown) demonstrate the solitons reflect as Rossby waves and equatorial Kelvin waves. This reflection process will be treated in future work.

4. Summary

The wind excitation of equatorial Rossby solitary waves during El Nino has been investigated numerically using a nonlinear, one-layer, reduced-gravity model of the tropical Pacific. The model basin, which is rectangular is shape, extends 15,000 km zonally and 1,500 km on either side of the equator. The northern and southern boundaries are open. The model is forced by patches of zonal wind stress representing equatorial trade wind events.

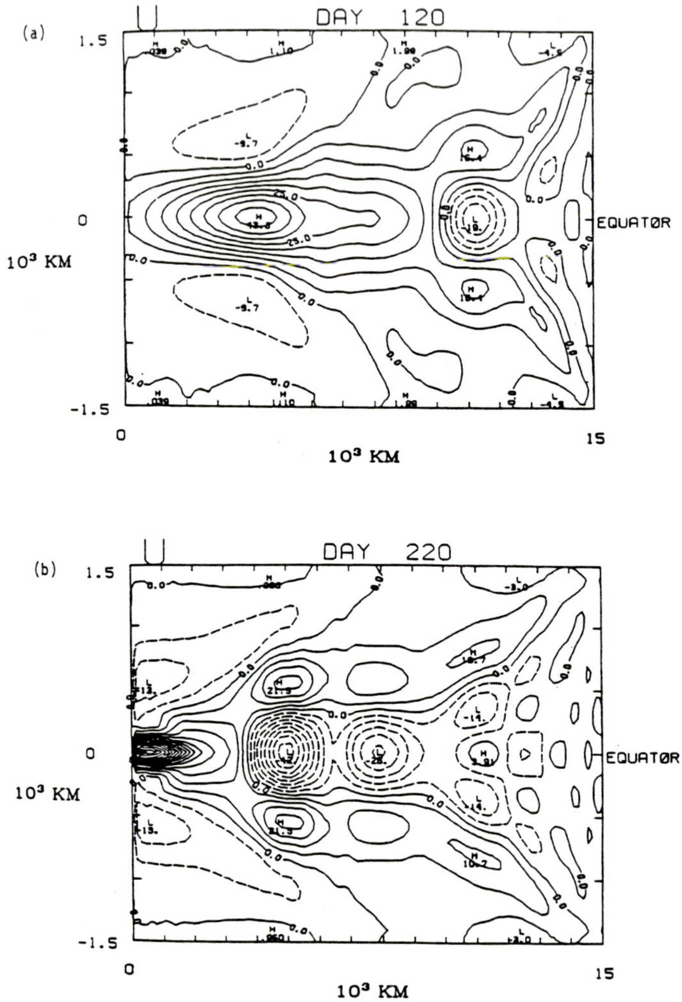

Figure 6. Same as Figure 5 except for nonlinear solution. Two solitons are evident in (b). (from Kindle, 1982)

a)

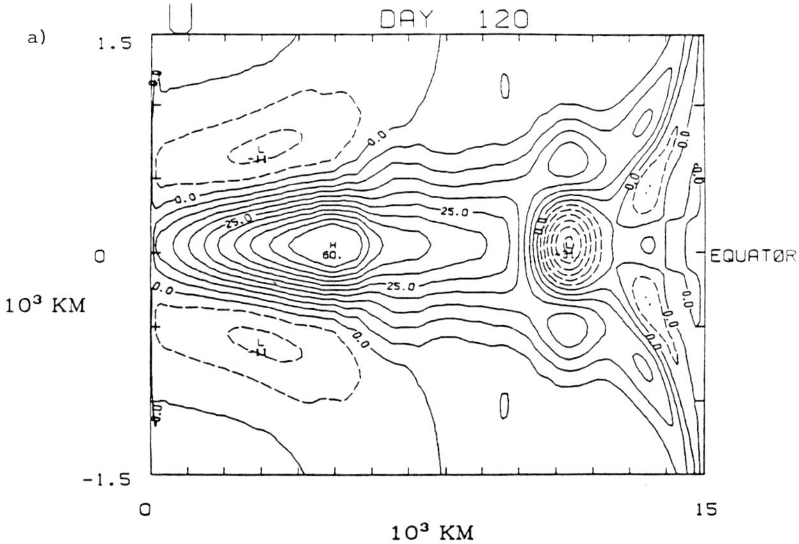

b)

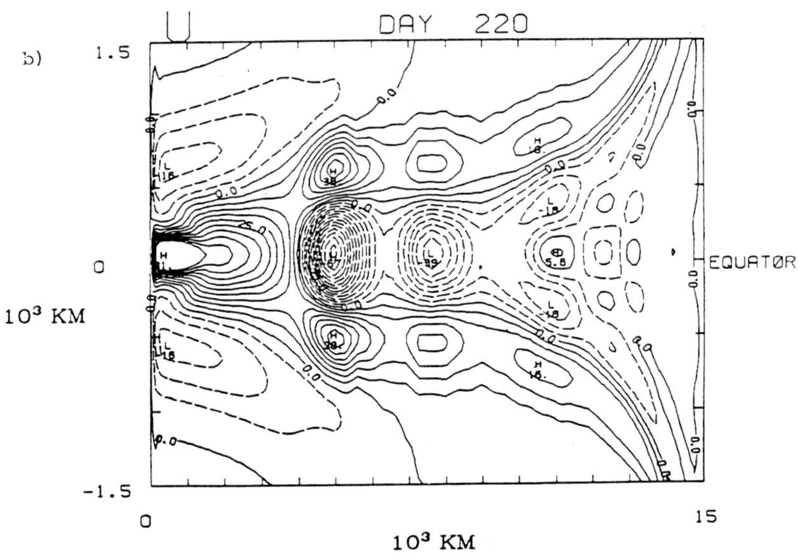

Figure 7. Zonal velocity component for large amplitude El Nino experiment (Case 7). a) Day 120; b) Day 220

a)

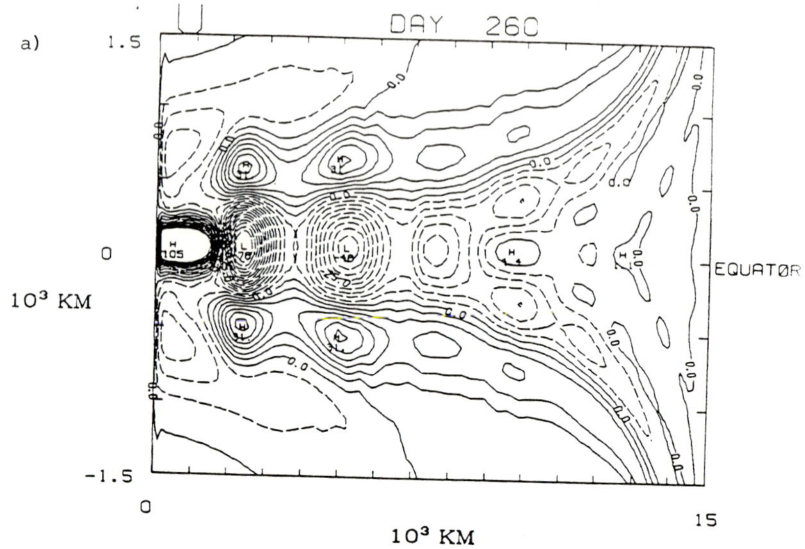

b)

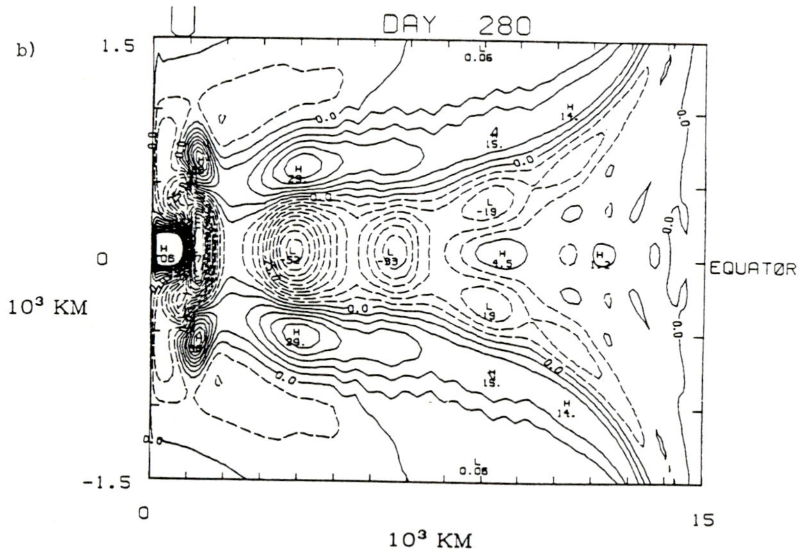

Figure 8. As in Figure 7, a) Day 260; b) Day 280. The third soliton is evident in a) at x = 7,500 km. By Day 280, it has intensified and propagated to x = 6,000 km.

A prolonged relaxation of the equatorial trades, such as that associated with El Nino, generates a soliton train subsequent to the reflection of the Kelvin wave front at eastern boundary. Even a very modest event in which the amplitude of the relaxation is .5 dynes/cm^2 over a 2,000 km patch generates two Rossby solitons which effect a thermocline depression of 30 meters. A major El Nino event would create a significantly larger response, including three or more first mode Rossby solitary waves. The ease with which the numerical model yields solitons for the prolonged wind events suggests that Rossby solitary waves may be most easily observed following a major El Nino.

5. Acknowledgements

I would like to thank Drs. John Boyd, Harley Hurlburt, and Dennis Moore for their helpful comments. Appreciation is extended to Charlene Parker for typing the manuscript. The computations were performed on the two-pipeline Texas Instruments Advanced Scientific Computer at the Naval Research Laboratory in Washington, D.C.

REFERENCES

Barnett, T. P., 1977: An attempt to verify some theories of El Nino. J. Phys. Oceanogr., 7, 633-647.

Barnett, T. P., 1981: Statistical relations between ocean-atmosphere fluctuations in the tropical Pacific. J. Phys. Oceanogr., 11 (8), 1043-1058.

Bjerknes, J., 1966: A possible response of the atmospheric Hadley circulation to equatorial anomalies of ocean temperature. Tellus, 18, 820-829.

Boyd, John P., 1980: Equatorial solitary waves. Part 1: Rossby solitons., J. Phys. Oceanogr., 10, 1699-1717.

Busalacchi, A. and J. J. O'Brien, 1981: Interannual variability of the equatorial Pacific in the 1960's. J. Geophys. Res., 86, 10901-10907.

Gardner, C. S., J. M. Greene, M. D. Kruskal and R. M. Miura, 1967: Method for solving Korteweg-de Vries equation. Phys. Rev. Lett., 19, 1095-1097.

Hickey, B., 1975: The relationship between fluctuations in sea level, wind stress and sea surface temperatures in the equatorial Pacific. J. Phys. Oceanogr., 5, 460-475.

Horel, J. D. and J. M. Wallace, 1981: Planetary scale atmospheric phenomena associated with the Southern Oscillation. Mon. Wea. Rev., 109, 813-829.

Hurlburt, H. E., J. C. Kindle and J. J. O'Brien, 1976: A numerical simulation of the onset of El Nino. J. Phys. Oceanogr., 6, 621-631.

Julian, P. R. and R. M. Chervin, 1978: A study of the Southern Oscillation and Walker Circulation phenomenon. J. Atmos. Sci., 106, 1433-1451.

Kindle, J. C., 1979: Equatorial Pacific Ocean Variability - seasonal and El Nino time scales. Ph.D. thesis, Florida State University, 134 pp.

Kindle, J. C., 1982: A numerical study of equatorial Rossby solitons, J. Phys. Oceanogr., In Press.

McCreary, J. P., 1976: Eastern tropical ocean response to changing wind systems: With application to El Nino. J. Phys. Oceanogr., 6, 632-645.

McCreary, J. P., 1977: Eastern ocean response to changing wind systems, Ph.D. Dissertation, Univ. of California, San Diego.

Mesinger, F., and A. Arakawa, 1976: Numerical methods used in atmospheric models. Garp. Publ. Ser., No. 17, WMO.

Moore, D., 1968: Planetary-gravity waves in an equatorial ocean. Ph.D. Thesis, Harvard University.

____________, and S. G. H. Philander, 1977: Modeling of the tropical oceanic circulation, The Sea, VI, ed. by E. D. Goldberg, I. N. McCave, J. J. O'Brien and J. H. Steele, 319-362, John Wiley & Sons, New York.

Namias, J., 1976: Some statistical and synoptic characteristics associated with El Nino. J. Phys. Oceanogr., 6, 130-138.

Rasmussen, E. M. and T. H. Carpenter, 1982: Variations in tropical sea surface temperature and surface wind fields associated with the Southern Oscillation/El Nino. Mon. Wea. Rev., 110, 354-384.

Weare, B. C., 1982: El Nino and tropical Pacific Ocean temperatures. J. Phys. Oceanogr., 12, 17-27.

Wright, P. B., 1979: A simple model for simulating regional short-term climatic changes. Mon. Wea. Rev., 107 (12), 1567-1580.

Wyrtki, K., 1975: El Nino - The dynamic response of the equatorial Pacific Ocean to atmospheric forcing. J. Phys. Oceanogr., 5, 572-584.

Wyrtki, K., 1977: Sea level during the 1972 El Nino. J. Phys. Oceanogr., 7, 779-787.